ENGINEERING OF CREATIVITY

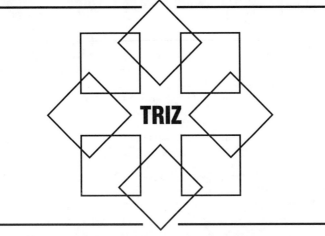

TRIZ

Introduction to TRIZ Methodology of Inventive Problem Solving

Semyon D. Savransky

CRC Press

Boca Raton   London   New York   Washington, D.C.

Library of Congress Cataloging-in-Publication Data

Savransky, Semyon D.
    Engineering of creativity : (introduction to TRIZ methodology of inventive problem solving / by Semyon D. Savransky.
        p. cm.
    Includes bibliographical references and index.
    ISBN 0-8493-2255-3
    1. Engineering—Methodology.   2. Problem solving—Methodology.   3. Creative thinking.   4. Technological innovations.   I. Title.
    TA153 .S28 2000
    620′.0028—dc21

99-053640

Preface

Any successful enterprise, be it a Fortune 500 company or a small shop, recognizes the importance of inventions and innovations to its business. The root of almost any invention or innovation is a solved technical problem, and this book describes a problem-solving methodology, TRIZ, that focuses on solving such problems efficiently, effectively, and creatively.

TRIZ (pronounced "treez") is the Russian abbreviation of what can be translated as "the theory of inventive problem solving." It was developed by Genrich Saulovich Altshuller (1926-1998) who was interested in basing creativity on science. He began developing TRIZ in 1946, but it was not until 1956 that his first paper on TRIZ (written with his friend Rafael B. Shapiro) was published in the USSR. Because of political tensions between Altshuller (and later his students) and the communist authorities, TRIZ development was hindered and at times blocked. After Gorbachev's perestroika, economic troubles slowed TRIZ development and research. Nevertheless, because of Altshuller's enthusiasm, and that of his colleagues, TRIZ became an extremely powerful methodology for creativity in the engineering fields. TRIZ was unknown outside the USSR until the 1990s, and its popularity in the US, Japan and the Pacific Rim, and Western Europe is now growing rapidly. Many Fortune 500 companies have cited a phenomenal increase in productivity, and they credit TRIZ for the breakthrough ideas and quality solutions to tough engineering problems as fueling that increase.

Many engineers in what is now the former Soviet Union have studied and successfully applied TRIZ. They have not only registered thousands of patents for the resulting inventions, but have also become what in effect is an unofficial, virtual TRIZ research and development laboratory. Unfortunately, the majority of these unofficial research results have not been published; instead, research papers have circulated usually as either typed or handwritten manuscripts among those interested in TRIZ. This book attempts to summarize these achievements, and it references publications where these results are formally reviewed. Many original sources are unavailable to the Western reader; consequently, these sources are not referenced. Moreover, it is often hard to track down many of these results, so I apologize to those authors and researchers whose names are not mentioned here.

It might seem that the most important methods for solving technical problems would be unique to each specialized area of engineering, but such is not the case. There is, in fact, a generic problem-solving method, illustrated, albeit much simplified, in Figure 1. There are many universal problem-solving concepts, heuristics, and instruments that work well with engineering and nontechnical problems; there are, of course, also specific methods that are applicable to only a single or limited number of engineering fields. TRIZ deals with both types of heuristics and methods but emphasizes general and universal instruments. There are six classes of inventive problems, which are divided according to whether they require an entirely new

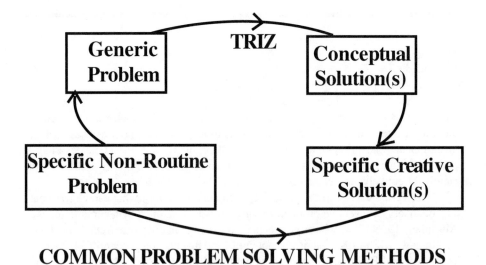

COMMON PROBLEM SOLVING METHODS

FIGURE 1. Simple schema of problem solving. Most problem-solving methods are based on the trial-and-error approach. Note that TRIZ works with technical problems in the same way that mathematics works with problems. The majority of other problem solving methodologies avoid generalizations.

solution or only a change in an existing technique. TRIZ is unique because, using the technique's relatively small number of easily understood concepts and heuristics (supported by effective knowledge databases), one can solve problems of any of the classes:

1. improvement or perfection of both quality and quantity (considered Contradiction Problems in TRIZ)
2. search for and prevention of shortcomings (Diagnostics)
3. cost reduction of the existing technique (Trimming)
4. new use of known processes and systems (Analogy)
5. generation of new "mixtures" of already existing elements (Synthesis)
6. creation of fundamentally new technique to fit a new need (Genesis)

It is inaccurate to say that TRIZ can resolve any technical problem today, but the many inventions created with TRIZ do confirm its power.

The ability to solve such problems is vitally important today. Consider the key economic principles and leaders during the twentieth century:

- beginning of the century — production in mass quantities (Germany, England)
- first half of the century — mass production at the lowest price (US)
- middle of the century — cheapest mass production with the maximum quality (Japan)
- end of the century — cheap and high-quality manufacturing of well-designed products (multinational companies with strong think tanks)

Perhaps in the twenty-first century an additional component — the maximum speed of development and introduction of next-generation products — will determine global economic leadership. If such is the case, then TRIZ takes on even more importance because it enables its practitioners to quickly obtain very high-quality and even breakthrough conceptual solutions and then to effectively remove technical obstacles in implementing the solution.

This book is primarily a practical guide to solving inventive or nonroutine technical problems within the framework of TRIZ, although some aspects of personal creativity development (another goal of TRIZ) are also briefly outlined. The discussion of TRIZ ideas, concepts, heuristics, and instruments is organized in a logical way that will help the reader understand, remember, and apply them.

Most of the TRIZ concepts presented here were proposed and developed by G. S. Altshuller. The book also includes the results of research over the last two decades and the ideas of many other TRIZniks (specialists in TRIZ and also considered as co-authors) who have developed TRIZ not only without financial support but often under negative pressure of Soviet authorities.

It is impossible to use TRIZ effectively without firmly understanding its concepts; therefore, the reader should not jump directly to the final part of this book. Moreover, the methodology of technical problem solving is far from being precise enough at present for this book to provide simple "cookbook" instructions for solving most problems. Teaching by example is an important and popular approach generally used in TRIZ, and although examples and case studies cannot replace solid proof of ideas, this book follows the convention of using examples. First, each TRIZ concept, heuristic, or instrument is discussed theoretically, and then it is applied to real problems, so the reader can see how to use it in actual practice. When you see an example, try to solve the problem yourseif. Then compare your solution with the solution presented in the book. Although the case studies are drawn from specific fields and consequently might not be fully understood by all readers, they do illustrate key concepts. They will undoubtedly help those readers with sufficient engineering background, but readers without such a background in the particular field can compensate by spending greater time on the theoretical discussions, which are geared toward a general audience. Regardless, understanding TRIZ can enhance your thinking in nontechnical fields.

This book is for anyone who is interested in solving engineering problems and developing his or her creativity. Although it might be difficult to learn the methods in a particular engineering field, at least those methods are taught in numerous schools and universities. However, general problem-solving methodologies are rarely, if ever, taught, though they are quite helpful for solving problems in every field of mathematics, science, and engineering. Since such courses generally do not exist, the reader can consider this book a handbook for a general methodology of solving technical problems. Becoming proficient in TRIZ concepts, heuristics, and instruments requires time, but it results in much more effective problem solving. As you gain extensive practice applying TRIZ, you will become so skilled in it that the problem-solving process will be less conscious and more automatic. You will also experience the joy of creativity and be able to solve problems in other fields. This ability to go beyond your own engineering field is also extremely significant today and is becoming even more important; it is clear that we need to acquire increasingly more knowledge and often switch our innovation skills and experience to the most promising directions. TRIZ can help us do that.

The Author

Semyon D. Savransky earned his M.S. and Ph.D. degrees in physics and materials science in Russia. His academic background is split among Novgorod State University in Russia, University Pais Vasco in Spain, and New York City University in the United States.

Dr. Savransky has applied TRIZ for several research and development projects in various industries and for pure scientific research. He is the author of more than 150 patents and scientific papers. He is the founder of the Research Center at Novgorod State University and "The TRIZ Experts," an international company with headquarters in Silicon Valley, CA. After becoming acquainted with TRIZ in 1981, Dr. Savransky became one of the most distinguished leading TRIZ researchers and summarizes the results of his development of TRIZ in this, his first book.

Dr. Savransky can be reached by e-mail at TRIZ_SDS@hotmail.com.

Contents

PART 3:   TRIZ Information

8   Inventions

9   Effects

PART 4:   Preparations to Problem Solving

10   Before Starting

11   Inventiveness

12   Su-Fields

PART 5:   TRIZ Heuristics and Instruments

13   Resolution of Technical (Pair) Contradictions

14   Physical Point Contradictions: Ontology and Resolution

Part 1

Problem-Solving Aids

1 How Do We Solve Problems?

This chapter discusses the process of problem solving and outlines various methods and aids for problems solving [1-27]. The chapter shows the weaknesses of these methods and presents requirements for an inventive problem-solving methodology and for dealing with nonroutine problems.

1.1 ROUTINE AND INVENTIVE PROBLEMS

Throughout life people face various problems; e.g., a child wonders how to dismantle a strange toy, while an adult wonders how to make a technological process more effective. They either try to solve the problems (Why do competitors have higher quality and production but also lower costs?) or refuse to (still thinking about whether to get married after ten years).

The ability to see and resolve problems is extremely important for engineers, managers, scientists, politicians, and others in our competitive world. Thus, the author hopes this book will be useful not only for engineers but also for anyone wanting to improve his or her problem-solving skills.

A *problem* is a gap between an initial (existing) situation and the desirable situation. *Problem solving* is a single- or multi-step transformation of the existing situation to the desirable situation or to a situation closer to the desirable one than is the initial situation. The steps in the transformation span the gap.

Problems are *routine* if all critical steps to a solution are known. A step is *critical* if a solver cannot resolve the problem without it. There are numerous methods for routine problem solving in specific fields, such as mathematics, marketing, and design; routine problems can be solved solely by standardized or automated procedures. Often computer programs or repetitive solutions that have worked in the past can be used to solve a current routine problem. A problem is *nonroutine* if at least one critical step to a solution is unknown. TRIZ defines technical problems for which at least one critical step to a solution as well as the solution itself is unknown as the *inventive problems*. The complexity of the initial situation, a poorly defined desirable situation, or hidden search directions can lead to inventive problems.

Inventive problems are often mistakenly considered to be the same as engineering problems. Actually, inventive problems are only a fraction of all engineering, technological, and design problems. The situation here is the same as with regular and creative artistic problems. A creative problem is one whose resolution is nonobvious. Inventive problems are a subclass of creative problems in fields of technique. They are ones where the input and output of the solving steps have not all been defined or where there is irrelevant, conflicting, and/or inadequate information provided. In addition, the solver may only poorly understand the information that is important for the problem. In short, an inventive problem is usually novel, elusive, and slightly out-of-focus in the sense that it is often ambiguous and poorly understood.

The difference between an inventive problem and a simple, routine problem is time-dependent because the *recognition* of the important steps grows with time. The process for solving a technical problem depends on how often the solver has faced similar problems and on the solver's ability to recognize the similarity.

Technical solutions must satisfy three requirements: They must be

- physically possible (corresponding to the laws of nature)
- technically possible (corresponding to resources and the scientific and technical ability of a society)
- economically profitable

In the statement of a problem, a solution model is formulated at the economic level and only partially at the technical level. However, the solution is sought first at the physical (or other adequate natural sciences) level, then at the technical level. All characteristics at any level can be determined for any technique (from its analysis). A complete set of purposes (which could be achieved with this technique) is among these characteristics. An inverse transition, from purposes and characteristics to a technique, is possible only for trivial problems. When there is a lack of congruence between the solution requirement levels, as well as an informational gap between them, the result is an inventive problem.

Problems can be classified according to criteria, such as difficulty, complexity, structure, and understanding of the problem. Some aspects of problem difficulty and complexity are discussed here to clarify the strength and weakness of various nonroutine problem-solving approaches.

1.2 DIFFICULTY OF A PROBLEM

Difficulty is defined in terms of the number of variables involved in the problem. Simple problems involve only a few variables and can be solved by an individual; complicated problems involve many variables and are usually solved by a team. The *problem space* is defined in terms of the original problem statement and ranges from known to unknown causes of the problem. Known causes contain all the information necessary to define the problem's objectives, constraints, variables, and assumptions. Consequently, the problem solvers do not need to make any assumptions in order to solve the problem (the cause of problem is known), but they may need to translate the problem statement into language more familiar to them. If causes are unknown, problem solvers may have too much or too little information. (The role of information in modern problem solving is discussed in Section 1.7 and in Chapter 10.) The *solution space* is defined in terms of the uniqueness of acceptable solutions. The solution space is considered *closed* if there is a finite number of correct solutions of an analytic problem formulation that will satisfy the requirements. Often only one acceptable solution exists for a problem. If several solutions are possible, the problem solution space is considered *open*. Since an open solution space accepts many alternative solutions, traditional optimizing procedures are not applicable for such problems, especially when the direction to search for solutions is unknown.

The following is a simple model for a problem of difficulty D that can be defined by the ratio

$$D = V/S, \tag{1.1}$$

where V is the number of possible variants (trial steps), and S is the number of possible steps that lead to acceptable solutions. The higher this ratio is, the more difficult is the problem and the longer is the time necessary to resolve the problem. It is assumed for simplicity's sake here that all trial steps are equal.

From this perspective, all problems can be represented in a simple two-dimensional scheme as shown in Figure 1.1.

In large part, the engineer's job is to solve, from all quarters shown in Figure 1.1, numerous problems during the design or production phase of a new product or when improving production or technology.

1.3 PSYCHOLOGICAL INERTIA

The problem-solving process itself depends on the ability of a solver. Two people with different knowledge will have different ideas about the necessary steps in solving the same problem. For example, a car that won't start may be a very complex problem for a rocket scientist but not for the corner mechanic. On the other hand, some rocket scientists may solve car problems that are too complex for the corner mechanic. In other words, one person's inventive problem is another's simple routine problem.

The amount of time needed to solve a technical problem should reflect the complexity and effort of determining the unknowns in the problem's possible causes and the steps in the solving process. For difficult problems, the solver rarely knows

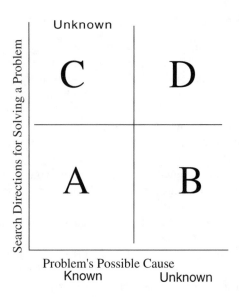

FIGURE 1.1   A simple classification of problems.

all possible variants and cannot perform all possible trial steps. Time can be measured in any convenient way, such as person-days. *Psychological inertia* (or psychological barrier) greatly affects the time needed to solve a problem. The term *psychological inertia* was introduced in creativity and innovation research and is analogous with physical inertia. It has been argued that most trouble in solving difficult (routine and nonroutine) problems is the solver's own psychological inertia [1].

Physical inertia is the effort made by a system to preserve the current (meta-)stable state or to resist change in that state. A system in the (meta-)stable state always resists transition to a new state if that new state does not coincide with the change tendency of the present or previous states. For example, because of our inertia when we are running at a high speed we cannot stop momentarily. Generally, rate of transition from one state to another cannot be changed. However, if another method is found to make transition, then the transition time may be shortened. For example, placing a metal spoon into a thin glass before pouring boiling water into it slows the temperature transition rate and usually prevents the glass from cracking. Therefore, knowledge plus invention is required to overcome the inertia. Systems inertia in itself may be positive or negative depending on the specific situation and person. In many cases, psychological inertia is useful because it prevents our brain from high concentration of its force during the routine operation. We learn how to walk or how to handle a spoon in our childhood so as adults we do not need to focus attention on these skills because of psychological inertia. But in nonroutine problem-solving, creativity, and innovation, psychological inertia is usually negative because it may strongly counteract the search for possible solutions.*

---

* It is interesting that "inertia" came to modern European languages from Latin where it means "lack of skill" rather than the current meaning of indisposition to motion or change.

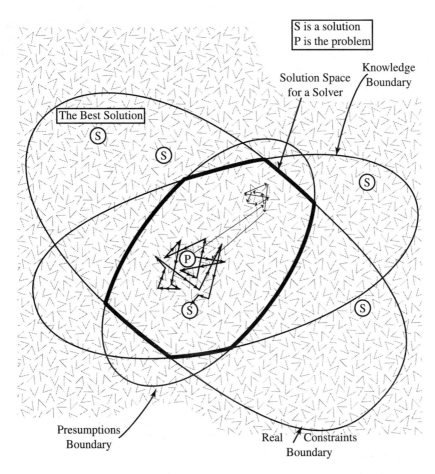

FIGURE 1.2   The trial-and-error method by a solver with high psychological inertia.

Psychological inertia often leads us away from the solution of the problem (see Figure 1.2), impedes problem recognition and clarification, creates barriers during the search for the steps to solution, complicates decision making, and makes other stages of the problem-solving process difficult.

1.4 TRIAL-AND-ERROR METHOD

Many surveys [2, 5, 8, 9, 17] have shown that the oldest and still most prevalent method for problem solving, regardless of the class of the problem, is trial and error, illustrated in Figure 1.2. In this method, the solutions are obtained after examination of various trial steps (the arrows in the figure represent these trials) through irregular search of a problem's solution space.

The trial-and-error method is suitable for simple, well-defined, routine closed problems. It works quite well for finite and small V values: one considers variant after variant until all have been considered or a solution is found. The trial-and-error method should be used for the open problems only when the possible directions of search are known. To do so, we would create a checklist (described in more detail in Section 1.5.1) of all possible steps, using the simplest and perhaps most immediately useful aid to design thinking. However, if a problem is not well defined or is too complicated, the trial-and-error method is inappropriate and more advanced methods should be applied to the problem.

Unfortunately, many problems have been tackled by the trial-and-error method, and most of such attempts have failed.

The disadvantages of the trial-and-error method include the following:

The trial-and-error method wastes much time, thinking activity, material means, and even human lives for problems with unknown parameters (see Figure 1.1) or high D number.

The method is not time effective: the number of ideas (trials) that succeed per unit of time is small, so this initial stage of the problem-solving process lasts a long period of time. In other words, productivity of generating ideas is low.

Psychological inertia focuses the mind on what is known, i.e., along the assumed search direction, thereby keeping the solver from the right solution. This inertia is useful only when the solution's direction is recognized correctly.

There is no mechanism to uncover all possible variants (although a morphological box or a good questionnaire can help; see Section 1.5). As a result, the solution beyond the trial steps will not be found.

Criteria for "right" or "wrong" variants are subjective, so a solver can miss appropriate solutions that reside in domains of engineering or scientific knowledge beyond the solver's ken.

There is no mechanism for directing the solver's thinking towards the solution: the solver is unable to define the direction in which the necessary solution might be found. This is the crucial disadvantage of the trial-and-error method.

1.5   METHODS OF CREATIVITY ACTIVATION

The weakness of the trial-and-error method was recognized in the middle of the twentieth century. Since then there have been several efforts to perfect the trial-and-error method and to avoid psychological inertia by (1) methods for activating creativity, such as Brainstorming, Synectics, Lateral Thinking, "Mind Machines," Neurolinguistic Programming, and Mind Mapping; (2) methods of expanding the search space, such as Morphological Analysis, Focal Objects, or Forced Analogy; and (3) decision aids such as T-charts and Probabilistic Decision Analysis [2-24].

Let us discuss some of the methods for activation of creativity [2-13] that have been developed in Western countries first. These methods are not presented in detail here because of the many publications devoted to them [2, 5, 11, 13] (Checklists/questionnaires and morphological boxes are briefly discussed in Chapter 11.)

Some researchers believe that creativity has roots in a subjective human psychology or biology of brain. There have been numerous attempts to study creativity "from the first hands," i.e., to find a relatively small number of "mental mechanisms" used by a large number of artists, scientists, and inventors or to "investigate in detail the structure and quality of creative thinking and imagination" from the drafts of artistic, design, and scientific works, as well as to teach creativity and problem solving on this basis [25, 26]. However, the fact that a person is very creative or inventive or has made discoveries does not mean that the person necessarily understands the creative or discovery process. It is no more reasonable to expect an artist, scientist, or inventor to give a full description of his thought activities than it is to install a lamp on a podium to deliver a speech about the physics of light. Much of what goes on during a person's thinking activities is unavailable to any conscious awareness of a researcher, even if he has devices to record brain signals or methods for articulating human activities. Researchers use terms such as "judgment," "left brain process," "intuition," "insight," and "lateral thinking" to label phenomena that occur without awareness, but labels, unfortunately, are not explanations. Therefore, such approaches to studying problem solving do not have a scientific base, depend on cultural background, and cannot be used by everyone [2].

Nevertheless, the best methods of creativity activation help reduce the second and the third disadvantages of the trial-and-error method and decrease the influence of the fourth and fifth disadvantages. Some of these methods are useful for suppressing the solver's psychological inertia.

A popular means for overcoming the trial-and-error method's shortcomings is based on ways of covering all search directions. Two useful methods in this realm, checklists and questionnaires, are considered in the next subsection.

1.5.1 CHECKLISTS AND QUESTIONNAIRES

The various checklists of control questions and objects or tasks are among the oldest methods for increasing the efficiency of problem solving. They are still one of the most popular problem-solving aids. (The terms "questionnaires" and "checklists" are used synonymously here and for sake of simplicity are referred to as "lists.") Usually lists are prepared on the assumption that requirements that have been overlooked before and will be overlooked again when new solvers are employed on routine problems. It seems a good idea to create a list for any problem that has more than a dozen trial steps, because a person rarely can keep such a number of possibilities in mind. Working with the lists does not require special training, but it does require experience in asking correct and exact questions and providing exact answers.

Usually a list represents a set of questions or recommendations for organization of a decision process. The consecutive answers to all questions raise the probability of achieving a successful solution to a problem, provided the criteria by which a solution is to be accepted or rejected are established. In general, it is possible to identify the following stages of a decision using a list:

1. specification of the formulation of a problem, allocation of object, specification of restrictions and requirements;
2. the preparation of the entries, in which a specialist enters questions or answers, taking into account specificity of the given problem;
3. analysis of the response, formulation of new questions, search for new answers, and construction of a set of the possible solutions.

TABLE 1.1
Difficulties in Problem Solving with Lists

Difficulties	Preclude Ways
The time taken to think about each possible trial recommendation in a long list can be greater than the time available for problem solution.	Pattern (tree or net) the list in such a way that a solver can quickly discern the more promising ways relevant to a particular kind of problem.
The list itself can be based on assumptions that direct the solver away rather than toward a new solution. *This danger is always present when using analogy.*	Compare the assumptions upon which the list appears to be based with those that the solver considers important. Evaluate the importance of different assumptions. See Section 1.6.
Formulating exact and reasonable questions; they are necessary for the "move" from a previously planned (consciously or unconsciously) area of possible decisions.	Combine a few lists for searching. Add questions to the existing list. Create new list. This is the most troublesome step.

Although checklists and questionnaires are quite simple aids, preparation of a good one is not an uncomplicated task. For example, Marsh Fisher and his coworkers spent 12 years (1977–1989) and 4 million dollars to build IdeaFisher software that consists of an interactive database of structured questions, idea words and phrases for brainstorming, and problem-solving activities in various fields [6].

The most famous lists were assembled by George Polya, well-known Hungarian mathematician [3]; Alex Osborn, American researcher in the field of creativity [4]; and Marsh Fisher, American entrepreneur [6].

These checklists and questionnaires are presented in Appendix 1 and Chapter 18. These tools are first concerned with the attitude of the solver toward the initial situation and possible problems. Unfortunately, they mix these very different aspects of the problem-solving process. Furthermore, they incorporate numerous recommendations on possible directions of the decision of problems. It is interesting that European and American specialists who prepared the checklist and questionnaires recognized the same aspects, such as resources, complexity, and causes, as Russian TRIZniks believe to be important to problems. Perhaps such similarity signals that a successful problem-solving methodology is independent of a person's cultural background and can be applicable anywhere. On the other hand, all recommendations in the lists are equally weighted, so it is not clear which of them can help solve a specific problem. The lists are offered just to disassemble numerous variants of changes of an object and not to forget any direction of possible solution. Therefore, some difficulties that can arise with the lists are described in Table 1.1.

The preliminary focus of such questionnaires and checklists is the elimination of existing objects or the search for new ideas. Thus, they can be used at the initial stage of problem solving. Many problems with relatively small D have a sufficient number of predictable requirements, so the art of correctly stating the question or appropriately balancing the value of answers is determined by the efficiency of the search for the solution.

1.5.2   MORPHOLOGICAL BOX

Based on the works of famous mathematician and philosopher Gottfried Wilhelm Leibnitz (1646–1716), in 1942 astronomer Fritz Zwicky [7] proposed morphological analysis. The goals of this method are

- to expand search space for a problem's solutions and
- to safeguard against overlooking novel solutions to a design problem.

This method combines parameters into new sequences for later review. The result of this analysis is the so-called morphological box or matrix or table.

The best known morphological box is the Chemical Periodic Table (created by Russian chemist D. I. Mendeleev) with the number of electrons at the outer shell along the X-axis and the number of electronic shells along the Y-axis. Atoms with X and Y numbers are located in the correspondent cells. It should be noted that indices at both axes of this matrix are arranged in increasing order. Ranging of matrix axes allows creation of an "efficiency vector" in the matrix: atoms in the matrix are arranged along the corresponding vectors of radioactivity, inertness, metalloidness, etc. Position of an atom in the matrix allows us to judge the comparative parameters of the corresponding substance, i.e., to perform the quick analysis of information about parameters of known atoms. Such a matrix can be used for forecasting properties; in fact, Mendeleev predicted many elements based on the Periodic Table.

The primary phases of morphological analysis are represented in Table 1.2.

Most technical systems can have more than two features, so it is possible to create several morphological boxes (with any size $N \times M$ where $N = M$ or $N \neq M$) or a morphological tensor for a system. Such boxes allow fairly easy estimation of values of attributes in empty cells.

The decimal matrix method for searching for technical solutions, proposed by Genrich Ya. Bush, a Latvian invention scientist, is based on morphological analysis [27]. His assumptions are the following:

A. All main parameters of techniques considered during a design are separated into ten groups:
   1. **Geometric** parameters: length, width, height, area occupied by a construction in plan, cross section areas, volume, shape.
   2. **Physico-mechanical** parameters: mass of a construction and its separate elements; consumption of materials; strength and other characteristics of materials, including new ones; resistance to corrosion, etc.
   3. **Power** parameters: drive energy type and power, efficiency, etc.
   4. **Design and technological** parameters: machine adaptability to production, transportability, simple or complex construction, etc.
   5. **Reliability and service life:** purely technological factors — technical reliability and service life, as well as such factors as protection from harmful effects of ambient medium; Bush considered all factors associated with human participation in the operation to be in another group.

TABLE 1.2
Primary Phases of Morphological Analysis

Phase	Comment
Define the problem or system as clearly as possible.	Clear parameters must be established that delineate the problem or system being analyzed. In the case of a complex issue, it may be appropriate to identify aspects of the problem that can be dealt with individually rather than to seek comprehensive coverage in a single analytical exercise.
Divide the system into subsystems or parts that can be considered singly.	Applicable models of the system may help to identify suitable subsystems for consideration. Define the functions that any acceptable design must be able to perform.[a]
Check connectivity of the subsystems.	Determine two principal features (X and Y) of the problem or system and check that X and Y do not strongly depend on each other (orthogonal).
Determine the possible solutions or approaches that can be applied to each subsystem and/or alternative sources for performing each function.	Create two lists with different variants of this feature (X_1, $X_2, \ldots X_i, \ldots X_n$; and $Y_1, Y_2, \ldots Y_j, \ldots Y_m$). Place all X_i along the horizontal axis and Y_j along the vertical. List solutions, options, or outcomes for each subsystem in cells of the morphological box.[b]
Develop possible outcome combinations.	The alternative outcomes are generated for each subsystem. Complete $N \times M$ cells of the box with entries.
Select an acceptable set of subsolutions, one for each function.	Evaluate the feasibility or practicality of each combination. See Section 1.6.

[a] The main difficulty is to identify a set of functions that are (a) essential to future solution(s); (b) independent of each other; and (c) inclusive to all parts of the problem. There should be enough functions to produce a box that can be filled and searched in a short time.

[b] The entry in cell X_k-Y_1 should have both peculiarities k and l of the features X and Y. This is the most difficult step, requiring divergent thinking and creativity.

6. **Performance:** productivity, accuracy and quality of operation, stability of parameters, degree of readiness for service, etc.
7. **Economic** parameters: cost of machines and parts, labor consumption for production and use, costs, losses, etc.
8. **Degree of standardization and unification.**
9. **Safety and ease of use:** all parameters connected with labor protection and accident prevention, ergonomics and psychology, ease of production and use, maintenance and repair, requirements for comfortable labor conditions, high culture of production.
10. **Exterior view:** all factors that affect the aesthetics of the machine dimensions, harmonious construction, proportionateness, etc.

B. The set of basic **heuristics** (i.e., procedures for the problem solving) is separated into ten groups:

1. **Neology** (originating from the Latin for "novelty" or "knowledge of new") consists of the new application of an established process, construction,

shape, material, their properties, etc. This process, etc. is not new but its application to the specific field is.

2. **Adaptation** includes fitting of known processes, constructions, shapes, materials, and their properties to specific conditions of labor.

3. **Multiplication** of system functions and parts and multiplied systems remain similar to each other, of the same type. Multiplication includes not only methods associated with enlargement of characteristics (hyperbolization) but also with their miniaturization.

4. **Differentiation** of system functions and elements: functional links between system elements weaken; elements of construction and working processes become spatially and temporarily separated.

5. **Integration** includes joining, combining, reducing in number, and simplifying of functions and forms of elements and the system as a whole: production and construction elements and working process become spatially and temporarily closer.

6. **Inversion** is a reversion of functions, shapes, and mutual position of elements and the system as a whole.

7. **Pulsation** encompasses the group of design and creative methods associated with changes in process continuousness. Pulses may repeat periodically or aperiodically; a pulse may also be single (method of overshoot).

8. **Dynamization** suggests which element parameters or techniques as a whole must be changeable and optimal at every stage of a process or in a new mode.

9. **Analogy** is using similarity or resemblance of some aspect of systems (objects, phenomena) which are otherwise different as a whole.

10. **Idealization** is presenting an ideal solution as an aim to be reached, to make a start from the best result.

C. Consecutive analysis of a chosen system should be performed. The work is aimed at filling every matrix cell (corresponding to a parameter or heuristic) with a new technical solution. Such simplified classification allows construction of the square morphological matrix 10×10 with qualitative parameters (A) in rows and groups of heuristics (B) for their change in columns along the horizontal and vertical axes of the matrix. The cells contain the data on technical solutions, in which specific parameters can be changed using the corresponding heuristic. A fragment of the decimal matrix for machine tool engineering is shown in Table 1.3.

Bush's method has the following stages:

1. Precisely formulate the problem to be solved. It is important at this stage to describe generally the system under study.

2. Describe the system's most important functions and subsystems (those required for the system's operation).

3. Reveal and fix all possible versions of realization of every function and every subsystem.

TABLE 1.3
Decimal Matrix for Machine Tool Engineering

Heuristics →	Adaptation	Multiplication	Differentiation
Parameters ↓			
Geometrical	Vertical configuration of a lathe ("put on a side")	Many-storied instrumental bedside-tables	Suspended remote controllers
Physico-mechanical	Oil smoke to cool processing of parts	Diamond processing of metals	Liquid polishing
Power	Electrical-insulation coatings of polymers	Use of laser for metals	Separated drives in a machine
Design and technological	Replacement of mechanical clamp of parts with the hydraulic one	Ball nut with a moving screw	Distributing shaft with cams

4. Reduce the set of obtained versions into the morphological matrix.
5. Choose the solutions corresponding to the sets of different attributes shown at axes or in columns of the morphological matrix.

This method is efficient in solving problems associated with change of a system's external view or design, as well as with fundamental changes in a technique (for example, search for new execution principles for its functions).

The ideal morphological analysis identifies *all* possible combinations of means to achieve a desired end. Of course, preparation and completion of the entire matrix require much time, especially if the morphological parameters are interchangeable. The number of cells rapidly becomes astronomically large. Even a simple box with 10×10 such parameters has 100 cells with variants of possible solutions. That is why the pre-prepared boxes published in several reviews of Morphological Analysis [7] are valuable.

After construction of a morphological box that, in principle, can contain all possible variants of a solution, a new obstacle appears: *How does one choose the right solution?* Morphological analysis does not offer an unequivocal answer to this question. It is the main drawback of the morphological approach and it perhaps cannot be solved in general in the framework of this approach.

Morphological analysis is a way to order all possible variants for the search of solutions for the given problem. It is most advantageous during the development and conceptual stages of new technique design or during the analysis of multifactor problems. But the absence of a precise tool for choosing the optimum decisions limits its sphere to problems with a relatively low number of possible variants.

Therefore, the main difficulty with using morphological analysis is the same as for using checklists and questionnaires: they all provide a time advantage when a person is seeking various forms of a solution, but they are also disadvantageous when it is time to evaluate the obtained versions. In a short amount of time, a solver can obtain thousands of solution versions, but their exhaustive analysis and evaluation can take several years.

1.6 DECISION AIDS

The methods of creativity activation and search expansion outlined in the previous section can generate many interesting ideas and solutions from which one must select the most promising. We used to think that analyzing the information would by itself produce solutions. However, we now know that analyses only allow us to select from ideas we already have and not generate new ones. In order to generate new solutions, however, we should be able to do some idea work in our minds with or without creativity activation before coming to the information. A few, almost trivial, methods for evaluating solutions or other statements [13] are discussed below: T-chart, the simple decision matrix, and the probabilistic decision tree.

TABLE 1.4
T-Chart

Positives	Negatives
1	a
2	b
3	c
...	...

The T-chart shown in Table 1.4 is a two-column table focusing on a solution's potential good/beneficial and bad/negative results. Blank cells encourage a solver to consider a new direction, and a list of negatives appreciably longer than the list of positives indicates the solution is probably not worth pursuing.

The decision matrix (Table 1.5) forces the solver to be specific about the qualities required of a solution, their relative priority, and, if possible, their relative value. These numbers can be obtained from firm data or quantitative estimate.

The probabilistic decision tree represents various options. Each junction is a decision point at which a solver must decide which step to take. For each step a value must be assigned in addition to the probability that the value will be achieved. These probabilities are what the solver considers most reasonable so they often are also subjective.

These aids do not provide additional information for a solver to analyze. Their value is that they encourage analytical thinking during decision making. They do not make decisions; they simply assure that a solver has consciously analyzed the situation. Once a solver chooses the most promising solutions, evaluation of them can begin. The steps in evaluation are listed in Table 1.6 with estimations of different proof parameters. The table also shows pros and cons of decision aids.

1.7 PROBLEM SOLVING AND INFORMATION

Many problem-solving and decision methods emphasize the importance of information [2, 8, 9]. Even a few years ago, the key was the collection of information followed

TABLE 1.5
Decision Matrix

Name of Object	Value 1	Value 2	...	Value N	Total
A					
B					
...					
Z					

TABLE 1.6

Evaluation Method → Criteria ↓	Judgment with Decision Aids	Mathematical or Physical Model or Simulation	Experiments with the Real System under the Working Conditions
Cost of prediction	Lowest	Medium	Highest
Error probability	Highest	Medium	Lowest
Required time	Lowest	Medium	Highest

by analyis. There was a time when solvers were very short of information, but today we usually have plenty of information, and with the Internet we now have even more. In fact, the amount of information we now have available makes it painful to sort out what is important. If decisions required only information, then often a computer could make the decision without the need for human intervention. Unfortunately, although computers can resolve sophisticated problems, they can solve only routine ones. According to Edward deBono [8], specialist in creativity, the relationship between information and a decision's value during problem solving can be represented by a bell curve. Initially, an increase in information leads to better solutions, but there comes a point at which more information has a diminishing effect. Information by itself does not produce solutions. We need the ability to work with the information effectively.

Mind mapping, developed by Tony Buzan for information representation [12], is a simple, attractive method for handling information. To make a mind map, one starts in the center of the page with the main idea, then, working outward in all directions, writes keywords and images, mapping connections. The result is a growing and organized structure of ideas. An example of a mind map is shown in Figure 6.1.

There is a strong tendency today to narrow specialization. Because of the exponential growth of information, we can afford (in terms of both economics and time) preparation of specialists in extremely narrow fields, the various branches of science and engineering having their own particular realms. As the knowledge in these fields grows deeper and broader, the individual's field of expertise has necessarily become narrower. One result is that handling information has become more difficult and even ineffective.

There is some value in specialization if the following assumptions will be valid forever:

The division of reality into these various fields of knowledge and/or experience is objective.
These fields are permanent.
The links between these fields are not important.
The whole of reality can be represented as an arithmetical sum of components, i.e., the knowledge and experience in these fields.

If these assumptions are true, specialists can successfully work in their fields throughout their active lifetime. But as we know, today progress is so quick that much of what is learned in even the best universities becomes obsolete in a few years. This persistent increase of information, diversity, and competition leads to an increase in the number of problems that have never been encountered before and that have a high D value. This fact devalues the above assumptions. As the philosopher Bertrand Russell noted, the limit of this narrowing process is the "perfect specialist's" abstruse knowledge about nothing. In contrast, over time a "universalist" knows less and less about an increasingly wide spectrum. In this case the limit of the process is zero cognitive penetration in everything. There are two processes of cognition: from the particular to the general and from the general to the particular. True cognition requires movement through these cycles, and each cycle advances human knowledge. Hence, a solver's education and a problem-solving approach must be knowledge-intensive and should not be limited to specific science or engineering fields. It should give a person who learned it an opportunity to become a virtual specialist-universalist centaur.

1.8 REQUIREMENTS

The effectiveness of a problem-solving methodology depends on quality or rationality of the solution, time spent on the problem resolution, and acceptance of a solution.

It is possible to formulate requirements for a problem-solving approach, and for a solver of inventive problems, based on experience with proposed methods and these criteria.

1.8.1 REQUIREMENTS FOR INVENTIVE PROBLEM SOLVING

There are three major requirements for a problem-solving methodology:

1. *It should have a mechanism for directing the solver to the most appropriate and strong solutions.*
   This requirement makes it possible to transfer a problem with a great number of possible variants to problem(s) having a few variants, i.e., to diminish the difficulty of the problems. In fact, this is the heart of the approach: it must reduce quickly the solution and problem spaces and eliminate a considerable number of barren trial steps, not coinciding with the strong solution's direction. This requirement resolves the key contradiction of the trial-and-error method.
2. *It should signal the most promising strategies.*
   During the solving process, this requirement leads to selecting steps based on a deep knowledge of peculiarities of the problem and the system under consideration, as well as each step itself. It is important that each step can be proven or at least that it has a statistically high probability of being the right step.

3. *It should provide access to important, well-organized, and necessary information at any step of the problem-solving process.*

   This requirement links the approach of solving inventive problems with knowledge-base research, systems that are conducted actively in the framework of Artificial Intelligence (a branch of computer science dealing with problem solving) [10]. Note that the current bottleneck is not information itself (see Section 1.7), but a presentation of information in a form useful for the problem solver.

1.8.2 NECESSARY QUALITIES FOR THE SOLVER OF NONROUTINE PROBLEMS

1. *A good problem solver must obtain very high quality solutions with a high level of recognition in a short time.*

   The solver needs to know an effective problem-solving methodology and be able to apply it to the problems he faces.

2. *A good problem solver has to know practically all relevant human knowledge.*

   Unfortunately, it is almost impossible for any one person to meet this requirement, so often a solver does not have the necessary knowledge for inventive or even difficult routine problems. In practice, this conflict is solved by teamwork. The minimal team includes a professional problem solver, a scientist (a physicist in about 90% of cases for inventive technical problems), and a professional engineer (or a few of them) in the problem's field. The features of team organization and management are discussed widely in the literature, so they are not considered in this book.

3. *A good problem solver should "turn off" his psychological inertia.*

   This requirement is a challenge for the cognitive sciences. Unfortunately, we are still far from understanding and finely regulating mental activities. Nevertheless, many methods to remove psychological inertia have been developed (see Chapter 11).

1.9 CONCLUSION

Inventive problems are a subclass of engineering problems and of creative nonroutine problems, and they cannot be resolved effectively by trial and error.

All known problem-solving methods consider the following critical steps:

1. Completely understand the problem.
2. Identify and evaluate all possible solutions.
3. Select the best solution.
4. Demonstrate that the best solution actually solves the problem and verify and validate the solution.
5. Document the problem-solving process.

It is proposed here to use the trial-and-error method to search for or select the best solution among others. All methods developed for nonroutine problems try to overcome the disadvantages of the trial-and-error method. These methods usually can help search for solutions of problems that are below the right bottom quarter C and the left upper quarter B of Figure 1.1. Although all methods for activation of creativity, search expansion, and decisions are not very powerful in their ability to solve non-routine, difficult, technical problems, it is nevertheless useful to know these methods.

Among the many methods of activation creativity, design theories, and problem solving developed around the globe during the twentieth century, only TRIZ is useful for solving most difficult problems with high D values from the right upper quarter of Figure 1.1. Although it is inaccurate to say that TRIZ can solve any technical problem today, the thousands of patents for technical problem solutions that used TRIZ confirm its power.

The knowledge of TRIZ is essential for any creative engineer and problem solver.

REFERENCES

1. Altshuller, G. S., *Creativity as an Exact Science: The Theory of the Solution of Inventive Problems,* Gordon and Breach Science Publishing, New York, 1984.
2. VanGundy, A. B., *Techniques of Structured Problem Solving,* Van Nostardam Reinhold Company, New York, 1992.
3. Polya, G., *How to Solve It: A New Aspect of Mathematical Method,* Princeton University Press, Princeton, 1973.
4. Osborn, A. F., *Applied Imagination: Principles and Procedures of Creative Problem Solving,* Scribner, New York, 1953, 1979.
5. Stein, M. I., *Stimulating Creativity: Individual Procedures* (vol. 1), *Group Procedures* (vol. 2), Academic Press, New York, 1974–1975.
6. Fisher, M., *The IdeaFisher: How to Land that Big Idea — and Other Secrets of Creativity in Business,* Peterson's/Pacesetter Books, Princeton, 1996.
7. Zwicky, F., *The Morphological Method of Analysis and Construction,* Wiley-Interscience, New York, 1948.
8. de Bono, E., *Serious Creativity: Using the Power of Lateral Thinking to Create New Ideas,* Harper Business, New York, 1993.
9. Wickelgren, W. A., *How to Solve Problems: Elements of a Theory of Problems and Problem Solving,* W. H. Freeman, San Francisco, 1974.
10. Russell, S. J. and Norvig, P., *Artificial Intelligence: A Modern Approach,* Prentice Hall, New York, 1995.
11. Higgins, J. L., *101 Creative Problem Solving Techniques: The Handbook of New Ideas for Business,* New Management Publishing Company, New York, 1994.
12. Buzan, T., *Use Both Sides of Your Brain,* Plume, New York, 1989.
13. Adams, J. L., *Conceptual Blockbusting: A Guide to Better Ideas,* Addison-Wesley Reading, 1986.
14. Jones, J. C., *Design Methods: Seeds of Human Futures,* Wiley-Interscience, London, 1976.
15. Kivenson, G., *The Art and Science of Inventing,* Van Nostrand Reinhold, New York, 1982.

16. Lumsdaine, E. and Lumsdaine, M., *Creative Problem Solving: Thinking Skills for a Changing World*, McGraw-Hill, New York, 1998.

17. Koberg, D. and Bagnall, J., *The Universal Traveler: A Soft-Systems Guide to Creativity, Problem-Solving, and the Process of Reaching Goals*, W. Kaufmann, Los Altos, 1978.

18. Kuecken, J. A., *Creativity Invention and Progress,* H. W. Sams, Indianapolis, 1969.

19. Buhl, H. R., *Creative Engineering Design,* Iowa State University Press, Ames, 1960.

20. Krick, E. V., *An Introduction to Engineering and Engineering Design*, Wiley, New York, 1969.

21. Sandler, B.-Z., *Creative Machine Design: Design Innovation and the Right Solutions,* Solomon Press, New York, 1985.

22. Helfman, J., *Analytic Inventive Thinking,* Open University of Israel, Tel Aviv, 1988.

23. Gordon, W. J. J., *Synectics, the Development of Creative Capacity*, Harper, New York, 1961.

24. Dewey, J., *How We Think*, D.C. Heath and Co., Boston, 1910.

25. Koestler, A., *The Act of Creation,* Macmillan, New York, 1964.

26. Mayer, Richard E., *Thinking, Problem Solving, Cognition*, W. H. Freeman, San Francisco, 1992.

27. Bush, G. Ya., *Creativity as the Dialog-Like Interaction,* PhD Thesis, Byelorussian State University, Minsk, 1989.

2 TRIZ Overview

2.1 WHAT IS TRIZ?

All known sciences (except mathematics and philosophy) can be classified according to three major groups: disciplines that study nature (such as physics, chemistry, biology, and geology), disciplines that study human behavior and society (such as psychology, economy, and sociology), and disciplines that study artificial objects (such as electron engineering, marine design, aerodynamics, and root cause analysis). The uniqueness of TRIZ is that it combines knowledge of all these groups:

- it uses some philosophical concepts of dialectics, materialism, and idealism for its roots,
- it uses the results of cognitive sciences for suppression of a solver's psychological inertia,
- it uses natural science effects and phenomena for improving artificial technical systems and technological processes, and
- it analyzes breakthroughs to recognize generic heuristics and design principles and to extract major trends in technique evolution.

As a science, TRIZ addresses the problem of determining and categorizing all regular features and aspects of technical systems and technological processes that need to be invented or improved, as well as of the inventive process itself. TRIZ is also concerned with deriving appropriate information from applied knowledge of the natural sciences and from practical experience (garnered from mostly patent records) in a form suitable for a user.

Any science passes through various stages during its development:

1. Description of phenomena.
2. Categorization in terms of apparently significant concepts.
3. Isolation and test of phenomena, with implied reproducibility by independent observers.
4. Quantification.

Even the first of these stages, known also as "natural history," is commonly acknowledged as "a science" when applied to a traditional area of study. Only the last of these stages, quantification, can lead to an "exact" science (e.g., physics), if it occurs with sufficient precision, repeatability, and "explainability," and if assumptions can be found to reduce the number of variables to be considered. In this sense, TRIZ will perhaps never reach the stage of an "exact" science. Hence, TRIZ is referred to as a *methodology* of inventive problem solving. Presently a major part of TRIZ is a set of descriptive and perspective statements (individual statements and their relationships) obtained from TRIZ information sources (see Chapter 8), mostly by induction and/or abduction:

- whose parts and elements have certain purposes,
- that are supported (substantiated) usually by patents examples,
- that each illuminate only certain aspects of inventing and problem solving,
- that are or can be learned from carrier of knowledge through practice.

After 50 years of development, TRIZ is still in its "childhood" because of political and economic situations that interfered with its progress. However, TRIZ is now a powerful methodology for technical problem solving that leads to enhancement of existing technique and strong acceleration of progress. Its power is evident by resolution in several thousand of difficult technical problems in the former Soviet Union and now Russia that it helped solve in the last decades, as well as those in the U.S., Europe, Japan, and other countries during the last few years.

The following sections define TRIZ, giving an overview of its sources, concepts, and heuristics, as well as the perspectives of TRIZ development. Unfortunately, there is no standard TRIZ terminology in either English or in Russian [1-3], so this book follows Altshuller's original terminology, replacing it with more accurate and common scientific terms when it seems appropriate.*

2.2 A DEFINITION OF TRIZ

Let us propose the following definition and then discuss its parts:

> TRIZ is a human-oriented knowledge-based systematic
> methodology of inventive problem solving.

Knowledge — TRIZ can be defined as a *knowledge-based* approach because

a) The knowledge about the generic problem-solving heuristics (i.e., rules for making steps during problem solving) is extracted from a vast number of patents worldwide in different engineering fields. TRIZ claims that it works with a relatively small number of objective heuristics that are based on the evolution trends of technique; this declaration is not proven and it

---

* One of the aims of this book is to establish a common scientific TRIZ terminology.

is based only on a statistical analysis of solutions represented in patents fund.

b) It uses knowledge of effects in the natural and engineering sciences. This large storehouse of information is summarized and reorganized for efficient use during problem solving.

c) It uses knowledge about the domain where the problem occurs. This knowledge includes information about the technique itself, as well as the similar or opposite systems and processes, technique environment, and its evolution or development.

Human-oriented — heuristics are oriented for use by a *human being*, not a machine. The TRIZ practice is based on dividing a technique into subsystems, distinguishing the useful and harmful functions of a technique, and so on. Such operations are arbitrary, because they depend on the problem itself and on socio-economic circumstances, so they cannot be performed by a computer. For the majority of problems we face, which are repeated again and again, it is reasonable to use computers. However, many problems occur only once (for example, during the conceptual design of a new technique), and for those it is ineffective to use computers; we would need to spend more time writing the computer program than would be required for a person to solve the technical problem. Therefore, we need to arm a human solver with an instrument for handling such problems.

Systematic — In the TRIZ defintion, *systematic* has two meanings:

1. Generic and detailed models of artificial systems and processes are considered within the framework of TRIZ special analysis, and the systematic knowledge about these systems and processes is important;
2. Procedures for problem solving and the heuristics are systematically structured in order to provide effective application of known solutions to new problems.

Inventive problems and solving — The previous chapter discussed some of the important aspects of problems, *inventive problems*, and *problem solving*. Primary TRIZ abstractions for inventive problem solving include:

- often the unknown step appears because of contradictory requirements for the system;
- often the unknown desirable situation can be replaced temporarily by an imaginary ideal solution;
- usually the ideal solution can be obtained due to resources from the environment or from inside the technique;
- usually the ideal solution can be projected from known trends of technique evolution.

The concepts of Contradiction, Evolution, Resources, and Ideal Solution are the main building blocks of TRIZ and as such are discussed in Part 2.

2.3   TRIZ SOURCES

The most important source for TRIZ has been patent and technical information. At
the moment TRIZ specialists have analyzed approximately 2 million patents worldwide that represent roughly 10% of all patents in the world. This extensive and
amazing work (I.V. Ilovajsky estimates that it took about 35 thousand man-years)
was possible because of the following factors:

- Engineers in the former Soviet Union were responsible to spend 8 hours
  at their work place but often had nothing to do (their regular salary did
  not depend on their effort, experience, or quantity and quality of work).
  Many of them, influenced by G. S. Altshuller and TRIZ, used this time
  to study patents. For example, Russian engineer and TRIZ specialist I. V.
  Vikent'ev went through all USSR patents (about 1 million patents at that
  time) collecting information for the list of geometrical effects. Altshuller
  himself studied more than 400,000 worldwide patents to create the Contradictions Matrix with 40 inventive principles.
- USSR patents have a simple structure: abstract, patent body, and at most
  a few claims (usually only one claim). Abstracts of all patents of developed
  countries (US, European Union, Japan, Germany, France, UK, etc.) and
  the former communist/socialist countries (Hungary, Poland, Czech and
  Slovak Republics, etc.) were translated into Russian.
- Several USSR federal departments, covering each branch of industry,
  published periodic reviews about the most important technical reports and
  scientific publications in numerous fields.
- The system of organizing engineering science simplified this task; in the
  former Soviet Union every person had almost unrestricted access to patent
  collections and the above mentioned reviews for free.
- This patent analysis followed a good procedure. The method for studying
  patents is described in Chapter 8, where classification of patents into
  different levels (from one, for obvious solutions, to five, for breakthrough
  achievements) is also presented.

The main result of these studies is the collection of TRIZ *heuristics* that helps
to solve the nonroutine problems. Often a TRIZ heuristic can help perform the most
important step in problem solving, filling the gap between the initial and desirable
situations. If a sequence of steps is needed to solve a problem, TRIZ offers quite
powerful instruments (e.g., ARIZ, Agents Method) to handle different aspects of
problem solving.*

During the analyses of patents from various engineering fields, Altshuller, and
later other TRIZ experts, discovered that very different technical systems and technological processes share similar peculiarities in their evolution. Western economists
extracted similar ideas from historic data. These studies of the history of technique
reveal general trends in the evolution of technical systems and technological processes. They have not only academic value, but they also help in the implementation

---

* TRIZ heuristics and instruments are often not distinguished because both perform the same function.

and timing of solutions of technical problems. The major results of these investigations are discussed in Chapter 7.

Another source for TRIZ is analysis of the problem-solving process itself. Such analysis shows wherein the problem-solving process psychological inertia has its greatest effect and also helps researchers to find ways to overcome it. Some methods for controlling and overcoming psychological inertia are presented in Chapter 11. A set of these methods, often called Creative Imagination Development, together with the Strategy for Creative Personal Growth, serve another TRIZ goal — development of creative and inventive individuals and groups [3, 4].

The last source for TRIZ to be discussed here is the body of all human knowledge about nature. The *Lists of Effects* are extracted from various scientific fields, such as physics, chemistry, biology, mathematics, and materials science, and prepared in an "engineering" format useful for developing new techniques. For example, TRIZ provides pointers for substances that can perform needed functions. The Lists of Effects (currently containing about 6,000 effects) play an intermediate role between TRIZ heuristics and knowledge-base sources.

2.4 MAIN TRIZ HEURISTICS AND INSTRUMENTS

The most attractive part of TRIZ is that it uses a relatively small number of heuristics (developed by Altshuller and other TRIZ experts [1-3]) for solving inventive technical problems with various degrees of difficulty (see Chapter 1 for discussion of difficulty).

According to Judea Pearl, an expert in Problem Solving and Artificial Intelligence:

> heuristics are criteria, methods, or principles for deciding which among several alternative courses of action promises to be the most effective in order to achieve some goal. They represent compromises between two requirements: the need to make such criteria simple and, at the same time, the desire to see them discriminate correctly between good and bad choices.
>
> A heuristic may be a **rule of thumb** that is used to guide one's actions. *For example, a popular method for choosing ripe cantaloupe involves pressing the spot on the candidate cantaloupe where it was attached to the plant, and then smelling the spot. If the spot smells like the inside of a cantaloupe, it is most probably ripe. This rule of thumb does not guarantee choosing only ripe cantaloupe, nor does it guarantee recognizing each ripe cantaloupe judged, but it is effective most of the time.*
>
> It is the nature of good heuristics both that they provide a sign which among several courses of action is to be preferred, and that they are not necessarily guaranteed to identify the most effective course of action, but do so sufficiently often. [5, emphasis added]

Later chapters present almost all of the TRIZ heuristics and instruments that are in the public domain and that had passed tests for applicability at the time this work was printing. Unfortunately, TRIZ heuristics for Creation, Genesis, Expanding and Convolution, and Trimming of Technical Systems and Technological Processes are not considered here because they required detailed Functional-Cost Analysis for the techniques and needs under consideration. Such analysis and instruments, together with Diagnostic Problem Solving (a set of templates and logical procedures for eliminating and preventing quality problems), demand more advanced examination of the technique and its functions and behavior than presented here. However, Structure-and-Energy Synthesis of Systems is outlined in Chapter 16 and some aspects of Expanding-Convolution of Technique are briefly discussed in Chapter 7. Complete description would require space comparable to the volume of this book. Readers interested in this Advanced TRIZ as well as who want to sharpen their knowledge and experience in TRIZ heuristics and instruments can find more information on the Internet: http://www.jps.net/triz/triz.html. Below, however, is a brief overview of the most popular TRIZ heuristics and instruments.

Preliminary Analyses can void trade-off solutions of problems containing contradictions and can help clarify important information about the technique and constraints of forthcoming solutions.

The *Contradiction Matrix* consists of technical contradictions between the characteristics to be improved and the characteristics that can be adversely affected. It also has a few inventive principles in each cell that may help resolve the contradictions.

Separations Principles help resolve the general physical contradictions between the opposite characteristics of a single subsystem.

Substance-Field (Su-Field) *Analysis* is a modeling approach based on a symbolic language that can record transformations of technical systems and technological processes.

The *Standard Approaches to Inventive Problems* (*Standards,* for short) is based on the observation that many inventive technical problems from various fields of engineering are solved by the same generic approaches. The Standards contain typical (from the TRIZ standpoint) classes of inventive problems and typical recommendations on their solutions, which usually can be presented in terms of Su-Field Analysis.

Algorithm for Inventive Problem Solving (ARIZ in its Russian acronym) is a set of sequential logical procedures for eliminating the contradictions causing the problem. ARIZ is considered one of the most powerful and elegant instruments of TRIZ. It includes the process of problem reformulation and reinterpretation until the precise definition is achieved, and the logical and disciplined process of solving the problem with iterative use of most of the TRIZ heuristics. It is very "solution neutral"; it removes preconceived solutions from the problem statement.

Agents Method is a graphical-logical procedure for implementing forward, backward, or bidirectional steps between the initial and desirable situations when they can, respectively, be presented as the correct statement of a problem and the Ideal Final Result.

As mentioned in Chapter 1, the most common way to solve problems is still the trial-and-error method. There have been several attempts to improve this approach with various methods of creativity activation. While studying TRIZ, these methods and analysis of science-fiction literature are used for developing inventiveness and creative imagination and for suppressing psychological inertia. TRIZ heuristics and instruments also have "built-in" mechanisms for overcoming psychological inertia.

David Pye, Professor of Design at the Royal College of Arts (England), wrote that "most design problems are essentially similar no matter what the subject of design is" [6]. This is the initial point from which TRIZ grew 50 years ago. Consequently, the solution process is supported in TRIZ by an expansive set of abstract problems and solutions that can be *Analogs* of the problem under consideration (usually such sets are created by each TRIZ expert; at the moment they are not organized). These effects and analogs can be beneficial for effective solutions of complex engineering problems because different inventions are usually transferable from one discipline to another. This means that about 95% of the inventive problems in any particular field have already been solved in another field. Thus, analogies are very important in TRIZ. The *structures of problems* help to navigate among various Analogs. In addition, techniques have been created in the framework of TRIZ for many operations that are always present in analysis and improvement of a technique, such as the statement of problems, search of resources, and reduction of contradictions topology.

The following three points should be stressed:

- Each TRIZ heuristic or instrument presented here has been tested numerous times; a few hundred solutions have been obtained with each heuristic during and after these tests.
- Each new heuristic or instrument must be tested very carefully.
- It is impossible to solve high degree of difficulty problems with only the knowledge of TRIZ heuristics and instruments without learning the whole TRIZ methodology and sharpening solver skills with real and educational problems.

Additional heuristics and instruments, such as operators, evolution trends and paths, and various analyses are under investigation now by TRIZ experts.*

TRIZ research is the process of discovering principles, and design or invention is the process of applying those principles. A TRIZ expert discovers a heuristic — a generalization — and an engineer or inventor prescribes a particular embodiment of this heuristic to suit the particular matter of concern. These heuristics and instruments, therefore, represent a system for handling different steps during problem solving. It is important to remember that these heuristics and instruments are "for thinking" and *not* tools to be used "instead of thinking."

2.5 TRIZ BRANCHES

Until the end of the twentieth century, TRIZ had been developed exclusively in the former Soviet Union by a couple of dozen TRIZniks (or TRIZovszev, as specialists in TRIZ usually name themselves in Russia). Many TRIZniks also teach TRIZ, so several thousand engineers have become familiar with this methodology in the

---

* The research cycle for a new TRIZ heuristic or instrument is relatively long compared with experimental and theoretical studies in the natural and technical sciences; consequently, TRIZ does not progress as quickly as other scientific and engineering disciplines. Its progress has also been slowed by the absence of support for TRIZ research.

former Soviet Union, where TRIZ was as popular as brainstorming was in Western countries. TRIZ was unknown outside the USSR because of linguistic and political barriers. However, during the last 50 years a few interesting methods have been proposed and implemented in the US, Japan, and Europe: Value Engineering, Design Theories, Quality Function Deployment (QFD), the Taguchi Method, and the Theory of Constraints are just a few [6-9]. Some TRIZ projects are based on creative re-engineering of particular ideas that have been developed in the framework of these methods. Such development led to the creation of new TRIZ branches, such as TRIZ-based Function-Cost Analysis or TRIZ+QFD. Most of these TRIZ branches developed recently, and it is still too early to review these activities thoroughly.

Toward the end of the 1980s, TRIZ became so powerful that it was applied to a wide range of nontechnical problems, such as management, education, journalism, public relations, and investment. Unfortunately, the scope of this book limits us to the technical applications of TRIZ. Moreover, these so-called "soft" applications have not been thoroughly evaluated and extensive research is still needed in this newer area of TRIZ development.

The primary area of TRIZ is still the technical problem that occurs in an existing technique or during development of a new technical system or technological process. This area of focus can be further divided according to the following categories:

A. Problem solving in an existing technique encompasses
   1. Improvements (Qualitative or Quantitative)
   2. Troubleshooting (such as Yield Enhancement or Quality)
B. Problem solving during development of something new encompasses
   1. A fresh mixture of subsystems that either exists or can be created (Creation, Synthesis)
   2. Genesis (theoretically from only knowledge of a human need)
   3. New use of an existing technique (Discovery)
   4. Prevention of malfunction of an existing technique (Diagnostic)

TRIZ works well with nonroutine problems that belong to these classes because of:

- contradiction overcoming first of all for A1 problems,
- diagnostics problem solving first of all for A2 and B4 problems,
- functional search of a new system or process first of all for B1 problems,
- functional analysis first of all for B3 and B2 problems.

2.6 FUTURE OF TRIZ

Hard (technical focus) and soft (nontechnical focus) TRIZ is an attempt to provide a general, comprehensive scientific methodology for creating novel ideas (thinking "outside the box"), breakthrough concepts, and innovative solutions. As such, it is a revolutionary methodology that should be developed as a science of innovation engineering.

Future developments of TRIZ can be categorized according to the following three attributes:

- area of validity (general, appropriate to any technique, vs. special, for a particular area of engineering),
- approach (theoretical conclusions obtained by induction and abduction vs. experimental studies of whole patent fund and other sources of technical information), and
- content (theoretical results vs. applied case studies).

Like any other science, TRIZ can be developed in one of two directions:

- by an empirical way of observing, describing, abstracting, generalizing, and formulating hypotheses and guidelines, or
- by postulating a set of axioms and postulates, proving theorems, modeling, refining, and formulating a theory.

Right now the first direction dominates because of the strong influence of Altshuller and the activity of his students (S. S. Litvin, V. M. Gerasimov, Yu. P. Salamatov, A. I. Gasanov, V. A. Mikhajlov, A. V. Podkatilin, A. B. Selyuczky, and M. I. Sharapov, among others), although some research along the second direction has been performed by S. D. Savransky, A. M. Pinyaev, G. A. Yezersky, V. N. Glazunov, N. N. Matvienko, and others. However, these approaches are not mutually exclusive, and a workable system can be obtained from research in both directions. I think that the recommended result achieved by degrading from a recognizable theory (i.e., by the second way) should deliver better results in TRIZ understanding and usage.

The following areas are where TRIZ application can be developed:

- a cornerstone for a theory of nonroutine problem solving;
- an undertstanding of theory and history of techniques and engineering;
- a core for coordinated and extended insights into the genesis of techniques;
- a starting point and coordinating concept for special theories of technical systems and technological processes;
- a source of knowledge for engineering design;
- a base for the psychological sciences dealing with creativity;
- an essence for the processes and procedures of inventive engineering;
- a starting point for new software: databases, algorithms, and artificial intelligence and expert knowledge-based systems;
- a kernel for an integrated educational approach for engineers;
- a base for communication and understanding between engineers from different fields, managers, and nontechnical persons.

REFERENCES

1. Altshuller, G. S., *Creativity As an Exact Science: The Theory of the Solution of Inventive Problems,* Gordon and Breach Science Publishing, New York, 1984.
2. Sklobovsky, K. A. and Sharipov, R. H., Eds., *Theory, Practice and Applications of the Inventive Problems Decision*, Protva-Prin, Obninsk, 1995 (in Russian).

3. Gasanov, A. I., Gohkman, B. M., Efimochkin, A. P., Kokin, S. M., and Sopel'nyak, A. G., *Birth of Invention,* Interpraks, Moscow, 1995 (in Russian).

4. Altshuller, G. S. and Vertkin, I. M., *How to Become a Genius. Life Strategy of a Creative Person,* Belarus, Minsk, 1994 (in Russian).

5. Pearl, J., *Heuristics. Intelligent Search Strategies for Computer Problem Solving,* Addison-Wesley, Reading, 1984.

6. Pye, D., *The Nature and Aesthetics of Design,* Herbert Press, London, 1983.

7. Wilson, P. F., Dell, L. D., and Anderson, G. F., *Root Cause Analysis: A Tool for Total Quality Management,* ASQC Quality Press, Milwaukee, 1993.

8. Dettmer, H. W., *Goldratt's Theory of Constraints,* Quality Press, Milwaukee, 1998.

9. Miles, L. D., *Techniques of Value Analysis and Engineering,* McGraw-Hill, New York, 1972.

Part 2

Main TRIZ Concepts

3   Technique: A Resumé

CONTENTS

3.1   INTRODUCTION

Human beings organize various natural and artificial objects and activities to bring about what is necessary or desirable. The limitations imposed by the laws of nature must be respected, but some natural processes and some properties of natural objects can be improved, degraded, increased, reduced, accelerated, or retarded. By making these changes we are creating artificial systems and processes. Any artificial object within an infinite diversity of articles, regardless of its nature or degree of complexity, can be considered a *technical system* (TS). Any artificial single action or consequences of procedures to perform an activity with assistance of a technical system or a natural object can be considered a *technological process* (TP). Technical systems participate in technological processes in order to satisfy the needs of human beings or another TS. On another hand, any technological processes occur because of the existence of one or several TS. Therefore, TS and TP are bound to each other — they supplement each other. We can also look at this as a set of parts with *links* in *space* (TS) and as a set of parts with *links* in *time* (TP). In a TS, the smallest part, or link in space, is often called an *element* of the system. In a TP the smallest part, or link in time, is often called an *operation*. Elements and operations forming a TS or TP are usually called *subsystems* in TRIZ.

The development of TS and TP is often similar, and they share a number of common properties, which allows us to consider them as a unified group — *technique*.

The systematic study of techniques (i.e., technical systems and technological processes) and their *functions* are the background and foundation of TRIZ. The *system* approach to technique was proposed by Russian philosopher Alexander A. Bogdanov (Malinovsky) at the beginning of the twentieth century [1] and developed under the influence of the systematic paradigm introduced in science by German researcher Ludwig von Bertalanffy [2] in the 1930s. The idea of functions was introduced by the English economist Adam Smith, who proposed to assign specific people to perform different functions. The concept of functions was then introduced in the world of technique by American purchase agent Lawrence D. Miles, who created value engineering in the middle of the twentieth century [3].

This chapter briefly characterizes the TS and TP based on results obtained in the frameworks of TRIZ, design theory, artificial intelligence, value engineering, and other disciplines [4–16]. We will also note some limitations of the current state of the art for description of technique in modern science.

3.2 INPUTS AND OUTPUTS — RAW OBJECTS AND PRODUCTS

All techniques have *inputs* and *outputs* from/to other techniques, humans, or the *environment*. Relationships between input and output of a technique can be static:

- combining: number of inputs > outputs,
- equal: number of inputs = outputs,
- dividing: number of inputs < outputs.

They can also be changeable:

- volatile: number of outputs depends on the number of inputs,
- fickle: the ratio of inputs to outputs fluctuates or depends on time.

The *input* and *output* also represent the relationship between the super-system or environment and the subsystem or technique under consideration.

The **raw object** (often also called raw material, although it can be a field) is the main input in a technique, while the **product** is the main output. Much research has distinguished three classes of raw objects and products [6]:

- *substances* — any matter with nonzero mass and that occupies nonzero space, for example, cup, car, piece of plastic, TV.
- *fields* — consist of many energy carriers regardless of their nature and mass, for example, radiation field, smell field, thermal field. During a

technological process, various types of fields are transformed into others, and/or their parameters are changed.

- *information* — commands (requests, desires, rules, normative statements) and data (verbal, graphical, and symbolic/numerical). The transformation concerns the form, quality, quantity, and location of information within the information carriers.

The following points should be noted:

- From a TRIZ viewpoint, information is not independent. It cannot be created and moved from one location to another without simultaneously moving a substance (at least in the form of subatomic particles) and/or a field (at least in the form of changes of field potential). Hence, classical TRIZ does not distinguish information as an independent raw object or product and works only with the first two classes of raw objects and products. Perhaps this is why TRIZ is not yet useful for software development.
- Both substances and fields are considered in TRIZ regardless of complexity. For example, both an airplane and a pencil can be viewed as substances, and both a gravitation field and a flow of multi-component chemically active liquid can be viewed as fields.
- In general, no one class can be changed without affecting another. This matter-field dualism is well known in physics and used in Su-Field Analysis in the framework of TRIZ [8]. Often it is useful to focus attention on the most important aspect and class of the product for a particular TP or TS and to give only passing regard to another class.
- Substances include those of biological origin (living human beings, animal or plant life-forms). Although their state (for example, sick versus healthy), properties (temperature, weight), or location can be transformed by TS or TP, modern TRIZ does not work with these objects because biological substances are often very specific. Hence, we consider only inanimate substances and fields in this book.
- The product of one TS or TP can be the raw object for other technical systems and/or technological processes.

Product and raw object are the generic terms used in TRIZ for all substances and fields that are changed during a TP by a TS. The transformation from the entry state of the raw object (at the input) to its desired state of the product (at the output) takes place during a period of time in a certain space and in some environment.* A raw object is the substance, field, or information to be changed, moved, measured, etc. The transformation may affect the raw object by altering its structure and internal properties (for example, material hardness), its external characteristics (for example, form or shape), its location (space, for example, by transportation), its time (for

---

* Environment is everything outside a technique. The environment for a technique most often is formed by other TS, TP, and nature that can be considered as external factors for the technique under consideration. The environment can be used for improvements of the technique or its parts, as shown later in this book.

example, by storage), and/or information about it (for example, by a measurement). A product, in TRIZ understanding, is not only finished goods at output, but also the changed raw object at all stages of its transformation into goods or a subsystem that acts under another subsystem of the same TS.

3.3   PARTS OF A TECHNIQUE

Before we characterize TS and TP, we describe TRIZ ideas about the most important *parts* — often referred to as subsystems and links — of a technique. The "from bottom to top" analysis used here corresponds to the inventive problems for diagnosis of a technique and its modification.

3.3.1   SUBSYSTEMS

As already noted, technique might consist of subsystems arranged in some way in space (systems, i.e., devices, machines, substances, and elements), and/or of subsystems interconnected in time (technological processes, i.e., working steps, methods, and operations). The subsystems are themselves systems, in turn, for subsystems of their own, and so on down to some granular level where the elements and operations can be labeled. TRIZ reasoning does not require that elements and operations be invisible. Various points of view with different granularity on the same technique can coexist.

An element (operation) can also be regarded as a system; likewise, a system can be regarded as an element within a larger system — a *super-system*. All techniques have subordinate subsystems and also serve as a part of a higher-level super-system (see Figure 3.1 below). Furthermore, the super-system or environment can itself be a technique for a subsystem of the technique under consideration. For example, the technique "TV production," as a TP, is part of the super-system "electronic industry," yet it also includes a series of subsystems: manufacturing of special cathodes, assembly of glass tubes for screens, and other processes. TV production is part of the TP television which, in turn, is part of a super-system of higher rank — information distribution, visual entertainment, or propaganda — and so forth. On the other hand, TV, as a TS, may be part of the super-system electromagnetic wave receivers, and it consists of its own subsystems: screen, amplifier, acoustic channel, plates, connectors, etc. One of these TS — the amplifier — has its own subsystems: transistors, wires, capacitors, etc. TV is a part of the super-system television that includes such TS as TV stations, satellites, cable companies, etc.

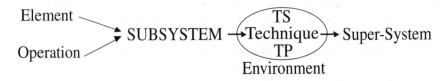

FIGURE 3.1   Technique in relation to its environment and sub- and super-systems.

Therefore, a technique (technical system or technological process), its subsystems, and its super-systems form a simple *hierarchy*, i.e., arrangement of parts in a series from lowest to highest in a multilevel structure. Any TS can be divided into subsystems of various granularity degrees that depend on the specific field of engineering. For a TV, we usually do not disjoin the system to the levels of its individual materials or even individual radio elements, such as transistors, resistors, etc., except to understand some failure. A material scientist, on the other hand, developing a new nano-technology will usually discriminate down to molecules and atoms.

Choosing the correct granularity degree is important in TRIZ analyses of TS and TP. The success of such decision-making depends on the skills of the problem solver. Often the decomposition into subsystems is fairly determined for a particular technique, but determination of its super-systems may depend on the viewer's perspective of the TS or TP.

In this book, subsystems are considered the central object of TRIZ. This selection reflects complexity of current technique that leads to the situation when an engineer (or a problem solver) does not deal with whole TS or TP. On the other hand, such selection allows us not to care about other vitally important subsystems of any technique, such as controls and energy sources.

3.3.2 LINKS

Links connect individual elements and operations, thus forming subsystems and then a technique. Almost every link can be considered an input or an output, for example, cables, pipes, wires. A link is the real physical channel for transition of fields, substances, and/or information and relation and interaction between subsystems. (As already noted, information cannot be nonmaterial; it always is a substance or field.) The main condition for technique operation is the gradient of field(s) and/or substance(s) between elements or subsystems (that can be viewed as a deviation from the thermodynamic equilibrium that is due to the Onsager principle, which is well known in physics). With a gradient, the moving force arises, causing the flow of substances or fields. Some examples are:

- temperature gradient causes heat flow (heat conductivity),
- gradient of concentration causes the flow of substance (diffusion),
- velocity gradient causes the flow of momentum,
- gradient of electrical field causes electric current.

Often it is required to organize the flow with the gradient of another field, for example, flow of substance at temperature gradient. A flow can be facilitated by a substance (pipe, shaft, gear), by a field (heat, electric), or by a substance-field mixture (smelled particles, magnetic fluid, information signals, luminescent gel).

The main types of links [5] are:

- one-way (directed)
- two-way (totally conducting)

- neutral (insensitive to direction)
- delaying (with delay in time)
- positive (increasing power at increasing "difference in potentials")
- negative (decreasing power at increasing "difference in potentials")
- zero
- reflexive (arising under effect of internal cause)
- selective (filtering out unnecessary flows)
- counter-link (proportionally dependent on the state of elements between which the link occurs)
- positive feedback (as power of one link increases the power of other also increases), the mechanism of mutual stimulation of functions, which leads to build-up of a process
- negative feedback (as power of one link increases the power of other decreases), the stabilizing mechanism, which leads to equilibrium or oscillations around the point of equilibrium
- double negative feedback or mutual oppression (as power of one link decreases power of other also decreases), which leads to unstable equilibrium resulting in strengthening of one side and suppression of the other

Main link characteristics are:

- type of substance or field transfer through a link
- transparency (intensity of substance or field flow)
- bandwidth (amount of information, substances, or fields that can pass through a link in the unit of time)
- accuracy (grade of substance or field change during transfer through a link)

In addition, desired as well as undesired links in a technique can be

- functionally necessary — required for execution of the technique's functions,
- auxiliary — to increase the reliability of the technique's function(s),
- harmful, surplus, excessive.

Depending on the type of connection, links can be in series, parallel, linear, circular, star-like, transit, branched, etc.

Links can also be distinguished according to the following:

- mechanical — rigid or soft for position/movement of a subsystem or element in space
- electrical — conductive connections or isolations
- thermal — heat transmitting or insulating
- chemical — aggressive or neutral, acidic or basic, oxidizing or reducing, whereby any aggressive link can lead to chemical action and corrosion
- magnetic — field-exciting or magnetically screening
- optical — visual, infrared, ultraviolet beams
- combined — all mentioned above plus biological

3.4   TECHNIQUE: TP AND TS

The purpose of any technique is to perform some function in order to transfer a raw object into a product. Attributes always connected with the term *technique* include goals, behavior, links, subsystems, input, output, structure, property, state, hierarchy, parameters, organization, etc.

All techniques meet the following conditions:

* are designed for a purpose, i.e., execute useful functions;
* have a set of characteristics or properties and their parameters, the values of which represent the state of technique;
* are organized in space and time (subsystems have relations with each other).

Absence of even one of the above conditions does not allow us to consider an object as a technique.

We describe technique here in a top-to-bottom scheme — we discuss first the TP, then TS, and then their peculiarities. This order is typical during technique synthesis and genesis, redesign, and analyses which correspond to the inventive problems for contradictions, trimming, new usage of known TP or TS, etc.

3.4.1   TECHNOLOGICAL PROCESSES

A TP is a uniquely determined sequence of individual steps (operations) and activities, from input to output, needed to manufacture a product from the particular raw objects during a finite period of time.

TP can be simple and consist of only one working operation, or it can be very complicated and consist of several thousand operations and interrelated assets. According to V. Hubka and W. Eder [6], most TP have an internal structure of operations that can be described by a rule, prescription, regulation, instruction, procedure, etc. A TP can be either *rigid* — its sequence of operations is determined in advance, once and for all times, or it can be a *flexible TP with adaptive control* — the sequence of operations depends on the results of a preceding operation. For a flexible TP, certain conditions from the preceding operations must be fulfilled before the next operation can be started. In addition to being connected by rules, TP operations are also connected by substance and field links, such as compatibility of outputs from one operation with inputs to a following one.

The number of known factors affecting a TP is very large; consequently, any technological process can be categorized from various viewpoints:

* technology applied (chemical, mechanical, etc.)
* the primary and other actions, which are based on physical, geological, biological, and other phenomena
* the principal classes of raw objects and products (substance, field, and information)
* the location and/or time

TP can also be categorized according to other characteristics, such as economical, ecological, etc. This list can be continued due to existence of numerous different and similar types of technological processes. One point should be stressed: time is as important for a TP as structure is for a TS. A TP can be continuous and discrete in time. The major time characteristics for a TP are intervals, frequency, and phenomena repetition that govern causality.

In order to realize a TP, inputs are needed in the form of technical or natural systems, raw objects, and various other assets, such as energy, humans, and auxiliary substances (for example, lubricants, catalysts). TP can involve many raw objects of diverse substances and fields. Similarly, outputs in addition to the desired product appear from the TP, such as waste substances (dust, smoke, exhausted liquids) and unwanted fields (heat, vibration, noise). These quantities are commonly termed desired and unwanted secondary inputs and secondary outputs.

3.4.2 TECHNICAL SYSTEMS

Various definitions of the system concept and particularly TS can be found in the literature. Most authors agree that a system is a finite set of parts (components, subsystems) collected to form a conceptual whole under certain well-defined rules, whereby certain definite relationships exist among the parts and with its environment. The following definition is accepted for technique in TRIZ [5, 8].

> A system is a set of orderly interacting subsystems intended for executing specific functions. It possesses behaviors and properties that cannot be reduced to the behaviors and properties of its separate subsystems.

The notion of system can also depend on what, if any, interest we have in the particular system feature. For example, the fragments of a crashed car would not be considered a system by most people, but it would be a system for the policemen investigating an accident. Generally, any system has characteristics not equal to a simple summation of its constitutive elements or subsystems. In some cases, the system exhibits properties that are even opposite to properties of its subsystems; i.e., the so-called *system gain* arises "suddenly" at formation of the system. (This idea goes back at least as far as to Aristotle, who stated that the whole is more than the sum of its parts, in regard to material objects and ideas.) For example, the TS can opener has the system gain of cutting metal of specific surfaces, whereas none of its elements taken separately exhibit such a cutting property. (System gain is discussed more fully later in this chapter.)

All TS can be classified as dynamic (TV, car, kitchen mixer, piano) or static (building, optical lens, furniture) depending on the TS state in time or behavior (and this is a relative statement with respect to the observed properties and human scale of time). Often a TS consists of both static and dynamic subsystems. Dynamic TS involve some kind of exchange with the environment and transformation of substances and fields. Static TS have no explicit energy- and mass-exchange with the environment, although they also affect the environment; they are characterized by accepting external actions (mostly energy) and transmitting (sometimes with transformation) them to the subsystem that has the role of casing (e.g., foundation for a house). The patterns of such external actions can conditionally be considered unchangeable in time ("frozen") flow of energy in a static TS. TRIZ and design theories development has been mostly based on studies of dynamic TS [4–10].

It is possible for a nontechnical system to contain isolated elements (i.e., elements with no relationships to other elements or subsystems — relationships equal zero). However, such a situation cannot occur for TS (as discussed fully later). Most TS consist of a few mandatory subsystems, as described below.

3.4.2.1   Major TS Subsystems

All TS include the *working tool* that performs the primary function. The main operation fulfills the role of the working tool in a TP. The working tool is the only part of a technique that is useful for a human being because it is what can satisfy the purpose of the technique. More precisely, the function, rather than its carrier, is what is useful for a human being.

As a rule, a working tool is useless without a source of energy and/or engine that enables the working tool to perform its functions.

Other important subsystems that are present in practically any TS are

- transmission — transfers the initial energy to the working tool, sometimes with energy transformations (for example, chemical into mechanical).
- controls — regulate and manage other subsystems; make it possible to change energy and substance flows between other subsystems of technique. These changes (discrete or continuous) may be preset or initiated by events. At least one subsystem is usually controllable in modern techniques.
- casing — protects TS and environment from each other, provides safety, furnishes aesthetics, and maintains the structure and shape of the TS. The shape of the casing is the most important for technique users for comparing with shapes of other major subsystems.

Table 3.1 illustrates the roles of basic subsystems in a TS.
Note that

- the set of different energy sources and kinds of engines is quite limited by mechanical, electrical, chemical, thermal, biological (human and animal), nuclear, and a few other generic members;
- the controls can be divided into human-regulated, automated, and computerized on the one hand, and into analog and digital on the other. The control by deviation of parameter is the most widespread and reliable mode of controlling functions performance.

If an energy source is a subsystem in a working tool, its control and planning functions are executed (partially or completely) by the corresponding subsystem of the working tool as a result of subsystem evolution. If a control subsystem is a part of a working tool or an energy source, its controlling functions are executed (partially or completely) by this part of the working tool or the energy source as a result of subsystem evolution. (Chapter 7 discusses technique evolution.)

TABLE 3.1
Basic Subsystem Roles

Basic subsystem	Type of TS where present	Function
Working Tool	Any TS	Directly interacts with a raw object in order to obtain the needed product
Engine	Any dynamic TS	Generates the initial energy usually in the form suitable for the working tool
Transmission	Any TS with the engine separated from the working tool	Carries the energy to the working tool from engine
Control	TS whose functioning is associated with changes of some characteristics or parameters	Gathers information, generates a control action, and transmits it to or from the working tool and other subsystems
Casing	Any multi-subsystem TS for which the mutual arrangement of subsystems is functionally necessary	Provides shape of TS and the mutual arrangement of its subsystems in space (for example, a car body)

3.5   ATTRIBUTES

Subsystems and their structures (spatial and temporal) and properties are the most significant attributes for the technique itself. However, the goals, purposes, cost, and performance are the most significant attributes for the society that uses this technique. The technique's functions bridge these two groups. (Other attributes also important for TRIZ are discussed later in this chapter.) We use here a bidirectional scheme for attributes description, which is also used for solving various inventive problems. A good solver should be quite flexible, use all schemes (top to bottom, bidirectional, bottom to top), and be able to shift among the different representations.

In order to solve the inventive problem, we should be able to operate with knowledge about the attributes of techniques and their purposes. Whole knowledge can be grouped into the following epistimological types [12]:

- subsystems knowledge that describes which elements or operations compose TS or TP and their properties;
- structural knowledge that describes the topology of a technique, i.e., how subsystems compose TS or TP and how they are linked to each other;
- behavioral knowledge that describes the activities of a technique, i.e., how its subsystems act and interact in term of attributes that characterize their state and the laws of nature that govern their acts;
- functional knowledge that describes the responsibilities of a technique and its subsystems to goals, purposes, and demands;
- teleological knowledge that describes the missions assigned to a technique by its designer under the influence of users' or social demands and the existence and operational conditions to its subsystems that facilitate fulfilling the purpose of the technique creation;

- empirical knowledge that describes the explicit representation of technique properties through hypothetical associations derived from observation and experimentation, as well as experience and subjective competence that human experts usually acquire through direct interaction with the technique.

The epistemological types defined above can be approximately grouped into three categories. Subsystems, structural, and behavioral knowledge are fundamental knowledge used for objective reasoning about a technique using the terms of natural sciences (physics, chemistry, etc.). TRIZniks use the fundamental knowledge of scientists (usually physicists) during problem solving. Functional and teleological knowledge are interpreted knowledge for subjective reasoning about a technique using the sociocultural terms of liberal arts (economy, management, etc.). For example, when we say "subsystem A works as... for ..." we express a function, i.e., a relationship between a subsystem's behavior (structure for static subsystems) and a goal, which is generally not valid for other subsystems of the same type in the same or other techniques. Finally, empirical knowledge is a separate category which concerns an explicit statement of technique properties and may refer to both fundamental and interpreted knowledge. TRIZniks use empirical knowledge of the particular industry's experts during problem solving.

3.5.1 GOALS

Every technique serves a particular purpose (intended output of a technique), which can be described with reference to a specific goal or set of goals. A common goal for TS and TP is transformation of a raw object into a required product. In the world of techniques, the engineer/designer formulates the goal as a hypothetical result that will satisfy a demand. A goal, therefore, is a state of affairs that is thought to be possible, and this state of affairs should preferably be realized. Some goals may be hierarchically related to one another in the form of "super-goals" and "subgoals" (or partial goals). A subgoal can concretize a super-goal, but usually it represents a means to realize that super-goal. The specification of the goals assigned to it by a designer under a user's influence is named the teleology of a technique. Generally, a goal may be decomposed in several different ways; i.e., there exist alternative decompositions of the goal into constituent subgoals. Therefore, the definition of a goal can easily be visualized through an AND/OR tree, where OR nodes represent alternative decompositions of a parent goal and AND nodes represent subgoals which are all needed to achieve the parent goal.

Often goals themselves stem from the understanding statements of the problem (also know as the demand, which is a thing to be obtained or done). Goals are differentiated from the function. A function is realization of a demand through a technique. A demand can be satisfied in different ways, i.e., by various techniques. Since there may exist alternative decompositions of a single goal into subgoals, the correspondence between goals and function is "many to many." In other words, the same function can be used to achieve different goals, while a single goal might be potentially achieved by different functions. Nevertheless, our society chooses only one or very few ways of demand realization by minimal efforts; i.e., a technique should have minimum mass, size, and energy consumption, because mankind always has limited resources.

In a world of modern techniques, the goals are assumed to be achieved by functions through physical phenomena or through information processing. The information level can be viewed as the result of an interpretation performed upon an underlying physical level. Thus, information goals always refer to appropriate physical goals which constitute their background. Nevertheless, it seems that TRIZ presently does not resolve problems involving the achievement of information goals (such as software development) or problems involving the fulfillment of physical goals (such as hardware development).

3.5.2 PROPERTIES AND STATE

Every technique and each of its elements, operations, subsystems, and links possess a set of particular constants and properties that define them. A *property*, or characteristic, p_i is any attribute that is held by an object and that describes or identifies that object. All elements, operations, and subsystems in the technique, as well as the technique itself, possess physical and pragmatic properties. Pragmatic properties are goal-oriented, such as suitability for manufacture and transportation, heat-insulating ability, stability, and corrosion resistance. Physical properties include characteristics and constants of substances and fields. They can be grouped as

- universal constants — independent of time and the subsystem itself (e.g., electron charge)
- general constants — depend on the subsystem and do not change in time (e.g., the speed of sound in pure water)
- fixed parameters — specific attributes of the subsystem (e.g., the number of cogs in a gear)
- variable parameters — current value of a property with reference to the physical phenomena in which it can participate (e.g., voltage drop across a resistor, speed of car)

The principal property of any technique is its ability to do something useful (to perform its functions) in order to satisfy the purpose of the technique. The ability of a subsystem to do something (to *behave* — discussed in the next section) can be expressed by quantitative and qualitative defining, theoretical, and phenomenological equations, ε_i, with constants, parameters, and properties of the subsystem and its environment. The defining equations stand for a new concept which can be specific to the subsystem or general for any technique. Theoretical equations do not involve specific knowledge about the subsystem and are valid regardless of the technique that has this subsystem (e.g., Kirchhoff principles of electrical circuits, equilibrium equations in mechanics). Phenomenological equations represent the subsystem's parameters' relations in the specific technique or similar techniques.

Usually a technique must have the particular values (or limits) for variable parameters in order to perform in accordance with all goals. The properties, general constants, and fixed parameters, as well as the qualitative measures of variable parameters (values) distinguish similar techniques from one another.

The subsystem's *state*, ϑ (p_1, p_2, ...,p_N), is the total set of all constants, fixed parameters, and measured parameter values of all properties of the subsystem (element, operation, raw object, product or whole technique) at a given time. A subsystem can usually have many states, and a transition from one state $\vartheta 1$ to another $\vartheta 2$ can occur in continuous or discrete steps. Sometimes the states of a subsystem can be mutually exclusive, for example, saturation and active operation regions for a transistor. Relations between different states usually can be described by a finite set of equations, ε_i. Any nonsignificant or uninteresting properties are usually neglected in practical descriptions of the subsystem states; hence, the state can be symbolized as ϑ (p_1, p_2, ...,p_M), where M < N, and described by a relatively small number of ε_j, where $\vartheta \leq I$.

In analogy with mathematics, the state of a technique in the simplest case can be regarded as a multidimension vector (or tensor) that has the individual properties as its subsystems. Two techniques are equivalent if there is no difference in the set of all their states.

3.5.3 BEHAVIORS AND FUNCTIONS

The most important feature of an artificial system or manmade process is its *functions* φ_i and behaviors.

A *behavior* is a change of properties, characteristics, and parameters between the input and output of the subsystem and its environment in time and space due to a chain of operations, actions, and events realized as a TP. It can be defined as a set of successively attained states ϑ of a subsystem, while a *function* can be defined as an interpretation of a subsystem behavior — more precisely, of the physical equations ε_i governing its behavior. In this case, "function" acts as an abstract or manifestation of behavior. From another perspective, it can be said that a function also acts as a bridge between behavior and goal. In this case, a subsystem's function can be understood as an interpretative relation between the technique goals and the actual behavior of the subsystem in the whole technique. Both points of view are important because usually economic requirements rather than physical and technical capabilities determine what is the most suitable mode of function performance (the selected behavior). For example, it is possible to transport an object by 1 m using a heat expansion effect or 13 miles using an airplane, but it is not economically worthwhile.

There are many attempts to distinguish behavior and function in the framework of philosophy, design theories, and artificial intelligence. Semantic formulations of a function or a behavior are very similar and consist of two parts: a noun for an object of the function/behavior and a verb for its action. The object is a raw object (substance or field), properties, and/or parameters which change either due to behavior or in the process of function execution. The action is aimed at changing the characteristics and/or parameters of the object into the product that satisfies the goals of the particular technique. Usually it is enough to understand that a function can have an emotional appeal, but a behavior is neutral and does not depend on a designer or user viewpoint. A very general approach to defining behavior and function states that behavior is "what the technique does" and function is "what the technique is for." Although this simple approach works for the existing technique, it fails for any designed technique undergoing synthesis. Even for an existing technique, people interested in the answer to "What does the technique do (or what should it do) in order to satisfy ... (or for)" need to know the functions of a technique. Moreover, it is practically impossible to separate behavior and function for static subsystems, and because many inventive problems arise from the last question, we often do not need a solid distinction between behavior and function. Therefore, here we use the terms *behavior* and *function* almost interchangeably.* Behavior is mixed with function not only in TRIZ but also in value engineering, design methodologies, and other disciplines oriented toward human usage. In contrast, specialists in artificial intelligence must differentiate and formalize each very small chuck of reality in order to obtain computable procedures, even for resolution of simple routine problems or for prediction of a simple subsystem behavior by numerical and/or qualitative simulations [11–16].

The association between functions/behaviors and physical equations results in an indirect structural assignment of functions to subsystems. Sometimes a technique with a given structure exhibits a single behavior predetermined by its structure. The observed behavior usually does not uniquely determine the structure that caused it; the same behavior can be realized by a number of different structures. On the other hand, several alternative behaviors can satisfy a given goal. Therefore, the association between functions and subsystems can be of two types:

1. One-to-one — when a subsystem is bound to a single function in a single view and behavior. For example, a pipe is viewed as a conduit in the hydraulic view.
2. Many-to-one — when one of the following three cases occurs:
   - A subsystem is bound to several coexisting functions in the same view and behavior. For example, a bulb is simultaneously a source of light and a source of heat.
   - A subsystem is bound to several coexisting functions in different views but in the same behavior. For example, a window is simultaneously a conduit, from an optical perspective, and a barrier, from a thermal one.
   - A subsystem is bound to several coexisting roles in the same view but in different behaviors. For example, from a hydraulic perspective a valve represents two behaviors: open and closed. In the open mode it is a conduit, and in the closed it is a barrier.

Note that a subsystem may dynamically change its behavior depending on the values of characteristics that determine the subsystem. Thus, a fuse is a conduit until the electrical current flowing through it is below a specified threshold of its properties; then it becomes a barrier. The function of technique, however, refers to something more stable than behavior, namely a desired ability to perform a set of actions (e.g.,

---

* It seems that it is usually preferable to speak about behaviors for inanimate natural phenomena and about functions of biological objects, especially in cases when the question of "why" is legitimate. Note that TRIZniks more often work with functions than with behavior.

TABLE 3.2
Arithmetic for Functions

X & Y

X → Y ↓	HF	NF	UF
HF	HF	HF	not defined
NF	HF	NF	UF
UF	not defined	UF	UF

X • Y

X → Y ↓	HF	NF	UF
HF	UF	NF	HF
NF	NF	NF	NF
UF	HF	NF	UF

to be able to exert a set of effects). This desired ability might not be ensured by the actual behavior of the technique; it can also behave incorrectly or badly. The term *behavior* can thus be applied in a broader sense than *function*; the latter is always connected with the desired actions. Under normal circumstances, the behavior of a technique is determinate and controllable — it can be adequately predicted and regulated by suitable control inputs — unless the technique is damaged in any way.

A technique can execute several functions, among which only one, the *primary* function, is the working function, the aim of technique's existence. Other functions are auxiliary, accompanying and lightening the execution of the primary function (PF). (PF is also known as the main, major. or principal function.) Each TS or TP is designed to provide one or several *useful functions* (UF). Unfortunately, each TS or TP also has one or more *harmful functions* (HF), the fulfillment of which is undesirable. For example, in order to perform its PF, a car also produces gas pollution, heat, vibration, and noise, which in TRIZ are considered harmful functions. G. Yezersky and G. Frenklach proposed to consider a *neutral function* (NF) in order to complete this nomenclature. It is possible to define the addition (&) and multiplication (•) operations for such functions in cases in which all functions are equally important (have the same "weight") as presented in Table 3.2.

Note that the terms UF, HF, and NF reflect only our interpretation of subsystems' behaviors. The primary function is always a UF from the perspective of the technique's creators.

Useful functions have a hierarchy in nonprimitive techniques. *Support UF* support PF performance and provide PF reliability; *secondary* ones reflect subsidiary goals of the system's creators; *auxiliary* ones ensure the accomplishment of the basic and secondary functions. These support and auxiliary functions are essential and serve to ensure that the primary or secondary function can be realized or its operation can be maintained. The most important such functions of TS and TP are

- delivering energy, instrument, and raw object
- removing waste

- driving or propelling
- regulating and controlling
- planning (mostly for the future technique)
- connecting and supporting
- production and materials storage
- repairing equipment

Usually different subsystems are distinguished in a technique to perform each of these functions. The energy creation functions often can be excluded from a technique, the regulation and control functions are usually presented in control subsystems of the technique, and the isolation and separation of a technique from the environment is performed by the casing subsystem, while connecting functions make the technique into a single entity. For example, a system TV turns electromagnetic power of electromagnetic signals sent from a remote source into a color picture that people can watch. At the same time, an aerial converts radio waves into alternative electric current, which an amplifier magnifies, and an electron gun transforms it into the flow of electrons which are turned into a visible picture on a luminescent screen. All these sets of operations are needed to achieve the PF of TV. Support functions supply electric power to parts of TV and scan the electron beam on the screen, among other things. The modern TV can also work as an alarm: one of the secondary functions is time count, and an auxiliary function for this secondary function is to generate a signal with frequency proportional to 1 Hz.

Sometimes a partial subsystem shares the support, auxiliary, and secondary functions and performs several of such.

If the primary and/or support functions are not carried out, the technique is useless in terms of its purpose, even if other secondary functions of a technique are extremely well developed. To use a car as an example, if the car does not run, it — the technique — is useless. The object — the car — could, however, still fulfill a different purpose — for example, it could be a museum exhibit. Primary, support, secondary, and auxiliary functions are inseparably linked. The auxiliary, secondary, and support functions are defined in the technique with respect to the PF, but not with respect to their importance. The PF of a given TS or TP is the fulfillment of requirements of the major super-system. Other requirements lower in the hierarchy have less effect on the given technique than those higher in the hierarchy. Although these functions are considered as auxiliary, secondary, or support with respect to the main subsystem (the working tool of TS), they are also the PF of the partial subsystem that performs their function. Both secondary UF and major HF of TS or TP often have auxiliary UF and HF of the subsystems.

Several techniques can have the same primary function or UF. All techniques that have the similar primary function can be classified into groups according to auxiliary, secondary, and support functions or the selected mode of PF performing (behavior). There are various techniques named *heater* and they can be classified by energy source: electricity, fuel (solid, liquid, or gas), etc. Electric heaters can in turn be subclassified according to how they convert electrical energy into heat, such as by resistance (direct or indirect), discharge (electric arc), and induction at different frequencies (microwave).

Table 3.3 can aid analysis of functions in a technique:

TABLE 3.3
Functions in a Technique

Technique	A multitude of interrelated subsystems (elements or processes) that possess the features not brought to the features of the separate subsystems.
Subsystem	Parts forming the technique (subsystems are the focus of TRIZ).
Element/Operation	The smallest part of a technique recognized for a problem.
Super-system	That which the technique is a part of.
Environment	All that is outside the technique.
Primary functions	Functions for which the technique was created.
Support functions	Functions assuring the execution of the PF.
Secondary functions	Functions reflecting subsidiary goals of the technique creators.
Auxiliary functions	Functions assuring the execution of the higher-level functions.
Harmful functions	Functions not intended for or desired of the technique and that have undesired results.

Three types of relations between functions can be identified for UF, HF, and NF:

- Interdependent — two functions φ_i and φ_j, which refer to the equations ε_i and ε_j, respectively, are directly interdependent if ε_i and ε_j share a variable characteristic.
- Mutually dependent — two functions φ_i and φ_j, which refer to the equations ε_i and ε_j, respectively, are mutually dependent if there exists an equation that links a variable characteristic of ε_i with a variable characteristic of ε_j.
- Influence — a function φ_i, which refers to the equations ε_i, influences a function φ_j, which refers to the equations ε_j, if a variable characteristic of ε_i is a parameter of ε_j.

Often the problem is how to break a relation between functions for better achievement of a goal. We discuss how TRIZ resolves such problems later in this book.

3.5.4 STRUCTURE

The spatial relations elements (subsystems) are important for a TS, and the temporal relations between operations are important for a TP. The term *structure* is usually applied to the internal arrangement, order, organization, breakdown, segmentation, conformation, constitution, or construction of a technique. Note that *structure* is usually used in one of three ways. The most comprehensive refers to the term as used here, the set of subsystems (operations and/or elements) and the set of their interrelations (links). Structure can also refer only to the set of relationships for a system in philosophy and is frequently used for the set of static subsystems of a TS that fulfills the partial functions of "connecting" and "supporting" in engineering.

The concept of structure is more important for TS than for TP. Only two basic structures of TP can exist: concurrent (parallel) operations and sequence (series) operations; all other TP structures are combinations of the two.

Structure of a technique is a set of subsystems (elements and/or operations) and links between them that are determined by the physical principles of executing a PF and secondary UF. An element (operation) is the smallest but a relatively integral part of a technique at the chosen granular level; it does not disappear when separating out the technique and it can execute a function and possess properties. However, properties of an element (operation) in the technique are not identical to properties of a separate element (operation). The sum of an element's properties in the system may be greater or smaller than the sum of its properties outside the system. Usually an element's properties are neutralized or even suppressed within the technique when they are included in subsystems.

If we agree that defining the PF (the goal of a technique) is subjective to some degree, then structure is the most objective; it depends only on the type and material content of elements used in the TS, as well as on general physical laws. To form a structure means to define the TS behavior for the purpose of obtaining the useful functions. The required UF and the chosen modes of their performance unambiguously set the structure.

The correspondence between function(s) and structure or subsystem(s) has been discussed for centuries in philosophy and biology. Surprising correspondences between functions executed by different organs of living organisms and the structure of the corresponding organ (construction and constructive parameters) have repeatedly been noticed and analyzed in biology. At the end of the twentieth century, this correspondence attracted the attention of some experts in artificial intelligence. Although many interesting and competing hypotheses about the relation between functions, behaviors, and structure have been proposed [11–16], as of yet they can be used for solving only a very few routine problems with the simplest TS and TP.

Extensive research of techniques in the framework of TRIZ and design theories has resulted in forming the following one-to-many hypothesis:

> A single function can be achieved in many different structures, and in turn any single subsystem (element, operation) can execute many different functions.

As an example of this hypothesis, consider a metal wire that can be viewed as a good electrical conductor, as a sensor in which resistance changes with temperature, as a ruler, or as a string for a musical instrument.

In the world of techniques, a variety of solutions can be presented by one functional structure. On the other hand, this variety is presented by a single and finite set of functional structures. The same one-to-many correspondence was noticed in various subsystems of diverse TS or TP by several researchers and reflects the priority of function over structure in the world of technique.

3.5.4.1 Shape

The TS's shape is the external manifestation of its structure (what we see when we look at a system), while the structure is internal. These two concepts are intimately related, and one might dictate the conditions of the other. For example, the shape of an airplane wing dictates or controls its structure. The logic of a structure's construction is mainly determined by the system's functions, whereas the shape usually depends on the requirements of the super-system (again, consider an airplane and its wing).

Shape is dictated by numerous requirements which have their own hierarchy of importance:

1. function (screw thread shape)
2. economics (simplest, cheapest tool handle)
3. technology (ease and convenience of production, processing, transportation)
4. raw object state (solid, liquid, or gas; field and its properties)
5. service (maintainability, stability, transportability, convenience for manu-facturing, storage, repair)
6. aesthetics and ergonomics (convenient design, beauty)

Even if the TS shape is largely determined by one set of requirements, the others must still be considered, especially economics.

3.5.4.2 Hierarchy and Organization

A modern technique usually has a hierarchical structure (for example, computer, car, VLSI semiconductor chip). Hierarchy of structure organization orders interactions between levels from the highest to the lowest. For example, a car is a system composed of subsystems (brake, motor, etc.). Each subsystem can be regarded as a system and consists of smaller elements (for example, bearings, bolts) and separate components (shaft, cover, housing). However, if instead of looking within the car system, following the hierarchy down, we follow the hierarchy up, we can consider the entire car an element in a factory that produces cars and this factory as a subsystem of an automobile company.

Hierarchy is relevant only for multilevel techniques. Every level acts as a control in relation to all lower levels and as a controlled subordinate in relation to the higher one. Each level is also specialized for executing a specific function (the PF of the level). Properties of a technique at any level are influenced by characteristics of the higher level — super-system — and lower level — subsystem — of the hierarchical structure. There is no absolutely strict hierarchy; some subsystems at lower levels are independent to a greater or lesser degree in relation to the higher levels. Within one level, members (elements, operations, subsystems) are equal to each other; they mutually complement each other, and some indices of self-organization are inherent in them (they are laid at formation of the structure). Sometimes the number of levels in the hierarchy of a technique reflects the technique's complexity.

A simple TS or TP does not require a hierarchy; subsystems interact directly by links between each other. In complex techniques, direct interactions between all subsystems are impossible (too many links are required); therefore, direct contacts survive only between subsystems (elements, operations) of the same level, while links between levels sharply decrease in number.

When a technique needs to be improved, it is usually accomplished by changing the technique's subsystems at the hierarchical level at which the need is felt. For techniques of medium and high complexity (i.e., the modern one), the only way to increase efficiency, reliability, and stability is by creating and developing hierarchical structures. Moreover, often to achieve the goal, the other systems and/or processes, or even super-system, also has to be changed. Effective development of a technique depends on one's skill to discern the subsystem being improved as a multilevel hierarchical structure, which is why considering the main properties of hierarchical systems is important for technical problem solvers and technique creators. These properties [5] are described below.

Duality of elements (operations) — When an element becomes a part of the technique, it loses its initial properties, and subsystem simplification takes place (a technique needs simple useful functions, rather than "complex"). On the other hand, when an element becomes a part of the technique, it acquires the system gain (see Section 3.6).

Note that this property of hierarchical systems causes a widespread type of inventor's psychological inertia: an inventor sees only one (usually the system) property of an element and does not see much of its individual properties. Any element simultaneously possesses both individual and system properties.

Dictate of levels — All controlling actions (signals) and power necessarily come to the working tool, inducing it to function in a strictly determined way. In this sense, the working tool is the most subordinate element of the system at the lowest level of the hierarchy. Its importance in synthesis of the technique is exactly opposite: it dictates the structure for executing the PF. Dictate of upper levels over lower ones is the main order of hierarchy.* Higher levels are insensitive to changes at the lower ones, but lower levels are sensitive to changes at the higher. The higher the level of hierarchy, the less strict are links between elements, and elements are easier to replace or reposition.

Filtering of useful functions — Due to the properly organized hierarchical structure, the useful functions reinforce each other at each transition from a lower to a higher level; and/or harmful functions at every next level are dampened or, at least, do not increase in number. The main contribution to PF is formed at the lower levels, starting at the working tool. At successive levels, the UF is enhanced. With the increasing number of levels, PF growth slows, which is why complex techniques with many levels are inefficient (expenses for energy consumption, mass, dimensions, etc. begin to exceed the gain in PF). The highest hierarchical level usually executes only the matching functions; a technique cannot have more than one such level. Therefore, the filtering of useful functions at different levels of the hierarchy determines the state of a technique.

Organization of a technique arises simultaneously with its structure and often has a hierarchical character. The organization arises when objectively regular, stable-in-time, links arise between elements; at this point, some properties of an element come to the fore or are enhanced, while others are suppressed. Such properties of elements are transformed in the process of organization into UF. Increase of the

---

* Often the dictate of the higher levels extends even beyond the working tool. This "tendency" of technology to change the environment for its own convenience seems mistaken from the ecological perspective.

degree of system organization directly depends on the number of links between subsystems. With an increasing number of links per subsystem, the number of UF of this subsystem usually increases. One of the main properties of organization is the possibility to control — to change or keep the state of subsystems (operations and elements) in the process of system functioning. The control as the sequence of information flux in time often goes through special links. Usually organization appears only when the links between subsystems and/or their properties exceed in force the links with nonsystem parts.

The degree of organization reflects the predictability of PF performance. Absolute predictability is either impossible or possible only for static or nonoperating dynamic techniques. Complete nonexistence of predictability (disorganization) means absence of the technique — that situation when we need to achieve a goal but absolutely do not know how to do it.

Complexity of the organization is characterized by the number and variety of subsystems, the number and variety of links, and the number of hierarchy levels. It grows as the technique expands and decreases at the technique's convolution (see Chapter 7). At expansion in usefully functional subsystems, principles of organization (conditions of interactions, links, and functions) are worked out, and then the organization comes to the micro-level (the function of subsystem is executed by a substance with smaller sizes in space).

Three groups of HF are among the factors destroying organization:

- External (from super-system, nature, human beings) — factors destroy links if their power exceeds the power of links in the technique.
- Internal (forcing or occasionally strengthening of harmful properties) — factors are present in the technique from the very beginning, but their number increases in time due to disruptions in the structure.
- Operational (self-destroying of elements due to limited lifetime) — factors are present in the technique during its functioning; they destroy subsystems (carry-over of some substance from or to a subsystem) and degenerate links because of entropy factors (for example, fatigue of springs, rust).

3.6   SYSTEM GAIN FORMATION

Subsystems (element or operation) of a technique possess many properties. When subsystems are included in a technique, the number of properties belonging to each subsystem is changed. Some of these properties are suppressed in formation of links, while others, in contrast, manifest themselves more markedly. According to the completeness rule,* the subsystems that compose a technique should not only be matched in inputs and outputs, shape, and some properties (so their joining is possible), but they should also complement each other, intensify each other, sum their positive properties, and neutralize harmful ones. In other words, some properties

---

* Detailed discussion in Chapter 7.

are augmented, while others are neutralized (see also the discussion above on filtering useful functions).

In addition, every technique has joint (integral) properties which are not equal to the sum of the characteristics of the subsystems entering into the technique. These properties "suddenly" arise when the technique is formed, and the sometimes unexpected addition is the main profit of the synthesis of the new technique*. *System gain*, then, is the appearance of a property, absent in all elements or operations before their inclusion in the technique. These *joint* properties are an integral part of the technique, and their parameters can be used to characterize it. Note that not the only the property is important, but also the form in which that property appears and its measurable value and/or quality.

The appearance of the system gain with the technique's creation can be manifested in three possible cases:

1. The positive properties add together and strengthen each other, while the negative characteristics are at least unchanged:

$$UF \ \& \ UF = UF \text{ and } UF \ \& \ HF = NF \text{ or } UF;$$

2. The positive characteristics add together, while the negative properties eliminate each other:

$$UF \ \& \ UF = UF \text{ and } HF \bullet HF = UF;$$

3. The negative properties are suppressed or converted to positive characteristics:

$$HF \bullet NF = NF \text{ and } NF \ \& \ UF = UF.$$

To determine the system gain of a given technique, we can use a simple approach: separate a technique into subsystems (elements and operations) and find what property (characteristic) disappears.

3.7   DESIGN SCENARIO

In this section we show how the ideas presented in this chapter apply to the conceptual design of a new technique.

The main reference point in the process of technique synthesis is the future system gain. The following sequence of steps is possible during the conceptual design:

1. Goals for a new technique are recognized and verified (teleological analysis).
2. Primary function (PF) is formulated.

---

* Apparently, the so-called law of transition from quantitative changes to qualitative changes, well-known in dialectical materialist philosophy, reflects only the content side of a perhaps more general law — the law of property or characteristic formation or change in parameter values due to system gain.

3. Principles of working tool effect upon a product are determined (behavior analysis).
4. Mode of PF performance is determined (physical and economical analyses).
5. Working tool is selected or synthesized (structural analysis).
6. Functional scheme* is constructed in the first approximation based on an idea about the major subsystems (see Section 3.3.1).
7. Transmission, engine, energy supply, and system control are selected and joined with the working tool into a casing. The parameters of all subsystems have to be adjusted in order to enable a technique to achieve its intended goal.

For example, a simple scheme for linear energy flow is:

Energy Source → Engine → Transmission → Working Tool → Product

More detailed schemes are created with regard to the hierarchy of subsystems, their characteristics and parameters, and type of fields and substances flow from one subsystem to another.

The scheme of structure construction is very important during the synthesis of the new technique. It is based on the rules of structure formation that include functionality, causality, subsystem completeness, and complementarity.

Moreover, the structure is determined by the decisions made prior to it in the following flow:

goal → function → behavior → structure → subsystem → element/operation

Each entry in the above string usually has more variants of realizations than the one preceding it, in accordance with the one-to-many rule. The main requirement for the structure is minimal loss of energy and unambiguous operation (exclusion of an error), i.e., high "conductivity" and reliable cause-and-effect chain. Likewise, the main requirement to the element (operation) is maximal Ideality of the technique (see Chapter 5). Every event in a technique has at least one cause; at the same time, this event itself is a cause for subsequent events. The mode of UF performance is based on what will best realize the causality rule. The working tool, which performs the technique's primary function, is what constructs the reliable chain of actions, from the final event to initial one. The principle of function performing unambiguously determines the possible structure of the technique and reflects goals-functions. Based on the chosen performing principle, the functional scheme of a technique can be drawn up (possibly in Su-Field form; see Chapter 12).

---

* The **completeness** and **transparency rules** (see Chapter 7 for details) can be taken as a basis for initial construction of the functional scheme. Any technique must provide energy transfer over all its subsystems and also substance transfer to all subsystems that require it. Furthermore, all techniques must provide all necessary transformations of fields and substances within the system that are necessary to its actions.

Of course, a technique synthesized according to this scenario will have certain values of PF and UF performance. After a period of time, the request to the technique will increase and the following cycle will occur:

1. PF growth — attempts to take more from the technique than it can give.
2. With the increase of PF, a property or parameter of the technique deteriorates, bringing a contradiction and a possibility to formulate an inventive problem.
3. The inventive problem is solved by TRIZ heuristics and instruments using scientific and technical knowledge.
4. Change of the technique in accordance with the solution obtained in the framework of TRIZ.
5. PF increases again — return to step 1.

Such a cycle, described by Yu. P. Salamatov [5], can be applied to any UF of a technical system or technological process.

3.8   CONCLUSION

Although this chapter might appear at first to be too theoretical, we believe that a TRIZnik should know the object he or she is dealing with as best as possible. For "hard" TRIZ, the object is *any* technical system or technological process and its attributes, especially functions. For any TRIZnik, it is *the* system/process and its useful, neutral, and harmful functions that he or she needs to improve or reduce.

Let us stress once again:

1. TP is impossible without a technical or natural system and a raw object. TS is a set of objects for generating purposeful products (that serve to satisfy various human needs) from raw materials during a TP. Technical systems and technological processes (i.e., any techniques) are designed to accomplish a specific series of useful functions to achieve specific goals. The technical system is only the carrier of the necessary functions and the desired behavior; it usually is not a purpose in itself. The technological process is only the set of necessary steps (operations) to the desired behavior of a product; it also is usually not a purpose in itself. Unfortunately, harmful functions also exist in almost all TS and TP.
2. Techniques are characterized by the following attributes:
   • have system gain; i.e., some properties of the technique are not equal to the sum of the appropriate qualities of its subsystems' characteristics
   • have a primary function, input, and/or output
   • have a structure; i.e., consist of subsystems connected by particular links
   • each of these subsystems can, in turn, be considered a system (down to some elements that can be elementary particles, such as electrons, or complicated mechanisms, such as aircraft, depending on the perspective on the initial system)
   • each is a subsystem of some super-system

At first glance, the different multidimensional classifications and nomenclatures of TS, TP, and their functions presented here might appear excessive for an engineer or TRIZnik. However, we need to remember that TS and TP as classes of objects are distinguished by an extremely high diversity of all elements and a multiplicity of functions, which have no analogies in any other field. As an example, the nomenclature in mechanical or electrical engineering includes many thousands of kinds of technical systems and many hundreds of technological processes, without even considering the variations of types. Existence of TP, TS, and their products are essential for human life, and their improvement is the main goal of a TRIZnik. Detailed studies of TS and TP help us to understand and improve nontechnical systems and processes through "soft" TRIZ, which uses a functional and systematic approach similar to the "hard" TRIZ approach discussed in this book.

REFERENCES

1. Bogdanov, A. A., *General Organization Science: Tektologia,* Kniga, Moscow-Leningrad, 1925 (in Russian).
2. von Bertalanffy, L., *General System Theory,* George Braziller Publ., New York, 1968.
3. Miles, L. D., *Techniques of Value Analysis and Engineering,* McGraw-Hill, New York, 1972.
4. Pahl, G., and Beitz, W., *Engineering Design,* The Design Council, London; Springer, Heidelberg, 1984.
5. Salamatov, Yu. P., A System of Laws of Engineering Evolution, in *Chance for Adventure,* A.B. Selutsky (Ed.), Petrozavodsk, Karelia, 1991, pp. 5–174 (in Russian).
6. Hubka, V., *Theorie technischer Systeme,* Springer-Verlag, Berlin, 1984. (English translation *Theory of Technical Systems: A Total Concept Theory for Engineering Design* by V. Hubka and W. E. Eder. Springer-Verlag, Berlin and New York, 1988).
7. Polovinkin, A. I., *Theory of New Technique Design: Laws of Technical Systems and Their Applications,* Informelektro, Moscow, 1991 (in Russian).
8. Altshuller, G. S., *To Find an Idea,* Nauka, Novosibirsk, 1986 (in Russian).
9. Suh, N., *The Principles of Design,* Oxford University Press, Oxford, 1993.
10. Koller, R., *Konstruktionsmethode für Maschinen-, Gerate- und Apparatebau,* Springer, Berlin/Heidelberg, 1979 (in German).
11. Chandrasekaran, B., Functional Representation and Causal Processes, *Advances in Computers,* Academic Press, 1994, Vol. 38, pp. 73–143.
12. Chittaro, L., Tasso C., and Toppano, E., Putting Functional Knowledge on Firmer Ground, *International Journal of Applied Artificial Intelligence,* 1994, vol. 8, pp. 239–258.
13. Kuipers, B., *Qualitative Reasoning: Modeling and Simulation with Incomplete Knowledge,* MIT Press, Cambridge, MA, 1994.
14. Chittaro, L., Tasso, C., and Toppano, E., Putting Functional Knowledge on Firmer Ground, *Applied Artificial Intelligence,* 1994, Vol. 8, No. 2, pp. 239–258.
15. Umeda, Y. and Tomiyama, T., Functional Reasoning in Design, *IEEE Expert, Intelligent Systems and Their Applications,* 1997, Vol. 12, No. 2, pp. 42–48.
16. Sasajima, M., Kitamura, Y., Ikeda, M., and Mizoguchi, R., A Representation Language for Behavior and Function: FBRL, *Journal of Expert Systems with Applications,* 1996, Vol. 10, No. 3/4, pp. 471–479.

4 Contradictions

4.1 INTRODUCTION

Throughout the history of human knowledge, there have been two conceptions concerning the law of development of the universe, the idealistic conception and the materialistic conception, which form two opposite world outlooks. TRIZ ideology is based on two major ideas; Contradiction and Ideality. Contradiction is the basic law of materialist dialectics, whereas ideality is the essence of idealism. These two opposite philosophic approaches are united in TRIZ, which uses their mutual cooperation. Perhaps this amalgam predetermines the unique power of TRIZ. The concepts of Ideality and/or Contradiction should be consciously included in any process of solving inventive problems. Hence this chapter and several following consider these main TRIZ concepts.

4.2 CONTRADICTIONS: ONTOLOGY

A contradiction literally means saying "No" but more generally refers to the propositions that assert apparently incompatible or opposite things. George Berkeley introduced the contradiction concept in his "A Treatise Concerning the Principles of Human Knowledge" in 1710. As the main point of critique of the formal logic developed by G.W.F. Hegel between 1812 and 1816, contradiction is the most popular concept for introducing dialectical ideas. For example, F. Engels wrote at the end of the XIXth century that the unity (interpenetration) of opposites is a basic law of dialectics, and V. I. Lenin said, "The splitting of a single whole and the cognition of its contradictory parts is the essence … of dialectics." Lenin draws attention to the fact that the contradiction is central not just to "logic" (as normally understood) but to cognition (analysis), and that the dialectical concept of contradiction is not the contradiction between two things external to one another, but the contradiction which is at the essence of a thing. According to dialectics, contradictoriness within a thing is the fundamental cause of its development, and contradictions have a universal presence in many fields. In his "On the Question of Dialectics," Lenin stresses that one contradiction cannot exist without the other, and he illustrates the universality of this mutually contradictory phenomenon by noting its presence in various fields:

mathematics: plus and minus, differential and integral;
physics: positive and negative electrical charges, mechanical action and reaction;
chemistry: the combination and dissociation of atoms;
war: offense and defense, victory and fault.

According to TRIZ, often the most effective inventive solution of a problem is the one that overcomes some contradictions. A contradiction shows where (in so-called operative zone) and when (in so-called operative time) a conflict happens. Contradictions occur when improving one parameter or characteristic of a technique negatively affects the same or other characteristics or parameters of the technique.

When a solver has extracted a contradiction from the problem that fits into the classes defined below, it becomes easy to find a variety of creative and effective solutions for the problem. Usually a problem is *not* solved if its contradiction is *not* overcome, as shown in the following example:

> A new and more powerful engine is installed in an airplane in order to increase its speed. The engine increases the total weight of the airplane, and the wings now cannot support the heavier airplane during takeoff. In an effort to solve this problem (remove the contradiction), the size of the wings is increased. Now, there is more drag which slows the airplane.

In the above example, the goal is not achieved because the central contradiction was not resolved. A solution should keep or strengthen the characteristic in this example (speed) in such a way that other properties (weight, wing size) are maintained or improved. This way is almost always nonobvious and requires some creativity on the part of a problem solver, or knowledge of and experience in TRIZ. In contrast to a routine design that leads to a smoothing of the contradiction (the trade-off dogma) or choosing one of the preferable combinations in the conflict (OR...OR), a design based on TRIZ aspires to permit and solve the contradiction, creating a system in which the improvement of one characteristic is not accompanied by deterioration of others (the AND...AND), so the so-called win–win principle can be achieved.

Altshuller and his coworkers [1] distinguished the following three types of contradictions*: administrative, technical, and physical:

Administrative contradictions — Something is required to make or receive some result, to avoid the undesirable phenomenon, but it is not known how to achieve the result. For example, we want to increase quality of production and decrease cost of raw materials. Such a form of a problem recalls an inventive situation. The administrative contradiction itself is provisional, has no heuristic value, and does not show a direction to the answer.

Technical contradictions — An action is simultaneously useful and harmful, or it causes UF and HF; the introduction or amplification of the useful action or the recession of the harmful effect leads to deterioration of some subsystems or the whole system; for example, it creates an inadmissible complexity of the system.

---

* Unfortunately, Altshuller's terminology of contradictions (administrative, technical, physical) is not perfect, but it is widely used in TRIZ. Also, the set of contradictions proposed by Altshuller [1] is not exhaustive for various problems outside engineering.

The technical contradiction represents a conflict between *two subsystems*. For example, we want to increase the penetration depth of ions into a semiconductor and decrease the electrical power (energy source) that is necessary for the ion implanter operation.

Such contradictions occur if

- creating or intensifying the useful function in one subsystem creates a new, harmful function or intensifies an existing harmful function in another subsystem;
- eliminating or reducing the harmful function in one subsystem deteriorates the useful function in another subsystem;
- intensifying the useful function or reducing the harmful function in one subsystem causes the unacceptable complication of other subsystems or the whole technique.

Physical contradictions — A given subsystem (element/operation) should have property A to execute a necessary function and property non-A or anti-A to satisfy the conditions of a problem. A physical contradiction implies inconsistent requirements to a physical condition *of the same* element of TS or operation of TP, i.e., the same key subsystem of a technique. For example, we want the insulator in semiconductor chips to have low dielectric constant **k** in order to reduce parasitic capacities, and we want that insulator also to have high dielectric constant **k** in order to store information better.

Physical contradictions also occur if

- intensifying the useful function in a subsystem simultaneously intensifies the existing harmful function in the same key subsystem;
- reducing the harmful function in a subsystem simultaneously reduces the useful function in the same key subsystem.

For example, if the gate bias voltage increases, a metal-oxide semiconductor transistor can have higher threshold voltage (good for power MOSFETs), but the transistor will operate at lower frequencies (bad); if the gate bias voltage decreases, a metal-oxide semiconductor transistor can operate at higher frequencies (good), but the high-speed changes of the gate voltage will lead to unanticipated triggering of the transistor (bad).

The physical contradictions as well as the technical contradictions usually appear during the special problem analysis that is described particularly in Chapter 18. Sometimes technical contradictions can be discovered by analysis of technique within the framework of Root Cause Analysis or Goldratt's Theory of Constraints [2, 3].

According to Altshuller [1], an inventive situation is usually inherent in some groups of technical and/or physical contradictions in a technique. Choosing a contradiction from the group means transition from an inventive situation to the beginning of problem solution. Usually, successful formulation of the physical contradiction shows the problem's nucleus. When the contradiction is extremely intensified, often the problem's solution will be straightforward. Below we illustrate the simplification of a problem by considering the contradictions, and introduce the names of logical operators for two of the most common contradictions in TRIZ.

> **Administrative contradiction:** It is necessary to detect the number of small (<0.3 micrometers) particles in a liquid with very high optical purity. The particles reflect light poorly even if we use a laser. What to do?
>
> **Technical contradiction:** If the particles are very small the liquid stays optically pure, but the particles are invisible. *XOR* if the particles are big they are detectable, but the liquid is not optically pure.
>
> **Physical contradiction:** The particle size must increase to be viewed *AND-NOT* increase to keep the optical purity of the liquid.*

After such transition we have reduced the difficulty of the problem.

In general, contradictions can be classified into three major groups:

- natural,
- social, and
- engineering.

A more detailed scheme of contradictions is shown in Figure 4.1. During solution of a problem in the framework of TRIZ, there is consecutive reformulation of contradictions generated by the problem. Each successive contradiction improves our understanding of the problem.

Let us discuss the first natural and social contradictions here; engineering contradictions are described in detail in Part 5 of this book.

Natural contradictions are divided into two groups:

> **Fundamental** contradictions — natural laws limit possible solutions of the problem. The impossibility to have a temperature below 0 degree K or to exceed the speed of light are examples of fundamental contradictions. Such contradictions perhaps represent only our current knowledge and some of them can be eliminated in the future.
>
> **Cosmological** — restrictions caused by Earth conditions. For example, it is impossible to keep any weight on a thin beam because of Earth's gravity, and a car cannot exhaust pure hydrogen because of the explosive interaction it has with oxygen in our atmosphere.

If a natural contradiction cannot possibly be overcome (at least currently for the cosmological contradictions), we can rather speak about constraints and trade-off solutions of the problem.

---

* It is possible to label the *physical* contradictions as *ANDNOT operators* and present the so-called *technical* contradictions through *XOR operators*, but these terms are not common in TRIZ research yet.

FIGURE 4.1   Types of contradictions. Difficulty of contradiction resolution usually increases from bottom to top, and from left to right. (A technically engineering contradiction would be easier to solve than would a fundamental natural contradiction.)

TABLE 4.1
Major Obstacles to Innovation

Individual	Managerial/Organizational	Cultural
Stereotyped thinking and/or lack of creativity (especially for elder specialists) and/or psychological inertia	Belief in some official functional method (often called a scientific)	The "holiness" of the current political/economic system
Risk of failure	Money constraints	Prejudice to change
Lack of knowledge and/or faulty memory	Decision-making and/or leadership styles	The occidental/oriental differing views of goals
Self-imposed constraints (e.g., taboos, fear of questioning)	Time restraints	Bias against (bigotry) recognition of the current paradigm limits

Social contradictions are divided into three groups (see Figure 4.1) according to the major restraints to innovation at different society levels (see Table 4.1). There is a simple hierarchy for the levels of social contradictions, and it is easier to break Individual and Managerial restraints than Cultural Ones. The knowledge of the society level to which a contradiction belongs as well as the types of restraints involved can help in choosing the strategy for the problem's solution.*

---

* Studies of social contradictions in TRIZ intersect with psychology, management, and other social disciplines but researchers often do not know results of studies in those fields.

At first glance, administrative contradictions (as defined by Altshuller) should be included in social contradictions, but more detailed analysis shows that the administrative contradictions can be often located between social and engineering contradictions in Figure 4.1.

It seems possible to resolve Managerial-Organizational contradictions within the framework of TRIZ methodology, and such research is an ongoing activity of some TRIZ experts. Often cultural and individual contradictions can be presented as the problem's constraints; however, human problems often do not have a contradiction. There are two opposite ways (trade-offs) for solving routine problems without contradictions:

Compromise   Strengthen a gain in one quality without losing considerably in another quality.

Radical   Keep or strengthen one quality at the expense of another quality if one of the requirements to the system is inconsiderable and if the problem does not lead to losses.

Contrary to solutions of natural and engineering contradictions, trade-off compromise solutions in human problems can frequently be good resolutions. Perhaps this difference is based on the difference in the types of systems: technical systems usually have a determined nature, while human systems are probable or stochastic.

Finding solutions for various contradictions is discussed in Part 5 of this book.

4.3   STRUCTURE OF A PROBLEM

Because of the diversity of TS and TP, a variety of problems can occur in any technique. Some relations exist inherently between useful (UF), neutral (NF), and harmful (HF) functions (see Chapter 3 for discussion of UF, NF and HF) in any TS or TP. On the other hand, any TS or TP can be presented as a set of subsystems. Initially the structure of TS or TP reveals almost nothing about how to solve the problem or the requirements for the solution. It simply describes the status quo of the technique. When conflicts between functions in a technique appear, they can be presented in the structures of problems that reflect the type of relations between UF, NF, and HF and the structure of the technique. In order to present such conflicts through the technical or physical contradictions, we consider here only UF and HF. Representation of contradictions together with the hierarchical structure of a technique allows us to figure out a structure (topology) of a problem. There are a few generic problem structures: point, pair, linear, network, triangle, and star.

Point problems either have physical or natural contradictions within a single subsystem or have no contradictions at all. Point problems involving a physical contradiction are usually hidden, but they are often the root cause of all other problems. A point problem is usually discovered after reduction of semantic links of more complicated problem structures or in careful analysis of problems set in a technique. The creative methods of idea generation, such as brainstorming, synectics, and lateral thinking,* are quite useful for the point problems without contradictions as well as for social contradictions.

Pair problems have a single technical contradiction between functions in two subsystems. They usually appear after reduction of an inventive situation or of more complicated structures to a set of independent technical contradictions.

Linear problems have chains of engineering contradictions. They usually can be presented as a sequence of dependent technical and physical contradictions in two or more different subsystems.

Network problems have a loop of several dependent contradictions (often so-called mathematical). They appear when l, m, and n subsystems depend on each other and/or can be dependent on k subsystem, resulting in several linked technical and/or physical contradictions. *Triangle* problems, the simplest case of a *network* problem, have three dependent engineering contradictions.

Star problems have a set of independent technical or mathematical contradictions with a common root which is usually a physical contradiction. They usually appear when one characteristic or subsystem k gets better, but other l, m, n, ... get worse. If contradictions A with B, C, and D are independent, the star problems can be reduced to pair problems. Star problems often have a hidden physical contradiction in the root, so they can be reduced to a point problem.

Beyond such generic (primitive) structures, more complex and even puzzling structures exist, such as a hierarchy (which usually reflects the structure of TS and TP). Such problems cannot be solved "as is" and should be reduced to one or a few generic structures. In the framework of TRIZ, in order to convert a problem's complex structure (e.g., nonsingle triangle or hierarchy) into a simple generic problem structure (e.g., point, pair, or triangle), it is essential to reformulate the problem in terms that distinguish subsystems, functions (primary, secondary), and resources, to crystallize a contradiction in the problem, and to systematically collect, organize, and document all important information related to the situation.

For simplicity, only problems in technical systems are discussed here, although the proposed ideas are also valid for technological processes as well. The different structures of problems can be presented using John Terninko's questionnaire, as the guide to semantics of the problems:

Useful Function

1. Is this useful function UF_n *required* for other useful functions UF_{n+1}?
2. Does this useful function UF_n *cause* any harmful effects HF_n?
3. Has this useful function UF_n been introduced to *eliminate* harmful effects HF_n?
4. Does this useful function UFn *require* other useful functions in order to perform UF_{n-1}?

---

* See references to Chapter 1.

Harmful Function

5. Does this harmful function HF_n *cause* other harmful functions HF_{n+1}?
6. Is this harmful function HF_{n-1} *caused by* other harmful functions HF_n?
7. Is this harmful function HF_n *caused by* useful functions UF_n?
8. Has a useful function UF_n been introduced to *eliminate* this harmful function HF_n?

SIGNS: ———▶ **cause,** – – – –▶ **require,** —✗— **eliminate**

It seems beneficial to present some aims of a solver in terms of simple structures of problems:

Point problem with the physical contradiction — Find a way to enhance or provide a UF_n that eliminates, reduces, prevents, or does not cause an HF_n within the same subsystem if UF_n causes HF_n and vice versa. For example, a high-k insulator improves the storage information in a semiconductor chip (UF) but it also increases the parasitic capacities in it, thereby slowing the speed of information exchange (HF) inside the same chip.

POINT

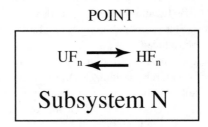

Pair problem — Find a way to resolve a single *technical* contradiction by which UF_1 should rise or eliminate HF_1 and does not cause an HF_2 or increase HF_2 in another subsystem, and/or UF_1 provides another UF_2 and does not increase or create HF_2 in another subsystem; i.e.,

UF_n causes and/or requires HF_k ($k \neq n$)

This pair problem is defined by two subsystems and at least one link.

PAIR

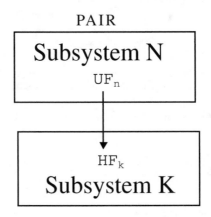

Consider the pair problem "UF_n requires HF_k" in TS ground transport (cars, buses, petroleum sources, gas stations, etc.). Primary UF of subsystem N (transport carriers) is "to transport objects" which requires expensive petroleum refinement into gas, the HF of subsystem K (energy suppliers and sources).

Triangle network problem — Find a way to solve a loop of *dependent* contradictions:

UF_n causes HF_k AND HF_k causes UF_l WHILE UF_l causes HF_n ($k \neq l \neq n$).

Note that the illustration of a network problem reflects a hierarchy between {N} subsystems and the subsystem M.

NETWORK
PROBLEM

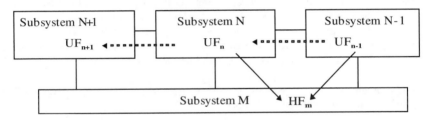

Linear problem — Find a way to benefit, eliminate, reduce, or prevent HF under the *sequence* condition of UF and/or to perform the *sequence* steps of a system improvement:

UF_{n-1} is required for UF_n and UF_n is required for UF_{n+1}

LINEAR
PROBLEM

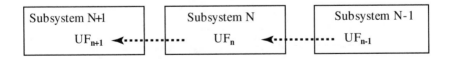

or

UF_n is introduced to eliminate HF_k, while UF_n causes HF_l ($k \neq n \neq l$)

or

UF_k is required for UF_n, while UF_n causes/increases HF_k ($k \neq n$).

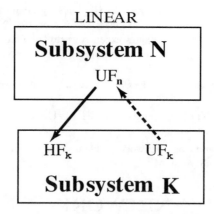

During operation of an electrically heated furnace for melting metals we can consider the walls of the furnace as a working tool (subsystem N), the electrical heater as an energy source (subsystem K), and solid metal as the raw object, and liquid metal as the product. A transition solid → liquid metal (the primary functions of the furnace) requires hot walls (UFn) and hence additional electrical power (UFk), but the high temperature of the walls destroys the electrical heater or its parts (HFk).

Star problem — Find a way to resolve a FEW technical contradictions by which UF should eliminate a few INDEPENDENT HF, i.e.

UF_n causes HF_k and HF_p, requires UF_l, and eliminates HF_m ($k \neq l \neq m \neq p$).

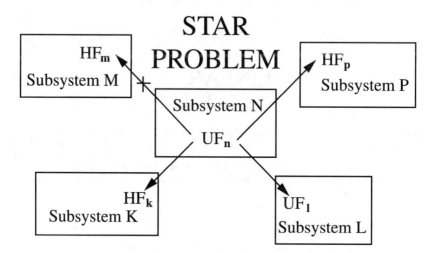

It is important to recognize the problem's structure because TRIZ heuristics work better on some problems than on others. Clarification of a problem's structure also provides the easiest way to use TRIZ knowledge databases. The problem's structure leads to a strategy and helps the solver choose appropriate information for solving inventive problems.

It seems reasonable to label the physical contradictions by the ANDNOT logical operator and present the technical contradiction through the XOR logical operator. Usually the simple logical operators AND and OR are used only for representations of TS or TP. "New" ANDNOT and XOR and traditional operators can be used for graphs of problems in terms of contradictions, subsystems, and functions of a technique that help resolve technical problems. It gives a strategy for solving the problem and helps choose appropriate information for solving the problem.

4.4 CONCLUSION

In contrast with other methods of solving technical problems, TRIZ emphasizes the contradictions and recommends solving them instead of making the usual engineering trade-offs. This approach leads to strong inventive solutions. This chapter reviewed the full set of contradictions known in TRIZ. The schematic presentation of the technique and problems in it reduces the numerous different objects to a few generic structures and paves the way for easily finding solutions by analogy with already solved problems that have similar structures.

TRIZ actively uses models of the problem that describe native contradictions in some primary functions of a technique. In this book we consider mostly point and pair technical problems with physical and technical contradictions correspondingly. A heuristic method for triangle problems, recent targets in TRIZ research, and network problems has been developed by the author, but it is out of the scope of this book. It is usually unknown in advance how to eliminate this contradiction in reality, but there is always the possibility of formulating an imaginary solution (known in TRIZ as the Ideal Final Result). This subject is discussed in the next chapter.

REFERENCES

1. Altshuller, G. S., *To Find an Idea,* Nauka, Novosibirsk, 1986 (In Russian).
2. Wilson, P. F., Dell, L. D., and Anderson, G. F., *Root Cause Analysis: A Tool for Total Quality Management*; ASQC Quality Press, Milwaukee, 1993.
3. Dettmer, H. W., *Goldratt's Theory of Constraints,* Quality Press, Milwaukee, 1998.

5 Ideality

5.1   INTRODUCTION

The concept of Ideality has its root in philosophy, where it refers to the status of ideas and pattern "per se" in metaphysics. The famous German philosopher Immanuel Kant discussed the Ideality of Space and Time in his work "Prolegomena to Any Future Metaphysics" (1783). Danish intellectual S. A. Kierkegaard and American philosopher E. A. Singer, Jr. more recently wrote of ideality. Modern metaphysics stresses that purposeful systems that can select between objectives can be ideal-seeking. Such systems can move toward Ideality by continuously changing to another objective once an objective has been achieved or the effort to accomplish it was unsuccessful.

Idealization is also fairly common in natural sciences. It is possible to define idealization as a mental act of creating abstract objects that cannot exist in reality and cannot be obtained as a result of any experiment. Such ideal objects represent a limit of the real objects. Ideal objects (such as a point or line in geometry, the absolute black body or ideal gas in physics) play an important role in axiomatic theories and in analysis of real objects. As such, ideality is an abstraction that represents the reflections of reality useful for the studies of various phenomena. It also serves as a powerful analytical and solution instrument of TRIZ which has been developed by B. I. Goldovsky, G. S. Altshuller, V. V. Mitrofanov, B. L. Zlotin, A. V. Zusman, G. I. Ivanov, S. I. Grigoriev, and other TRIZniks.

5.2   IDEAL TECHNIQUE, IDEAL METHOD, ..., IDEALITY

In TRIZ, Ideality applications include the ideal system, ideal process, ideal resources, ideal solution, ideal method, ideal machine, and ideal substance. Ideality in TRIZ has been described [1-4] in the following way:

- the ideal machine which has no mass or volume but accomplishes the required work
- the ideal method which expends no energy or time but obtains the necessary effect in a self-regulating manner
- the ideal process which actually is only the process result without the process itself: momentary obtaining of a result

- the ideal substance which is actually no substance (a vacuum), but whose function is performed
- the ideal technique which occupies no space, has no weight, requires no labor or maintenance, and delivers benefit without harm, etc., and "does it itself," without any additional energy, mechanisms, cost, or raw materials.

Such declarative descriptions represent Ideality primarily as a bias mental signal against psychological inertia of a solver (see Chapters 1 and 11). A technique aspiring to Ideality has "nothing extra" and can be characterized as follows:

- the primary and secondary functions perform only there and only then, where and when necessary, at any time and in any place
- the technique performs only what is necessary, i.e., only PF and secondary UF
- the technique consists of only what is necessary, i.e., only those subsystems and interactions required for the PF and secondary UF
- the technique does not have harmful and neutral functions (the subsystem for the useful functions must work against HF and achieve performance of NF function for "free") and also does not have support and auxiliary functions (need for these functions is eliminated).

It is necessary to remember that the native technique information (see Chapter 3) should be kept for the functioning of the technique. The ideal technique is an ultimate result which, unfortunately, cannot be reached, because a system or process that could perform only UF is impossible as perpetual-motion without an NF (or HF in some cases) of energy dissipation.

Although this initial approach represents Ideality as an impossibility and a nonentity, a solver needs to imagine the Ideal TS or the Ideal TP. Consequently, more instrumental description of Ideality has been proposed in order to satisfy engineering needs to link the concept of Ideality with ideas about functions of a technique. Any ideal system or process is an absent system or process, but the concept of Ideality can be practically realized. Since the primary function has to be performed, some materialistic system or process must exist to accomplish it. V. V. Mitrofanov has argued that the idealized technique is the limiting state of a real technique possessing the best combination of parameters and the optimal number of functions. Ideality can be considered only in the interaction of the technique with the environment and super-system while functioning. The degree of Ideality is, in essence, the "difference" between the real and idealized techniques in the number of functions and/or the values of parameters (see Figure 5.1).

Thus, a technique can be called ideal if it is not absent (really, physically) but possesses the set of parameters ideal for functioning. Mitrofanov's formulation is the sequential approach of the real technique to the ideal one in every undesirable effect or analyzed parameter. The digression from the absolutely ideal technique to reality occurs in the following steps:

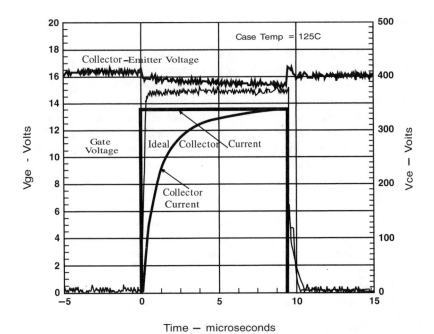

Time — microseconds

FIGURE 5.1   Difference between ideal and real collector currents in a semiconductor transistor. The rectangular power pulse, shown as a bold curve in the figure, is ideal in essence because the real current pulse has buildup and decay fronts as well as noise (more visible for voltages). In this example, the measure of the Ideality degree is the tendency to zero of the area between two pulses.

1. A useful result from a TS or TP is to be obtained without this TS or TP ("free of charge");
2. At every time and at every point of space, TS or TP must have only such properties and interactions that are needed for obtaining the useful result ("nothing unnecessary");
3. All available properties and interactions of subsystems and the environment must be used maximally; losses and waste must be eliminated ("maximal benefit from unnecessary");
4. Continuous working toward better time, speed, mass, efficiency, energy, etc. favors approaching Ideality.

The first step orients inventors not toward creating a technique, but to using something already in existence. This step can be implemented by using resources of the environment or super-system (see Chapter 6). When this step is implemented, the number of relations between TS and the super-system increases.

The second step orients inventors to develop TS or TP which has only what is essential. This step can be implemented due to specialization of subsystems in space and time, which leads to an increase in number and heterogeneity of subsystems and interactions between them. Since a change of operational condition of TS is rather a rule than exception, transition from "static" systems to "dynamic" ones is met most often. Higher optimality of distribution of properties and interactions of subsystems in space and time often require higher degree of variability of sub-systems. The subsystems become less rigid (transition to powders, liquid, gas); easily controlled types of energy (magnetic, electric) come into use.

The third step orients to maximal use of all technique resources. It is applied most often, because there are no absolutely essentials — only TS or TP. On the whole, implementation of this step leads to increased versatility of subsystems. This step is closely related to the first, which can be considered as application of the third step at the level of the super-system. This step can be predominantly implemented in the following forms:

1. Use of waste or otherwise unused products of techniques, subsystems, and elements.
2. Use of hidden properties (side-properties) of TS subsystems; load maxi-mally the subsystems.
3. In spite of introducing new elements, change existing subsystems or the environment due to additions, external fields, changes in the shape, etc.

For example, consider USSR Patent No. 1364508. A hand wheel is set on a shaft of an electric engine feeding from an accumulator battery, and the accumulator battery itself is used.

The fourth step corresponds to increased efficiency of processes running in a technique. This step can be predominantly implemented in the following forms:

1. Maximal use of the space due to increased TS dimensionality (in partic-ular, perform the transition "point — line — surface — volume") and due to use of the rear surface of objects.
2. Maximal use of time due to an increase in continuity of processes, idle and intermediate moves and pauses; i.e., time expenses for obtaining the useful result must be minimized.
3. Lower energy losses, due in particular to decrease in the number of energy transformations in the technique.

The main way for its implementation, in spite of process intensification, is a decrease in the number of operations and their combining in space and time. Com-bining transportation with processing is one of the basic directions of development of transport techniques.

It seems reasonable to distinguish *local* Ideality for different representatives of technique with the same principles of performing PF and *global* Ideality between various techniques with all possible technique principles of performing the same PF. The processes of local Ideality growth can take place through two pairs of mutual modes:

- **Reaching** — improving the performance of UF by means of optimization and various types of small enhancements; for example, using improved materials, introducing additional regulations, selecting optimum sizes, intermediate correlation, etc.

- **Squeezing** — the reduction of payoff factors and the elimination of losses and waste by means of full or partial compensation of harmful functions and/or the optimization of neutral functions; for example, using cheaper material, standardized elements, excluding ample excess, increasing workability, etc.
- **Universalization** — increasing the amount of useful functions accomplished by the multipurpose technique. Most often the secondary functions are transferred to a specific technique belonging to the super-system of this specific technique; for example, a modern multimedia computer that performs the functions of a TV, telephone, fax machine, music center, etc. and weighs and costs thousands of times less than the original mainframe computers.
- **Specialization** — the distinction of quality performing primary function by refusal of other secondary useful functions. Consider an example from the semiconductor industry: Some time ago a company designed IC chips, designed and made equipment for their production, produced IC chips, and used them in electronics fabricated by the same company. Now there are several specialized companies that design only a certain class or part of IC chips, corporations that make only some types of equipment for the semiconductor process, foundries for various technological operations which are necessary for IC chip creation, consulting firms for optimizing IC chips applications, companies that use these IC chips in boards, and assembly shops that produce the final electronic product from these boards.

The processes of global Ideality growth can take place through three modes:

- **Function Exclusion** — disregarding the neutral and auxiliary functions (together with the subsystems associated with their performance) without deteriorating PF performance. For example, painting metal parts with conventional paint releases dangerous fumes from the paint solvents. An electrostatic field can be used to coat metal parts with powdered paint. After the powder is applied, the part is heated and the powder melts. A finished coat of paint is thus produced without solvent.
- **Subsystems Exclusion** — delegating the functions of subsystems to available resources. The cost of a technique often decreases if auxiliary subsystems are excluded. For example, the high vacuum of the moon eliminates the need for the glass cover of light bulbs on the lunar vehicle. The function (protect the light filament from oxidation) was satisfied without a subsystem (cover).
- **Principle Change** — modifying the basic operating principles in order to simplify a system or to make a process more convenient. Hot, soft-sheet glass (used to manufacture plate glass) tends to sag between the rollers as the sheet moves on a conveyor. If the rollers were smaller, the sag would decrease. The ideal final result is eliminating any sagging. A solution is to convey the hot sheet and keep it flat by floating the glass on a pool of molten tin whose atoms work as the smallest roller. Another example is the transition from regular mail to e-mail in order to increase the speed and ease of information exchange.

The first mode, function exclusion, occurs due to the delegation of the system's function to another technique, which can be an existing subsystem, super-system, or even higher level system, or an outside system of the same hierarchical level. If a few systems perform similar PFs, the one that requires fewest expenses to build and maintain is the best. In all these cases, a solver has to use resources first to solve the problem (see Chapter 6 concerning resources). Some TRIZ instruments have been developed for such problems but they are beyond the scope of this book.

Suppose an initial technique that executes a set of primary and secondary functions at some level needs to be improved, to achieve a higher level of system performance, or to execute another, usually wider, set of functions. It may require adding a new subsystem. However, it is usually worthwhile to explore the possibility of using an ideal (sub)system to perform a new set of PF and UF by the existing technique. For example, it can be accomplished by some modifications to the existing technique, if, during the new system design the engineer

- eliminates the deficiencies of the original technique,
- preserves the advantages of the original technique,
- does not make the new technique more complicated than the original, and
- downsizes the HF of the original technique and/or does not introduce new HF.

Therefore, a problem solver should choose the target (local or global Ideality) before he or she starts the solving process (discussed also in the next section). Usually, another technique should be introduced only if all attempts to find the ideal system and/or to improve the existing technique have failed. The search for the local Ideality is preferable.

The technique, even the ideal one, is not itself the goal; it is needed only to execute useful functions to serve a human being or another technique. A variety of techniques may be capable of performing the same function. In fact, a technique is a "fee" for realization of the required function, because a technique always has some harmful effects. Therefore, we can write:

$$\textbf{IDEALITY} = \Sigma \textbf{UF} / \Sigma \textbf{HF}, \qquad (5.1)$$

where $\Sigma\textbf{UF}$ is the sum of all useful functions and $\Sigma\textbf{HF}$ the sum of all harmful functions.*

---

* Equation 5.1 is known in economics as "cost-benefit ratio," in management as "effectiveness," and in Value Analysis/Engineering as "value." In all these fields it is an evaluation indicator of processes or systems. Such indicators are roughly equivalent to Ideality as used in TRIZ. For example, Peter Drucker, famous American management guru, proposed two major indicators of efficiency and effectiveness that should be evaluated for any TP (and TS). Efficiency is "doing things right," and effectiveness is "doing the right things." The efficiency of a process is the ratio of the amount of a single quantity (e.g., of energy) appearing in the useful output to the quantity of the same kind entering the input to that process. The effectiveness gives an aggregate of benefits (yield, income) and expenditures (cost, human factors), usually expressed in some common measure, such as money. Such indicators must be interpreted with care because they easily omit, hide, summarize, simplify, or ignore many features that may be important. Analysis of pros and cons of the indicators mentioned above and concepts of Ideality should be performed.

Changes in the system that lead to some combination of increased numerator and decreased denominator bring the system closer to Ideality. Such combinations include

- increase the numerator at a faster rate than the denominator — **dUF/dt > dHF/dt > 0**
- combine the subsystems for several functions into a single system in order to decrease the denominator — **dUF/dt > 0, dHF/dt < 0**
- remove unnecessary functions in order to reduce the denominator — **dUF/df = 0, dHF/df < 0**
- increase the numerator by adding functions or by improving the performance of the more important functions — **dUF/df > 0, dHF/df = 0**

Ideality can be formulated from Equation 5.1 as a simple statement: *some subsystems remove the harmful (unnecessary, superfluous) effects themselves, while other subsystems increase their capacity to carry out the useful function(s).*

In many cases it is useful to separate all harmful effects in Equation 5.1 into two groups: expenses and harms. Expenses include the costs of raw materials, the space the technique occupies, the energy it consumes, and the noise it emits. Harms in the equation include all types of waste and pollution. For example, in the furnace for production of semiconductor silicon ingots, the electrical power expenses and the possibility of injury from high temperature or electrical power are parts of the denominator. The ability to melt materials because of the high temperature and the small amount of pollution from special quartz crucibles (compared with metal crucibles) are parts of the numerator in the equation for furnace Ideality,

Therefore, Ideality can be written as

$$\text{Ideality} = \Sigma\ \text{Benefits}/(\Sigma\ \text{Expenses} + \Sigma\ \text{Harms}) \tag{5.2}$$

The equation states that in order to expand Ideality of a system one should increase benefits and/or decrease expenses and/or decrease harm. It also implies that to increase the Ideality of a technique, one should try to

- increase the numerator at a faster rate than the denominator,
- increase the numerator by adding functions or by improving the performance of the more important functions,
- remove unnecessary functions in order to reduce the denominator, and
- combine the subsystems for several functions into a single system in order to decrease the denominator.

5.3   IDEAL FINAL RESULT

The Ideal Final Result (IFR) is the absolutely best solution of a problem for the given conditions. Although IFR was proposed by Altshuller and Shapiro in the 1950s [1] and it is well known for proving theorems in mathematics, it was rarely used in engineering outside TRIZ. The goals of formulating the Ideal Final Result are to

- eliminate rework (solve the right problem the first time) by addressing the root cause of the problem or customer need;
- think of how to get the IFR with the least work, without reconstruction or big changes in the technique;
- determine resources in the technique, environment, and super-system (discussed in Chapter 6);
- sketch an implementation-free description of the situation after the problem has been solved;
- recognize better the constraints' appearance in the problem.

It is quite natural and logical for an engineer to begin the problem-solving process by using his or her knowledge and experience to think of an improved subsystem that can be developed. Although this "forward" method does work, it can be slow and expensive. IFR provides the opportunity to apply the so-called "backward" method to the technical problem solving.

As we know, a problem is a barrier or gap between the existing situation and the desirable situation or goal, and problem solving is a search for the steps that transform the existing situation into the goal (see Chapter 1). Why should we want to work in the reverse direction from the desirable to the initial situation rather than use the quite normal method from existing situation to the goal? When is this backward method more appropriate than working forward, for what problems, and why? Working backward is likely to be useful if a problem satisfies one of the following three criteria [5, 6].

I. The problem has a clearly specified goal, as is the case for most technical problems. The IFR is the unique specified goal stated in the technical problem, thus the possibility exists of working backward. This approach is particularly true if, in contrast to the single goal statement, there are many statements about the initial situation and the information about it is uncertain for problem solving. Newell and Simon, researchers in the field of problem solving, have stated that the advantage to working backward in such problems is that there is no ambiguity as to what statement to start with, whereas such ambiguity is considerable when working forward [5]. As they noted, working forward in such problems is analogous to looking for a needle in a haystack, whereas working backward is analogous to the needle's finding its way out of the haystack. You can start from any of many places outside the haystack to try to find the single location of the needle. By contrast, the needle starts in a single location and can solve the problem of getting out of the haystack by going to an extremely large number of alternative locations outside the haystack.

Therefore, working backward is preferable to working forward if in working backward the number of different sequences of steps that have to be considered is considerably smaller than the number when working forward [5].

II. For many technical problems (especially during synthesis and/or genesis of new technique) the goal can be achieved if a problem solver starts from different existing techniques and their subsystems. In such problems, the initial different existing techniques or sets of subsystems are independent from one another. That is, to solve the problem by the forward method, a problem solver needs to organize and make steps from any one of these techniques or set of subsystems to the goal. Usually a problem solver does not know and cannot determine *a priori* which set of subsystems is preferable. Using Ideality and working backward from the goal, a problem solver can save much time if a choice between different initial sets of subsystems is not predetermined. Therefore, the method of working backward is frequently very useful in such problems because the unique starting point frequently directs you to only those aspects of the given information that are relevant to the solution [6].

III. Three cases of steps sequence can occur during the solving of a technical or other problem:
   - One-to-one step — possible to determine which input statements produce the output statement.
   - Many-to-one step — several different input statements are needed to produce the single output statement.
   - One-to-many steps — one input statement can produce several different, independent output statements.

   The determinism of each step can be completely adequate if one given input statement leads to one output statement and inadequate if one output statement can be produced by several input statements. The kind of steps needed to fill the gap predetermines the choice between forward and backward methods.

   In the "one-to-one" case, there is no preference for the backward or forward method. When steps are not one-to-one, working in one direction may lead to a more rapidly growing number of possible sequences than will working in the opposite direction. In such cases, it generally is preferable to work in the direction that produces more certain information about the subsequent steps. The backward method works well for the one-to-many steps case because in such cases it is necessary to find only one of the several input statements when the output statement is known.

The most general criterion for the applicability of working backward is that the method would produce fewer steps in problem solving than would be produced by working forward. Another powerful approach — the bi-direction solving of the technical problems — has been developed by the author and other TRIZ experts and is discussed in Chapter 17. Bi-directional methods work well when both one-to-many and many-to-one cases occur in a single problem during the transformation between the initial and desirable situations.

In problems where IFR is not so clearly and completely specified and there is, in fact, a variety of possible alternative goals, the advantages of working backward are often diminished. Even the incomplete formulation of the IFR helps a solver consider what constraints are required by the laws of nature, what constraints are required by human society, and what constraints were introduced in the system during previous design. A solver should distinguish all constraints in terms of importance. The IFR describes the solution of a technical problem, independent of many constraints of the original problem. Based on the IFR concept, an engineer can make "a step back from Ideality." Stating the IFR and retreating from it as little as possible offers strong technical solutions because of the possibility of designing the system that works almost ideally. Often backward solving is possible only after this "step back from Ideality."

However, if the mentioned criteria are not satisfied by a problem, working backward will likely be inferior to working forward.

5.4   CONCLUSION

The interconnection and unity of the two major TRIZ concepts, Ideality and Con-tradictions, provide a possibility for solving inventive problems. Usually purification of a problem (i.e., establishing correct statement of a problem), clarification of the contradictions (i.e., detecting the problem's roots), and imagining the best solution (i.e., discovery of Ideality) are powerful steps during problem solving. One of the major criteria of a technique's quality is its Ideality, and by affinity a major criterion of the quality of problem solving is its ability to decide about the main contradiction of the particular problem.

Usually engineers agree to "pay" for the effect required — in terms of machines, expenditure of time, energy, space, etc. This need for "payment" seems self-evident, and the engineer is concerned only that the "fee" is not too high. In contrast, Ideality signifies a possibility to move away from the "expenses" for the solution.

Another purpose of IFR is to present an orientation point for a problem solver and to stimulate a "backward process" during problem solving. It is impossible to make the first step to Ideality without knowledge of the features of techniques that were discussed in Chapter 3. Additionally it is often impossible to improve a tech-nique without knowledge of its resources, which are discussed in Chapter 6. The concept of Ideality is also linked strongly with another TRIZ concept, the evolution of a technique, outlined in Chapter 7.

REFERENCES

1. Altshuller, G. S. and Shapiro, R. B., About the Psychology of Inventiveness, *Problems of Psychology*, 6, 37, 1956 (Electronic copy at http://www.jps.net/triz/triz0000.htm).
2. Gasanov, A. I., Gohkman, B. M., Efimochkin, A. P., Kokin, S. M., and Sopel'nyak, A. G., *Birth of Invention,* Interpraks, Moscow, 1995 (in Russian).
3. Ivanov, G. I., *Equations of Creativity or How to Learn to Invent*, Prosvechenie, Moskow, 1984 (in Russian).

4. Altshuller, G. S., Zlotin, B. L., Zusman, A. V., and Filatov, V.I., *Search of New Ideas,* Kartya Moldovenyaske, Kishinev, 1989 (in Russian).
5. Newell, A. and Simon, H. A., *Human Problem Solving*, Prentice-Hall, Englewood Cliffs, 1972.
6. Wickelgren, W. A., *How to Solve Problems: Elements of a Theory of Problems and Problem Solving,* W. H. Freeman, San Francisco, 1974.

6 Substance-Field Resources

6.1 INTRODUCTION

Resources play an important role in the solution of problems that are close to the IFR. Resources in technical systems are discussed here from different points of view presented by Lev Kh. Pevzner, Svetlana V. Vishnepolskaya, Zinovy Royzen, Rashid H. Sharipov, and other TRIZniks [1, 2]. Any technique that has not reached Ideality should have some substance or field resources available.

6.2 RESOURCES OVERVIEW

As already mentioned, any technique is part of a super-system and a part of nature. It exists in space and time, consists of and/or uses substances and fields, and performs functions. Resources can consequently be grouped in accordance with the following descriptions.

Natural or environmental resources — any material or field that exists in nature (the world around the technique).

Example:
- Solar cells that use a natural energy resource.

Time resources — time intervals before the start, after the finish, and between cycles of a technological process, that are partially or completely unused:

- alteration of a subsystem's preliminary placement, application of pauses
- use of concurrent operations
- elimination of idling motion

Example:
- Simultaneously cooking several different foods for a dinner.

Space resources — positions, locations, and order of subsystems, the technique itself, and super-system:

- voids and holes in the technique
- distance between subsystems
- mutual location of subsystems
- symmetry/antisymmetry

Example:

- Posting ads on food packages.

System resources — new, useful technique properties or new functions obtained when changing connections between subsystems or when joining independent techniques in a new super-system (e.g., a few new TS become one new TP).

Example:

- A scanner and a printer used together as a photocopy machine.

Resources, such as by-products and wastes, may be found in the technique itself, in neighboring techniques and environment, or in the super-system.

Example:

- When producing pile foundations, soil extracted during drilling is mixed with binding agents, and this mixture is used for pile formation.

Substance resources — any materials composing or producing the technique and its environment.

Example:

- Exhaust gases of a snow-removing car are directed to formed snow rolls, thus compacting them.

Energy/field resources — any field or energy flows which are existing or produced in the technique and its environment or which can replace subsystems.*

Example:

- The difference in electric potentials between the ionosphere and the ground produces a low-capacitance electric field of approximately 100 V/m that can be used to control airplanes flying at low altitude.

Information resources — any signals that exist or can be produced in the technique. Note: information cannot exist without a carrier, and Altshuller chose to use fields for representation of information, although sometimes the information carrier is a substance. Information has both form and content.

Example:

- Reflective material on bicycling clothes.

Functional resources — the capability of a technique or its environment to perform secondary and auxiliary functions: application of existing neutral functions and/or harmful functions.

---

* Fields and energies are often not distinguished in TRIZ.

Example:
- Software for task scheduling relies on a built-in internal clock in a personal computer.

The application of resources in solving technical problems often leads to almost Ideal solutions. An additional and unexpected benefit often arises as a result of a problem solution when some (usually functional) resources are used. For example, it is expected that exercising one arm will increase bone mass. The benefit is that, to a lesser extent, bone mass in the other arm also increases. Hence, it is useful to discuss resources in more detail.

6.3   DETAIL ANALYSIS OF RESOURCES

Resources can also be categorized as internal and external. **Internal resources** are things, substances, and fields reachable in the conflict area or operative zone during or before conflict time or operative period. **External resources** are things, substances, and fields in the neighborhood of the conflict area or phenomena that occur before conflict time. For example, ash from heat electric stations is used as soil deoxidizer, plant growth stimulant, and filler for plastic products and concrete.

On the other hand, internal and external resources can be readily available, derivative, and differential. Chapter 18 shows such classification is quite useful when a problem is solved with ARIZ.

6.3.1   Readily Available and Derivative Resources

In many cases the resources necessary for solving the problem are available in the technique in the form that fits the application; these are *ready resources*. However, quite often the available resources can be used only after some sort of preparation, such as accumulation or variation. Such resources are called *derivatives*.

Let us first discuss both these types in general, then provide more systematic description of various resources, and list some beneficial resources.

Readily available resources can be used in their existing state. Substance, field, functional space, and time resources are elements available to most techniques. Since substance resources include all materials from which the technique and its environment are composed, any technique that has not reached Ideality should have substance resources available.

For example, to prevent overheating of machine components (such as bearings), a temperature control system is installed that typically includes thermocouples positioned where overheating is likely to occur. Sliding bearings often include an electro-conductive insert in an iron ring fixed within the component body. Overheating can be prevented if the contact between the iron ring and the body is used as a thermocouple. That is, the component is switched off if the thermocouple detects a temperature above a certain value.

Derived resources, or derivatives, are resources that can be used after some kind of transformation. These derivative resources can be raw materials, products, waste, and subsystems of the technique (including water, air, etc.) that may not be useful in their existing state but might be transformed or modified to become a resource. For example, warm water from an electrical power station can be used in a greenhouse.

Often phase transitions (in the sense of physics) and chemical reactions can be used for creation of derivative resources. An example is heat produced during dissolving of CaO in water.

More specific readily available and derivative resources are listed below, along with examples.

Ready available matter resources — any material, subsystem, product (final and intermediate), remnants, etc. consisting of anything that can be used directly.

Example:

- A factory produces very pure quartz for the semiconductor industry. Quartz is used in crucibles for melting ultra-pure metals at the same factory.

Derivative matter resources — substances and materials obtained either through transformation of the ready substance resources and under any action on the ready substance resources.

Example:

- A thin film of silicon dioxide is grown on the surface of semiconductor wafers for protecting semiconductor chips from impurities.

Ready available energy resources — any nonutilized amounts of energy existing in the technique or in the environment.

Example:

- Electrical energy available while a car is running powers the car's radio.

Derivative energy resources — energy received either through transformation of ready energy resources into other kinds of energy or through changing the direction, intensity, and other characteristics of their action.

Example:

- A thermocouple transfers thermal energy into an electrical signal that can be measured precisely.

Readily available field resources — any unused fields that are available in the technique or in the environment.

Example:

- The gravity and magnetic field of our planet.

Derivative field resources — fields received either through transformation of readily available field resources into other kinds of fields or through changing the direction, intensity, and other characteristics of their action.

Example:

- The electrical potential (electrostatic field) between a cloud and the Earth can be transformed into light (electromagnetic field).

Readily available information resources — information about the technique that can be received with the help of the fields or the substances coming through or out of the technique.

Example:

- Oil and other particles in a car's exhaust provide information about the functioning of the car engine.

Often a technique has a surplus of information that can be reduced without harm.

Example:

- Computer programs such as Pkzip.exe, Arj.exe, and WinZip for compressing files.

Derivative information resources — making irrelevant information relevant. *Timeliness and accuracy are important for information resources!*

Example:

- Small changes in the Earth's magnetic field can be used in searching for minerals.

Readily available space resources — free, unoccupied space available in a technique.

Examples:

- Compartment door in a passenger train slides into interwall space.
- Test patterns are placed in the scribing lines of semiconductor wafers. Tests are conducted on the whole wafer before individual semiconductor chips are separated along the scribing lines.

Derivative space resources — additional space obtained from the expanse of different geometric effects.

Example:

- Effective length of ring elements, such as magnetic tape or band saw, can be doubled by folding as a Mobius strip.

Readily available time resources — unused or only partially used time intervals during, before, or after, a technological process, or between operations.

Example:

- Having the oil in your car changed in a shop and having lunch in a neighboring restaurant at the same time.

Derivative time resources — time intervals obtained as a result of process acceleration, deceleration, or interruption.

Example:

- Transmission of information in compressed form.

Readily available functional resources — abilities of a technique or subsystem to perform functions in addition to those for which it was specifically designed.

Example:

- To stand on a table (to use its height) to change a light bulb on the ceiling.

Derivative functional resources — abilities of a technique to perform additional functions after some changes.

Example:

- Compression mold for producing parts of thermoplastics has figured gating channels, and gates become finished products, such as letters for an alphabet.

6.3.2   DIFFERENTIAL RESOURCES

Often the difference in the properties of a substance or field is a resource that forms a technique. This differential resource can usually be used at the expense of a difference in structure and properties of the technique's substances or fields.

Typical Differential Resources of a Substance

Structural difference — Anisotropy is the difference in physical properties of a substance in different directions. It is characteristic, first of all, for crystals (solid or liquid). Anisotropy can be used for the execution of the primary function and/or the increase of UF performances.

Examples:

- Optical properties — diamond's brilliance and other optical properties manifest themselves only when it is faceted along its symmetry planes.
- Electrical properties — quartz plates possess piezoelectric properties only if a crystal is cut out in a certain direction.
- Acoustic properties — the difference in acoustic properties of a part's sections with a different structure (defects) forms the base for ultrasonic defectoscopy.
- Mechanical properties — sharpest and cheapest scalpels are produced by splitting a piece of volcanic, noncrystal glass (obsidian) along slip planes. Even ancient Aztecs used such blades. The art of chopping wood is based on using the wood's anisotropic properties.
- Chemical properties — in etching hot-rolled steel, first the scale is dissolved, allowing preparation of the piece for rolling. Crystal etching first takes place at defect points, dislocations, and boundaries of microcrystal blocks, allowing production of samples for microscopic research into crystal structure.
- Geometrical properties — only the good spherical pellets roll down an inclined sorting table into the receiving bunker. Bad pellets, those with deviations in shape, remain on the table and then are rejected.

Difference in material properties — Variations in material properties can be used for executing the primary function or increasing UF performance.

Examples:

- Silicon epitaxy films with different electrical resistance are used to build semiconductor devices with various values of a specific property, such as different breakdown or threshold voltages or recovery time.

- A mixture of steel cutoff pieces can be separated using magnetic separation by heating the mixture, step-by-step, to temperatures corresponding to the Curie point of different alloys.

Typical Field Differential Resources

Field inhomogeneity in different parts of a technique allows the use of difference in potentials of this field.

Use a field gradient

Example:

- Difference in atmospheric pressure near the ground and at the altitude of 3200 m gives rise to draught in furnace with the help of a chimney.

Use a field inhomogeneity in space

Examples:

- Rail cross section can be considered an example of rational use of materials in only the loaded parts of a construction.
- In order to improve working conditions, workplaces are located in the region of acoustic shade.

Use a deviation of a field value from standard

Examples:

- Analyzing how the patient's pulse differs from the pulse of a healthy person, Tibetan medicine diagnoses more than 200 diseases.
- Heat vision is based on difference in the heat radiation of objects.
- Magnetic mines float to the surface and explode under the object having the strongest magnetic field among neighboring objects (usually a ship).

To perform the useful function, interacting substances and fields must have minimal difference in order to avoid harmful flows in the technique; however, to eliminate the harmful function or unwanted effect, the difference must be maximal in order to provide useful flows. This can be reached by controlling the aggregate states of matter of given substances or by the fields' interference. A particular case of the difference resource is the disappearance or appearance of some materials' properties in defective subsystems compared with normal subsystems. Such a difference forms the base for quality control and from this point of view this difference should be big enough.

6.4   RESOURCE USAGE

During the analysis of resources the following questions arise: how do we choose resources during problem solving? in which sequence should we look for resources? how do we use resources in a more preferable manner? The following two checklists can help in answering these questions.

Resource Estimations	
Quantitative	Insufficient
	Sufficient
	Unlimited
Qualitative	Useful
	Neutral
	Harmful

Availability of Resources	
Degree of readiness	Ready
to application	Derivative
	Differential
Arrangement	In an operative zone
	In an operative period
	In the technique
	In subsystem(s)
	In the super-system
Value	Expensive
	Cheap
	Free

The usual order of a resource search that allows for the maximum result at the minimum cost is

1. collateral resources; in particular, waste resources;
2. external environment;
3. the tool;
4. other subsystems of the technique;
5. the raw object and/or the product, if there is no interdiction for their changes.

During this analysis of the product's resources, we should keep in mind that the product itself is usually considered an unchangeable element of the technique. The only exceptions arise when the product can be changed itself or can allow the following:

an expenditure of any part of it (usually when much of the product is available);
a transition in super-system;
a usage of subsystems;
a connection with emptiness (void);
a temporary change in time.

It is easier to use resources that are available in an unlimited quantity. As a rule, such is possible with resources from the environment, i.e., air, water, their temperature, solar and wind energy, etc. If it is necessary to use resources that are not present in the environment, consider resources that exist in an adequate amount in the technique itself. Frequently, such resources are those connected with the UF and NF of the technique or adjacent techniques, such as matter or energy produced or consumed by the technique or the technique's free space. It is inconvenient to use resources that are not available in sufficient quantities, and such use usually requires additional effort to accumulate such resources.

We can also examine resources based on the degree of their utility in the following order:

1. harmful resources (especially manufacturing waste, filth, unused energy),
2. readily available,
3. derivative,
4. differential.

Such an approach increases the Ideality of the technique and allows improvement of the ecological parameters of manufacturing. Note that any transformation of a simple resource to a derivative or differential one requires complication of the technique, additional energy, and money, thereby distancing the decision from ideal.

These search directions are statistically preferable. Of course, that does not mean this order is the best for all techniques and problems. Sometimes subsystems, technique energy, technique behavior, and technique functions can provide resources.

6.5 OVERCOMING RESOURCE LIMITATIONS

Unfortunately, resources are often hidden or unavailable. The following general recommendations are therefore useful:

1. concentrate the resource on the most important actions and subsystems;
2. turn to rational and effective usage of the resource (exclude any loss, waste, pauses);
3. concentrate resources in space or in time;
4. use resources wasted or lost in other processes;
5. share the useful resources of other subsystems and dynamically regulate these subsystems;
6. use other resources according to the hidden properties of the subsystems;
7. use other resources, modifying them for the needs of the process (thus use the environmental resources).

Special features of different resources can also help to overcome their limits:

Space

- select the most important elements from the subsystems and put the rest into less valuable regions of space
- make the layout denser, use spare space (clearances, cavities), place one object inside another
- use surfaces of neighboring subsystems or the inverse side of a surface
- use another space dimension (point, line, surface, volume)
- use more compact geometric layouts (e.g., spiral)
- use temporally the spare space within another object, making shapes change in dynamics. (If two incompatible objects (processes) should be placed in one point, they should be cyclic so that one can continue during the interruption in the other)
- while solving information problems, data (including a graphical image) should be excluded from the subsystem

Time
- implement only the most valuable part of the process that determines the result to the highest degree
- minimize time waste during the most valuable period, implementing some stages in advance, making the subsystem responsive to the process
- make the action more intensive, eliminate pauses, exclude stops and idling, make the process continuous
- move from sequential execution to parallel

Material
- use thin films, powders (dust), vapors to spread a small amount of substance in a large volume
- use matter from the environment in mixture with the given substance (use the substance given as an addition, foam, inflatable shell, or a coating filled with liquid)
- use matter from the environment transformed for the needs of the process (as ice, gas, or jet fluid)

Energy
- implement only the most valuable parts of the process that determine the result to the highest degree
- limit usage of the most valuable type of energy and use cheaper ones instead
- if lack of power exists, concentrate energy in space (add several less powerful streams) and in time (with an impulse action or accumulating several impulses)
- use energy resources of nearby techniques
- if intensity of the flow for some kind of energy is limited,
    use this flow to manipulate more intensive flow
    add other types of flow using hidden properties of the technique
    transform other types of energy flow
- utilize energy wastes of the given and nearby techniques
- use environmental resources

Another way to analyze resources is a two-dimensional morphological box, in which the same resources are listed on the horizontal and the vertical axis (for tools, collateral products, external environment, emptiness, etc.). Let us stress the necessity to include all resources in this box. Often problem solvers overlook

- unused, specific properties and characteristics of the technique;
- existing, but incompletely used, subsystems;
- arrangements, mutual orientation, and communication between subsystems;
- subsystems that can be easily built in the technique;
- new, unexpected properties or features of the technique appearing after small changes.

Examples:

- To use the weight of a (sub)system or periodically arising effort to promote an additional effect.
- To compensate for excessive outlays of energy by any additional positive effect.

- To choose and to ensure (supply) the optimum parameters (temperature, humidity, illumination, etc.) and/or to specify settlement pressure (voltage) in (sub)systems on the basis of exact mathematical models and computer simulations and to use excesses.
- To exclude selection, alignment, and adjustment of subsystems during assembly of a technique in order to reduce manufacturing time.
- To use or to accumulate brake and other passive, received energy.
- To use for auxiliary purposes a (sub)system that carried out its goal or is becoming unnecessary.
- After improvement of any (sub)system, to define how other (sub)systems should be changed so that the efficiency of the technique overall has increased even more.

Such a morphological box allows us to see consistently not only simple resources, but also their combinations, at least pairs. Sometimes at this stage, three-fold or even more complex combinations from resources can arise. In this case more sophisticated morphological boxes can be created.

If, despite these methods, necessary resources are not found, try the following recommendations:

- Combine two or more different resources such as information and substrate.
- Proceed to a higher level of technique hierarchy.
- Consider whether the resource is necessary, and reformulate your search.
- Proceed to other physical principles for the primary technique's function with cheaper or accessible sources of energy or higher efficiency.
- Instead of the action required by the problem's conditions, carry out an opposite action (for example, not to cool a subsystem but to heat it).

6.6 CONCLUSION

Considering resources allows a solver to "see outside the box," which is especially important when a solver concentrates his attention to the particular subsystem(s), operative zone, and period during resolution of a contradiction in an inventive problem. Especially efficient are solutions that use empty space or voids, waste products, byproducts, or harmful substances and fields, as well as available and very cheap energies and substances. In short, resources are everything that remains idle in the technique and its environment. They often exist in the technique or super-system, but sometimes they are hidden from a problem solver. The mind-map in Figure 6.1 should help a solver navigate among various resources.

REFERENCES

1. Pevzner, L.Kh., *ABC of Invention,* Sredne-Ural'skoe Publishing House, Ekaterinburg, 1992 (in Russian).
2. Sklobovsky, K. A. and Sharipov, R. H., Eds., *Theory, Practice and Applications of the Inventive Problems Decision,* Protva-Prin, Obninsk, 1995 (in Russian).

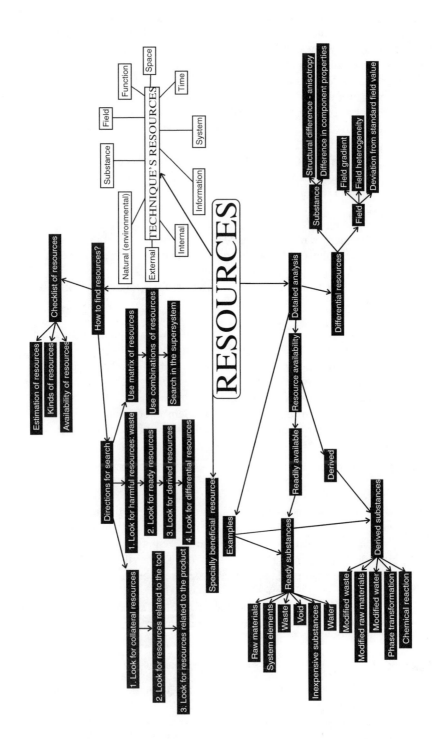

FIGURE 6.1 Mind-map of resources.

7 Evolution of Technique

7.1   INTRODUCTION

If we consider a family (or series) of techniques which are intended to fulfill the same need or PF and that were produced over a relatively long period of time, we can readily observe many changes that occurred in this family. Nevertheless, each technique has essentially remained true to its main purpose although properties and/or parameters of the technical system (TS) or technological process (TP) have been altered, as a consequence of increased perceived needs, advances in knowledge, and changes in the means of fulfilling the needs. Well-known examples are cars, personal computers, bread production, and processing of semiconductor silicon wafers.

Much research in various areas of science and engineering has shown that the general character of system development is essentially the same for any system — biological, technical, information, social, etc. While research on the evolution in biology and in economical systems is well established, similar studies on techniques are only now beginning. Most technical systems and processes change over time through gradual — consequent — or breakthrough — revolutionary — developments by engineers. Consequent development occurs when engineers need to reach new parameters of the existing technique and they, as a rule, face a problem with only a small **D** degree. Revolutionary development occurs, as a rule, when engineers face a problem with large values of **D** degree or need to create a new technique. These long-term changes of technique are recognized in TRIZ as the axiom on the existence of an *evolution of technique* (ET) that occurred because of human activities in research, design, and development. During such evolution, the construction of a TS or sequence of operations of TP changes according to particular regularities and trends (a stable, recurrent relationship between events). TRIZniks have shown that the trends revealed for ET in one engineering field can be transferred to other kinds of artificial systems.

G. S. Altshuller, Yu. S. Melechenko, and A. I. Polovinkin initiated studies of ET in the framework of TRIZ in the 1970s and 1980s, influenced by papers of Karl Marx. Such studies are still conducted by Yu. P. Salamatov, G. I. Ivanov, B. I. Goldovsky, V. A. Korolyev, N. N. Matvienko, A. P. Khvostov, I. M. Kondakov, S. D. Savransky, A. V. Netchiporenko, G. Zainiev, B. L. Zlotin, A. N. Zakharov, S. V. Strizhak, and other Russian TRIZniks. Because ET is very important for technological and economical forecasting, it is the object of intensive research in Western countries, mostly by economists. Although the styles of these investigations are quite different, their major results and conclusions are the same. This chapter presents primarily results obtained by TRIZ experts [1-4] because their ET research is virtually unknown to English-speaking readers.

The majority of TRIZniks believe that all techniques develop according to objectively existing laws. These laws can be recognized and revealed and then used for teleological perfection of techniques. At our current stage of knowledge, we deal with formulating and justifying hypotheses about the possible laws of TS and TP evolution. Although these hypotheses are often called "laws," it is more accurate to use the term "trends" because they are usually obtained by induction, or even abduction, from empirical correlations based on analysis of a huge number of patents and other TS and TP information rather than from formal logic. Sufficiently justified and generally accepted ET laws are absent yet, and a closed proof-resistance set or system of such laws, even hypothetical, is also not present.* Justification of evolution trends and construction of such a system of laws are among the most important directions of modern TRIZ research because of their potential use in technical problem solving, forecasting, and design. Assigning evolution trends to a specific technique class allows us to determine the most promising properties and characteristics of the next-generation TS and TP for this technique class and to recognize paths of evolution for it. Although currently several trends and paths of TS evolution are recognized and studied in TRIZ, this book presents only the most popular and commonly accepted. In addition, TRIZ uses some regularities known as the laws of dialectical materialist philosophy, such as "Unity and Struggle of Opposites," "Quantity becomes Quality," and "Negation of the Negation." As we will see, these regularities work during ET but their original formulations are too general for practical use in engineering. The following requirements allow distinction between somewhat stable and recurrent relations and an innumerable multitude of various ones:

1. The trends and paths must reflect real development of the technique and, consequently, should be revealed, established, and *proven* on the grounds

---

* The existence of a system for evolutionary laws is very important for engineering forecasting and for the application of such laws as a filter for solution of technical problems.

of sufficiently substantial patent and technical information and profound study of different techniques' development history.

Note

- The trend or path must be proven by many high Level (not lower than third) inventions because lower Level inventions do not alter the original technique and do not actually develop it. (Detailed discussion about the Level of inventions is presented in Chapter 8.)
- Each revealed trend or path should permit and make possible checking it in practice with the works of the patent fund while solving practical tasks and problems.

2. The trends and paths making theoretical grounds of TRIZ must be *instrumental*, also; i.e., they should assist in discovering new concrete heuristics and/or instruments to solve problems, to forecast and direct the development of technique, etc. Hence, the trends and paths provide a basis for deriving concrete conclusions and recommendations about forthcoming techniques.

3. Revealed trends and paths should be of an *open type*, i.e., allow further perfection as the engineering is being developed and new patent materials are being accumulated.

7.2   POSTULATES AND COROLLARIES OF TECHNIQUE EVOLUTION

There are two postulates and several corollaries of ET that will be described in this section. The cornerstone for the postulates is the following axiom of human society:

Both the quantity and quality of human needs, as well as requirements for humans, increase with time.

The first half of this statement is well known in political economics, and the second half is well known in social sciences where they are formulated at the qualitative level. Increased quality of our lives is a manifestation of the first half of the axiom, while specialization of our work is a manifestation of the second half. The postulates, corollaries, and paths about ET proposed in TRIZ, and discussed in the remainder of this chapter, are applied to the needs and requirements, which are satisfied with the help of techniques only.

Pioneering breakthroughs are often initiated by new needs which require new techniques (or more precisely for our day, the genesis of new subsystems). With sufficient scientific and engineering potential and when it is socially and economically expedient, a newly arising need is satisfied with the help of a newly created technique. In such a situation, a new primary function arises. This function may then exist as long as it is useful for people. V. Hubka and W. Eder [5] noted that the assortment of available techniques for each single primary function is continuing to expand, and their quality is increasing. Usually the lifetime of this function is longer than the lifetime of the technique that was created.

Such qualitatively and quantitatively different needs and functions belonging to the world of technique are steadily increasing in number. In this connection, invention of new needs or discovery of existing needs are a very important step in innovation. Methods for synthesis of new needs do not exist yet, although TRIZ enhancement of Quality Function Deployment (QFD) [6] should help to recognize needs better (discussed in this chapter).

Here is a brief review of the history of a very popular TS — the bicycle — that should help understanding of the theoretical aspects of ET presented below:

> The first bicycle, the Wooden Horse, was invented in 1817. This bicycle consisted of a frame, wooden wheels, no handlebars, and was powered by the rider's feet. Several engineering deficiencies existed: it was uncomfortable, impossible to steer, and hard to propel. In 1861 a newer generation of the same basic bicycle design, the Velocipede, had become very popular, but it had the same insufficiencies that had existed for more than 40 years. The Ariel was designed in 1870 to resolve a few of the problems — the front wheel was attached to a vertical shaft for steering, and a stomach rest was added to facilitate pushing with greater ease. Although some improvements were made, the vehicle was still unsafe, uncomfortable, and hard to propel.
>
> In 1879, after 9 years, pedals were introduced and bicycle speed increased. But there were no brakes on the bicycle! Only 11 years later, in 1888, did brakes appear. Higher speed was achieved by increasing the diameter of the front wheel, but speed escalation was further restrained by low durability of the wheel material. The appropriate material for the wheels was introduced only in the twentieth century; in the intervening time, about 10,000 patents were issued for different improvements of the bicycle.

It should be noted that the corollaries discussed below have been obtained by logical but nonmathematical conclusions from analyzing the patent fund and history of technique in accordance with the requirements described in the first section of this chapter.

A reader is encouraged to apply the following postulates and corollaries to the technique evolution with which he or she is familiar in order to learn these aspects of TRIZ better.

7.2.1 DIRECTION POSTULATE

Everything on our planet exists in 3-dimensional space and 1-dimensional time. A time arrow directs from the past through the present to the future. However, in various branches of science, different notations co-exist for the location of an object in space. The most popular in engineering is the Euclid one that includes 3 orthogonal coordinates, X, Y, and Z, which are crossed at a predetermined or random point. Frequently it is more convenient to operate with multidimensional space of some parameters, such as wave-vectors or impulses, instead of 3-dimensional space as it is widely accepted in physics. Such an approach seems suitable also for studying a technique, which exists not only in time and in regular space, but whose state can be presented also through a set of characteristics and/or parameters of different functions in multidimensional space. Position in multidimensional parametric space characterizes the state-of-the-art for the technique at each moment of time. Any

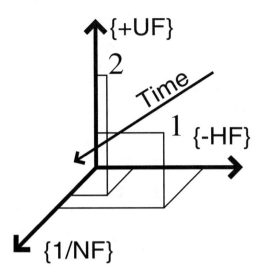

FIGURE 7.1 Two states-of-the-art for a technique (1 and 2) in the functional coordinates of multidimensional parametric space.

changes in a technique during its development can be shown as a transition from one point to another in the multidimensional parametric space.

It seems practical to group all parameters into 3 categories, each of which corresponds to a specific type of function, UF, NF, and HF, introduced in Chapter 3. It allows us to reduce the multidimensional parametric space to quasi-four-dimensional, where one axis coincides with the time arrow and each category of functions is shown along a single axis (see Figure 7.1). Therefore, the state-of-the-art for any technique can be shown as a point with coordinates {UF, HF, NF, t}.

Any technique keeps a status quo of the current state-of-the-art until it is suitable for the environment and society, i.e., until it satisfies the principal requirements listed in the order of importance:

1. Primary function (PF) and other useful functions (UF) must correspond to the human needs.
2. Cost for TS or TP manufacture and functioning must correspond to society's capabilities (technique must be efficient).
3. Reliable execution of UF must be guaranteed at some acceptable level.
4. The technique's harmful functions (HF) must be acceptable for the society and environment.

These requirements reflect the functioning of features of society, and they govern development of the technique as well. A change in a technical system or a technological process favoring satisfaction of the requirements listed above is considered a desired effect, while any change hampering satisfaction of them is considered an undesirable effect. Any change of a technique during its evolution occurs when one or more of these requirements is violated or because of the necessity to overcome current limits for performing parameters of the PF or a UF. These violations of the requirements are the driving force of ET. The criteria of technique development vary as a rule during long periods of time, because the requirements and their importance change with time as well.

Often the change in one category of the functions leads to reorganization in other categories of functions. Simple arithmetic (the rules of summation and multiplication given in Chapter 3) works for these three categories of functions.

> The positive direction of the axis for UF reflects the increase of parameter values for these UF and/or the number of UF with the fixed parameter values, shown as {+UF} in Figure 7.1.
>
> The positive direction of the axis for HF reflects the decreased parameter values for these HF and/or the number of such HF with the fixed parameter values, shown as {−HF} in Figure 7.1.
>
> The positive direction of the axis for NF reflects the decreased number of NF with fixed parameter values, shown as {1/NF} in Figure 7.1.

Choosing the positive directions agrees with the definition of Ideality of a technique (see Chapter 5). Consequently, it is possible to express the postulate about the direction for a technique's evolution through the concept of Ideality as follows:

Any technique's evolution brings the increase of its Ideality.

This postulate occurs because of numerous attempts to increase performance of a technique by engineers and other technique creators. Such an increase of Ideality occurs when it is worthwhile from the social and economic points of view and when it is allowed by the current scientific and technical level. Increase of Ideality can be described as follows:

(α) The cycle with unchanged mode of PF performing. Transitions to a more rational technique take place until approaching "local" Ideality inside the parameter values for the given paradigm for the mode of PF performing. During this cycle, parameters of TS or TP are improved (i.e., some or all UF rise and/or some or all HF and expenses fall) due to routine and inventive (but relatively low Level) solutions.

(β) A changed mode of PF performing step. Once the potentialities of cycle (α) are exhausted, the transition to a more rational mode of PF performing occurs due to inventive high Level solutions. Then the development of this next generation of technique again follows the cycle (α) for this new mode of PF performing.

TRIZ postulates that the direction of ET is common for all techniques and unchangeable. According to this postulate, ET occurs when the state-of-the-art for a technique changes from the point $\{UF_1, HF_1, NF_1, t_1\}$ to the point $\{UF_2, HF_2, NF_2, t_2\}$ in the multi-dimensional parametric space in such a way that for any $t_2 > t_1$ the following relations are correct (see also Figure 7.1):

$$\{+UF_2\} > \{+UF_1\} \text{ and/or } \{-HF_2\} > \{-HF_1\} \text{ and/or } \{1/NF_1\} > \{1/NF_2\}$$

The gradual change (α) and revolutionary breakthrough (β) in the technique correlate with the character of violations of the requirements revealed in the preceding generation of TS or TP. Among all possible changes, the one implemented first is the one that gives the necessary or considerable elimination of the most or even all violations of the requirements with minimal economical, industrial, commercial, and intellectual expenses. The postulate acts if the following condition is met: the transition to a new technical solution provides a gain in technique performance, significantly exceeding additional intellectual and industrial expenses. Transition to a next generation TS or TP performing the same primary function is caused by elimination of the revealed main violations of the requirements associated, as a rule, with rising technique performance criteria. Progress in technique can be imagined as two mutually supplementing cycles:

- in every functional niche, a mode of PF performing is inevitably replaced by a technique with a more promising one;
- a newly discovered mode of PF performing tends to fill all suitable functional niches.

The cycle (α) and the step (β) alternate until achieving "global" Ideality between all modes of PF performing for the set of known physical, chemical, biological, and other effects. Of course, during some periods of time a few types of technique can coexist although performing the same PF on the basis of different structures (material and functional).

Considering all transitions between generations of TS or TP, i.e., the whole history of the constructive evolution of a technique, we can reveal the following regularities in the hierarchical exhaustion of potentialities of technical solutions at the three levels. At the first, the parameters of the used technique are being improved, as a rule, without facing a contradiction. When changes of the parameters become inefficient, the process comes to the second level by transition to a more efficient technical solution without changing the performing mode of PF by the technique. Then, once the parameters are exhausted and/or contradictions between subsystems or within a single subsystem are raised, a new, more promising technique with the same performing mode of PF is implemented. Cycle (α) at the first and second levels repeats within the framework of the used technique until no new performing mode of PF can be found. Then the revolutionary change takes place at the third level if the scientific and technical potential is sufficient. This is the transition to the new, progressive performing mode of PF. Cycle (α) at the third level repeats within the framework of the new technique, and so on. Sometimes step (β) may occur in TS or TP before exhaustion of potentialities of a preceding technical solution for the performing mode of PF or UF. In the future, such abrupt transitions will apparently occur more often because of the use of computers (and TRIZ) for fast analysis of the cycles (α) and the possible steps (β). When developing new generations of TS or TP, it makes sense to study purposefully the evolution of other classes of TS or TP that have similar functions and are, as well, at higher stages of development.

7.2.1.1 Corollaries

The direction postulate leads to several particular corollaries and trends during technique progressive evolution. The most important and widely accepted corollaries for ET can be grouped as follows.

Group I

- Multiplication: trend toward the transition from a mono-system to a bi-system or to a poly-system with economy of support and auxiliary functions.
- Trimming: trend toward decreased number of subsystems for neutral and auxiliary functions in the existing technique.
- Poly-functionality: trend toward increased quantity of useful functions of a technique by adding new subsystems.
- Aggregation: trend toward achieving many functions within or by one subsystem.

Note: The last two trends reflect the fact that a class of techniques is developed first toward increased complexity or expansion, and then toward simplification or convolution. The aim of Accumulation or Integration of TS or TP functions is to decrease mass, size, and energy loss by a technique (see the next group).

Group II

- Dehumanization: trend to exclude people from performing a noncreative work through the use of mechanization, automation, and computerization.
- Resource: trend to achieve the low cost of a new technique by use of various resources during technique improvements.
- Minimization: trend to achieve the smallest or optimal dimensions (e.g., minimum size), smallest or optimal weight, and/or smallest energy consumption of any subsystem for a given function.
- New materials: trend to substitute current materials with new ones with advantageous properties, some of which can provide useful functions.

Note: The last two trends reflect the fact that a technique tends to evolve from a set of macroscopic subsystems to a set of microscopic subsystems. During this transition, different types of fields and substances are used to achieve better technique performance along various paths (see the next group).

Group III

- Encapsulation: trend toward putting replaceable subsystems into a more easily handled cassette or cartridge.
- Modular construction: trend toward connecting the construction elements into suitable groups from which many variants of techniques can be assembled.

- Standardization: trend toward unifying the dimensions, forms, and other properties and values of subsystems.
- Typification: trend toward establishing the optimum number of variants and sizes of selected characteristic properties of a particular kind of technique.
- Reuse: trend toward using subsystems that have already been designed and tested for a previous application as some of the subsystems of the new technical system or technological process.
- Recycling: trend to use elements of "dead" technique samples as raw materials in manufacturing of new samples of technique which can be the same or different, compared with "dead" samples, in order to overcome shortages of natural raw materials and to avoid ecological problems.

Some of these corollaries, which belong to Groups I and II, are discussed in detail later. The trends of the first group serve as a base for TRIZ heuristics for genesis of a new technique at step (β) and development of existing TS and TP. The trends of Group II serve as a base for design optimization and improvement during the cycle (α) of technique evolution. The trends of Group III serve as a base for cost reduction during the cycle (α) of ET (see details in [5]). Of course, there are no strong boundaries between these groups; for example, trimming allows improved economical aspects of a technique.

7.2.2   TIME POSTULATE

The word "evolution" itself implies that something happens in time. TRIZ emphasizes that the time of ET is the important factor for any technical system, technological process, their subsystems, or their super-systems.

The time postulate for an evolution of technique is formulated as follows:

Each technique and each subsystem has its own representative time of evolution.

The representative time τ_E of evolution can be calculated from the simple equation

$$\text{Subsystem performance } (\tau_0 + \tau_E)/\text{Subsystem performance } (\tau_E) = \vartheta.$$

Here the subsystem can be considered as a whole technique or as its element or operation, τ_0 is some initial moment, and ϑ is a constant that can be any number above one (for example 2, e, 1.1).* Actually, the exact value of the constant ϑ is unimportant for theoretical TRIZ and it is fixed by an agreement. Moreover, the time τ_E can be different for various characteristics of the technique or subsystem performance. A single TS, TP, or its subsystem can have as many times τ_E as it has secondary, support, and auxiliary UF.

---

* A technique (subsystem) becomes twice as good if $\vartheta = 2$ and a technique performance increases at 10% if $\vartheta = 1.1$ during the time τ_E.

It seems possible to distinguish at least three time scales for subsystems performing various functions (see Chapter 3) listed in the order of their longevity:

1. Working tool for the PF is rapidly developing, reflected in developments in processing and manufacturing technologies based on scientific principles.
2. The subsystems for the UF (especially the secondary and support) have seen intensive developments. The subsystems that fulfill control UF have continuing tendencies toward automation and computerization among the main directions of these developments. The electrical, pneumatic, and hydraulic subsystems that fulfill energy UF are being continuously improved in both technical and economic aspects.
   On the other hand, the subsystems linked with the HF have seen intensive reduction. The subsystems executing HF are complemented with new elements, replaced, or even discarded.
3. Subsystems for the neutral functions seem not to have been affected by any great changes during time scales for the UF and HF subsystems.

Any modern technique has, as a rule, many subsystems, and for each of these $\tau_E (PF) < \tau_E (UF) \sim \tau_E (HF) < \tau_E (NF)$. Within any new technique, or even in most existing technical systems and technological processes, the subsystems have not been evenly developed. Generally speaking, the more complicated a system or process is, the more uneven is the development of its subsystems. Different subsystems usually evolve according to their own schedule and reach their inherent limits at various times. The subsystem that reaches its limit first or is underdeveloped is "holding back" the overall performance of a technique.

The following predicate is logically derivative from the postulate about time for ET:

> **Any technique with nonsingle subsystems has a few**
> **time scales during technique development.**

This predicate about nonuniform evolution of TS or TP is widely used in TRIZ because the progress of various functions of technique depends on developing their subsystems and vice versa. Therefore, nonuniform ET is bound with the idea about correspondence between functions and subsystems as was considered in Chapter 3.

7.2.2.1 Corollaries

The first corollary of the time postulate and the derivative about nonuniform development of technique manifest the origin of inventive problems. The second and third corollaries deal with the extreme situations that should be considered during conceptual design of a new technique. Note that the history of technique shows cases in which technical systems proved unable to operate just "due" to violation of the following corollaries, e.g., *insufficient controllability*.

Corollary i

The most important corollary of the derivative of nonuniform evolution that has been verified in many studies is the following:

The uneven development of subsystems leads to existence of contradictions in a technique.

Many TRIZ heuristics and instruments have been established to resolve such contradictions in technical systems and technological processes (see Chapter 4 and Part 5).

Corollary ii

As we know, practically any technical system includes a few basic subsystems. Any TS should have a working tool. Most TS have or are connected with a source of energy and/or engine; usually TS also has a transmission, a control subsystem, and a casing (see Chapter 3). The system will not work if one important subsystem is missing or does not perform well. This corollary is often named as the rule of TS completeness in TRIZ. It corresponds to the extreme situation of totally nonuniform development of technique when one or a few technique subsystems are not created yet and, therefore, the time τ_E is infinite.

Corollary iii

As we know, links carry out relationships between subsystems of a technique. All TS and TP must possess the capability for easy energy (field) and/or substance transfer throughout a system or process, at least from an energy source and/or engine to the working tool (and then to the product). Any technique that requires a control must be capable of information transfer. The technique will not work if one of the energy/substance/information links is missing or does not perform well. This corollary is often named as the rule of energy/substance/information transparency through TS or TP in TRIZ. It corresponds to the extreme situation of totally nonuniform development of a technique when one or more of a technique's transfer links between subsystems is not created yet and, therefore, the time τ_E is infinite.

One of the manifestations of the last two corollaries is the requirement of correspondence between variability of functioning conditions and controllability of a technique. Either a technique must be insensitive to variation in external conditions and purposes of functioning without undesired consequences, or it must have changeable (controllable) subsystems which allow the technique to adjust itself to these changes. For each PF there exists some minimum of controllability, which necessarily should be exceeded to ensure operation of the technique. Similar minimums of controllability can be for UF.

TRIZ postulates and corollaries of ET are helpful at the initial stages of designing new generations of technique and for understanding the paths of technique evolution. Such paths or lines of evolution have been actively studied by TRIZniks and are described below.

7.3 PATHS OF EVOLUTION

Several paths of ET coexist, some of which are presented in this section. Although the order of the steps and guidance of each of these paths are not absolute for any technique, many successful developments of TS and TP have been based on them. The knowledge of these paths can also be useful for filtering and selecting solutions of technical problems. The following section briefly discusses the most important of these single- and bi-directional and adverse paths.

7.3.1 SINGLE-DIRECTIONAL

1. **Technique physical states** — A technique "moves" toward Ideality as its subsystems become more mobile, e.g., by changing the physical or phase state of a subsystem along the direction

$$\textbf{solid} \rightarrow \textbf{liquid} \rightarrow \textbf{gas} \rightarrow \textbf{plasma} \rightarrow \textbf{field} \rightarrow \textbf{vacuum}$$

2. **Interactions in technique** — A technique moves toward Ideality as interactions between two or more subsystems or between a subsystem and a product become more precise, from being continuous, to vibrating, to vibrating at the resonant frequency, to standing waves. Changing the interactions along the direction

$$\textbf{continuous} \rightarrow \textbf{vibrating} \rightarrow \textbf{resonant}$$

   often allows decreased energy expenses inside the technique and increased performance and productivity (see also the path Rhythms Coordination $\Leftrightarrow$ De-Coordination in the next subsection).
3. **Degree of dimensionality** — A technique moves toward Ideality when its dimensionality changes from a point (zero-D) situation to a line, then to a plane, and later to a volume (3-D). Changing the dimensionality along the direction

$$\textbf{zero-D} \rightarrow \textbf{1-D} \rightarrow \textbf{2-D} \rightarrow \textbf{3-D}$$

   of technique or its subsystems usually improves the useful function's performance.
4. **Adaptability** — A technique moves toward Ideality as its subsystems become more adaptable (flexible) to changeable (including opposite) requirements. Adaptability means going from a rigid, immovable, immobile subsystem to a subsystem that is more dynamic, with one and then many hinges or joints, followed by an elastic (i.e., infinitely jointed) and then a soft subsystem, or even beyond that to an elegant flexibility made possible through the use of fields (e.g., electromagnetic, thermal, etc.) instead of substances. Changing the adaptability along the direction

rigid → dynamic → multi-hinges → elastic → soft → field flexibility

is accompanied sometimes by a change of the physical state of a sub-system (see above).

5. **Degree of voidness** — A technique moves toward Ideality as its sub-systems' "voidness" increases. The lowest level subsystem has no void: it is a monolith — one solid piece. The next step in its evolution would be the inclusion of a large void and then many smaller voids, followed by a porous subsystem, or even beyond that to micro-voids in structure (e.g., zeolites, xerogels). Changing the "voidness" along the direction

monolith → inhomogenous solid → solid with void(s) → solid with capillaries → porous solid → solid with dispersed micro-voids

leads to reduction of subsystem mass and to its poly-functionality.

6. **Degree of human involvement** — A technique moves toward Ideality by exclusion of human participation in its functions' performance. The tendency of decreasing human involvement in the operation of techniques began many centuries ago when humans, as a source of energy, were replaced by animals or inanimate subsystems. Now humans are "losing" the control and planning functions as they are replaced by "smart" subsystems. Various TS with a processing function applied to a material subject of labor have three levels of development. These levels are connected with consecutive implementation, by technical means, of the four functions and the corresponding consecutive exclusion of man-executed functions from the technique:

1. TS executes only the primary function performed previously by a human being

    ↓

2. TS, along with the PF, also executes some of the useful functions as well:
    a. usually, initially the energy supply function, then

       ↓

    b. the control and support secondary function over the processing, and finally

       ↓

    c. the auxiliary functions

       ↓

3. TS additionally performs the planning function for arranging the amount and quality of products resulting from processing a subject of labor.

These grades reflect in ET during the twentieth century:

- Mechanization: allocating the propelling functions to the technique
  ↓

- Automation: allocating regulation and control functions to the technique
  ↓

- Computerization: allocating routine decision functions and performance monitoring tasks to the technique

Transition to every next grade occurs on exhaustion of natural human abilities in improving execution of the PF and/or UF with the aim to increase productivity of the technique or quality of products. This transition occurs only when it is possible from the social and economic point of view as well as if a sufficient scientific and technical ability exists.

Within the next couple of decades, the newly arising, pioneering TS or TP for satisfaction of new needs will probably often perform simultaneously the three basic functions (primary, energy, and control) without humans. As a result, human intervention will be fully excluded from the technological process, except at the higher levels of planning. Later in the 21st century, the planning function will probably also be performed by a prospective technique that will utilize nanotechnology, results of Artificial Intelligence, and other breakthroughs. It seems that ET will continue along the following direction:

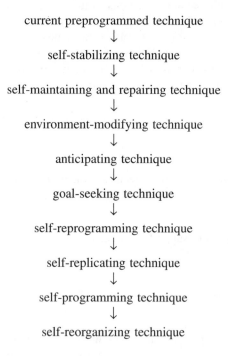

current preprogrammed technique
↓
self-stabilizing technique
↓
self-maintaining and repairing technique
↓
environment-modifying technique
↓
anticipating technique
↓
goal-seeking technique
↓
self-reprogramming technique
↓
self-replicating technique
↓
self-programming technique
↓
self-reorganizing technique

Of course, this direction corresponds to reduction of human participation in technological processes and even in the creation of a technique.

7.3.2   Bi-Directional and Adverse

7.3.2.1   Expansion ⇔ Convolution

Techniques begin from a single element or operation intended only for execution of the PF (as a rule, from the working tool). As the PF grows, there is increase or enhancement of some element properties. Then an element differentiates; i.e., it is divided into zones with different properties. The technique grows further at the expense of complication of its elements, which form subsystems. Apart from primary functional elements and subsystems, which include such elements, support subsystems appear in the technique. The technique becomes highly specialized when it begins to take over functions of other "neighboring" techniques by itself, so secondary and auxiliary subsystems are added to it, and so on. Usually such secondary functions can be considered supplementary to PF and often they were the primary functions of the neighboring techniques. Soon afterward the technique begins to take the functions of other TS or TP without increasing the number of its subsystems (elements and operations). At this point the technique becomes more and more versatile at the same time as it decreases the number of subsystems.

The chronology of this type of development includes the following stages:

1. Expansion — the number of subsystems (elements, operations, or links) increases with time with the increasing PF (usually PF grows faster than the number of subsystems).
2. Convolution — the number of subsystems decreases over time while PF grows or remains constant.
3. Reduction — the number of subsystems begins to decrease as does PF (usually PF grows more slowly than the number of subsystems).
4. Degradation — PF decreases with decreasing links, power, and efficiency.

This bi-directional process can be presented as shown in Figure 7.2.

Let us consider both sides of this bi-directional process of technique evolution in detail.

Expansion

As the number of systems and processes interacting with the technique increases, its functions begin to change. These quantitative changes in accordance with the dialectics

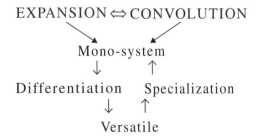

FIGURE 7.2   Bi-directional technique evolution.

law lead to qualitative changes, and these differences grow with time and expand the technique. The apparent randomness of such evolution is caused by the abruptness of the appearance and aggravation of contradictions in some subsystems; i.e., the process obeys the time postulate about inhomogeneous development of subsystems. Sometimes it is useful that the technique has the secondary functions while at others the straight concentration on PF is more beneficial.

The technique expansion process can be illustrated by the evolution of a monosystem that consists of only one element made from a single material; in this case the factors hampering the increase of UF (including PF) are most pronounced. These factors are

- absence of required properties in the element, and/or
- inability to use the hidden characteristics and parameters of the element (resources).

The simplest solution is to increase the necessary characteristics of this element to provide the required UF. This attempt most often faces a contradiction — other properties deteriorate. Thus, one needs to improve the element properties but cannot. It is usually impossible to replace the material or element immediately because it serves well in achieving the level of UF performance. Therefore, the next step focuses on the needed property and suppression of harmful by-properties. Thus, numerous modifications of the same material arise for different elements and different operating conditions. For example, about 3000 types of steel, and more than 100 different classes of silicon transistors, are produced. Such focusing is required because it is impossible to make the single material satisfy all needs. It is possible to obtain micro-gains in PF by enhancing some properties or parameters of the material. The process of evolution usually leads to separation of the element of the mono-subsystem into parts (areas, layers, regions, etc.) or functional zones, and the process of their specialization starts — each part executes only one function. The specialization by executed functions results finally in separation of the homogeneous element into parts and replacement of the "old" materials in some parts with "new" materials that provide higher UF performance. The specialization, accompanied by the transition to new materials for each zone, facilitates the growth of UF of the element and, hence, of the mono-subsystem or technique as a whole. The cause is simple: at a usual attempt to increase UF it becomes clear that the property on which this increase depends should be inherent not in the whole element but only its part (operative zone). It is easier to enhance the property in a zone by a new material and keep the "old" and still useful material in the rest of the element.

The evolution of element or the mono-subsystem inside the technique can be summarized by the following stages:

1. attempts to improve (separate out) the needed property of the element
↓

2. separation of a homogeneous single material element into functional zones
↓

3. specialization of zones by functions, transition to an inhomogeneous material

↓

4. compound material of specialized ingredients with high values of UF

↓

5. expansion of the compound material into individual elements

↓

and usually the evolution goes forward to

↓

6. convolution of the compound material into the ideal material and the individual elements into a single "new" element that is better than the "old" ones

Convolution

After the period of expansion, a technique enters a new period of transformations, which involve its structure, organization, and system properties. This process, called *convolution*, corresponds to increased Ideality due to decreased technique mass and space, and often spending energy and number of auxiliary subsystems at simultaneous increase of PF. For example, the structural steel and power transistors have progressively changed in quality during the twentieth century so that the user gets greater value per mass unit (for steel) and per volume and frequency units (for transistors).

Once a technique reaches the point of maximum expansion, a few options of convolution are principally possible:

1. Elimination of some subsystems from the technique and combining of them into new specialized techniques within the super-system;
2. Evolution of subsystems within the technique (miniaturization, accuracy of functioning, decreased power requirements, disappearance of some harmful factors, self-organization);
3. Package of technique into a subsystem, element, or single material.

All these options lead to the same result — the new technique executing the same PF as the initial technique with the same or even higher performance. Convolution often includes the randomly mixed steps in all the options above. While the second option has been outlined in other sections of this chapter, the first and third should be explained here.

In the first option, the number of elements and/or subsystems in the technique decreases to a tool or main element of a secondary function, while the technique's PF increases because

- the technique becomes "lighter" (it does not need to be versatile), its structure and organization become easier, it takes less space, time, and energy, and the functioning is improved;
- the function of the excluded subsystem is replaced with the same function from the super-system performing with the same or higher quality;
- the subsystem becomes a specialized technique in the super-system.

The higher PF performance is obtained by moving subsystems into the super-system, and the number of functions performed by the super-system increases because of these new elements in it.

In the third option, the number of elements and/or subsystems in a technique decreases because of the merging of two and then more subsystems into a single one; during this merging, other "excluded" subsystems "disappear," transferring their functions to the "preserved" versatile subsystem. This versatile subsystem can take the additional functions from the discarded subsystems. If the preserved subsystem is a tool (as often happens) then it remains PF and continues to be improved. The next step is the convolution of this preserved subsystem into a single universal material (a subsystem is absent, but its function is executed by a substance) or idealization of the versatile substance up to the level of technique (the substance, becoming more complicated, takes the increasing number of functions up to the function of the whole technique). Of course, transformations of other subsystems into the universal and ideal substance can occur before those excluded in the future subsystem "gather" to the tool or other preserved in the future subsystem. In all cases, the main properties of ideal substances are a high value of the executed PF or UF and an independent response to changes in the environment.

The third and first options cross: the subsystems excluded from different but similar techniques move into a super-system and execute the same or similar functions (informational, power, transport) as they did in the initial techniques. This creates conditions for their integration (the super-system packaging) during which individual subsystems and elements also disappear. The "collectivized" technique becomes a super-system of common use, to which all techniques refer continuously or periodically for service, control, repair, etc. However, before integration these subsystems often evolve outside the initial techniques, so their functions transform; these subsystems take execution of additional functions, i.e., follow the second option. All three options of convolution are interrelated. Moreover, different levels of the technique (element, substance, subsystem, super-system) can be on different paths of development (expansion and convolution) at the same time in accordance with the second ET postulate.

Techniques are not only a set of substances and energy fields but also the native technique information — a set of instructions or signals, determining a sequence and kind of interaction between the technique's elements with raw objects and the environment, and among themselves. At convolution, when one of the systems transfers function "idealized," the substances and energy fields disappear. Information parts of the disappearing technique (the ways determining application of the old technique to a new purpose and/or assignment) can remain (being transferred from "excluded" subsystems to "preserved" subsystems), change (because they don't directly apply to the new conditions), or are generated again. Any new technique, regardless of its potential opportunities to carry out new functions, cannot execute them without new instructions on how to perform UF. It is completely mistaken to represent the convoluted technique as absent, because its native information is not destroyed. Idealization of the technique occurs usually at the native technique information account.

7.3.2.2 Super-System ⇔ Micro-Level

A technique "moves" toward Ideality as its subsystems merge and divide. Two directions, internal improvements and external development, coexist because different subsystems have various representative times τ_E. At any stage of technique development, these two directions lead to new techniques due to a transition to a super-system (a kind of expansion) and a transition from a macro-level to a micro-level (a kind of convolution). Note that the level here is a conventional concept, which reflects only the peculiarities of human thinking and perception of the environment; it is always related to objects comparable with it and directly perceived properties of these objects.

The tendency to use properties of existing subsystems from a macro-level to a micro-level or the transition to micro-level occurs when a subsystem or its element is replaced with a material or field, capable of performing a required UF. There are many micro-levels in materials (domains, molecules, atoms, etc.); consequently, there are many different transitions to micro-level, as well as many transitions from one micro-level to another, lower one. Usually such transitions occur due to applications of physical, chemical, and other effects (see Chapter 9).

The tendency to merge subsystems occurs when the advantages for the technique outweigh the expenses for formation of bi- and poly-systems:

- some functions are delegated to a super-system;
- some subsystems are withdrawn from the technique, are consolidated into a subsystem, and become part of a super-system;
- upon their integration into a super-system, the constituent systems begin to display new properties and functions.

This bi-directional, reciprocal ET trend can be graphically represented as follows:

Substance/Field ⇔ Element ⇔ Subsystem ⇔ Technique ⇔ Super-system

where a subsystem is considered the main object here.

Let us consider both sides of this bi-directional process of technique evolution in detail.

Micro-Level

The transition to the micro-level usually starts from differentiation of properties, zones, and functions of the tool material. The contradictory requirements for properties of the same substance arising in the process of technique evolution are resolved by transition to the micro-level due to

- increase in the degree of separating the substance and joining of the parts into a new system,
- increase in the degree of substance mixture with the vacuum (transition to capillary-porous materials),
- replacement of the substantial part of a system with the field (transition to the "field plus substance" action or only the field).

Many peculiarities of the transition to the micro-level are similar to the features of the convolution, so some TRIZniks do not distinguish these paths.

Super-System

Evolution of a technique that has achieved its limit can continue at the level of the super-system. A technique or its subsystems can be joined with other subsystems of other techniques into a super-system with new characteristics and parameters. Such transition into a super-system is profitable for the technique because

- some functions pass to the super-system,
- some subsystems come out of various techniques, join into a single new technique, and become a part of the super-system,
- subsystems joined in the super-system acquire new functions and properties.

One way for such a transition to occur is that techniques are joined with formation of bi- and poly-systems. The initial single subsystem or element (mono-system) is doubled with formation of the bi-system or poly-system, if several mono-systems are joined. In the simplest case, to construct bi-systems and poly-systems, two or several similar substances or two or several identical fields are integrated. One of the mechanisms of a new bi-system or poly-system creation when joining two equivalent elements is the conservation of borders between these elements. This border conservation requires introducing a new boundary substance (it can even be emptiness). It results in the creation of an inhomogeneous quasi-poly-system with emptiness as a second, boundary substance. Joining can be applied not only to identical subsystems and to similar systems (often named *nonsystems* in TRIZ) with slightly different characteristics, but also to *concurrent or alternative techniques* with the complementary functions or properties, as well as to different techniques (with different functions) including the *anti-systems* (systems with opposite primary functions or properties). In all cases, the joining of systems follows the same stages. The sequence of formation and development of a bi- or poly-system may be the following:

Mono-Systems → Crossing → Combination → Subordination → Convolution

At formation of a bi- or poly-system, a new system property arises, which is inherent in only the joined system. The system property can arise from combination (cooperation) of earlier hidden or neutral properties of elements.

Joining is more efficient for dissimilar mono-systems, especially for alternative techniques than for identical or similar ones. Similar bi-systems, always execute a sole function, while dissimilar ones execute two functions. However, joining dissimilar elements into a single technique not always gives a system gain, so joining of mono-systems is justified only when it results in new properties and/or system gains. Increased efficiency of synthesized bi-systems and poly-systems can be reached, first of all, by developing links between elements and/or elements in these systems. Newly-formed bi-systems and poly-systems often have a "zero link" if they are simply a "heap" of elements. Consequently, development goes in the direction of strengthening the links between elements. On the other hand, elements in such systems sometimes are joined with solid links, and then link flexibility increases (see the single-direction path *adaptability* described in the previous subsection).

Mono-systems should be joined so that properties of elements interact in two ways: some properties amplify each other while combining (this is a dominant system property of the future new technique), and other properties suppress or neutralize each other. As a result, the UF begins to prevail in the new technique. The main meaning of formation of bi- and poly-systems (mono–bi–poly transition) is in the qualitative and quantitative changes of behaviors, properties, and parameters. In partially convoluted bi-systems some elements are replaced with others. In fully convoluted bi-systems, one mono-system (or even substance) executes the function of the whole system.

Bi-systems are not necessarily formed from two mono-systems; sometimes it is easier and more profitable to transform a mono-system into a bi-system by dividing the mono-system into two identical subsystems and then rejoining them in a new way to achieve a UF and/or exclude some HF. According to the dialectic law "Unity of Opposites," integration and disintegration give the same result: bi-systems and poly-systems arise. The gain is the same: appearance of new properties, UF enhancement, and exclusion of harmful properties and HF.

Details of the transition to a super-system are shown in Figure 7.3, which presents results obtained by Yu. P. Salamatov [2]. Duplication of the initial mono-system results in a bi-system, or, with more than two systems involved, in a poly-system. Integration is observed not only among identical (homogeneous) systems, but also among similar systems with slightly different characteristics, nonsystems (in which another technique has the same primary function but performs it differently), and anti-systems (systems with opposite functions). New properties in bi- or poly-systems arise as the technique evolves toward increased difference between elements. Therefore, the integration and merging of mono-systems pass through the same stages (see Figure 7.3) that often are similar to the evolution path for technique expansion.

7.3.2.3   Rhythms Coordination ⇔ De-Coordination

Coordination (adjustment) or purposeful de-coordination (de-adjustment) of vibration frequencies of all elements in technical systems (dynamic and static) or periodicity of operation of technological processes is important for increase of technique Ideality. From this point, all TS/TP and their elements can be divided into those that have "proper" rhythm and those that vibrate "improperly."

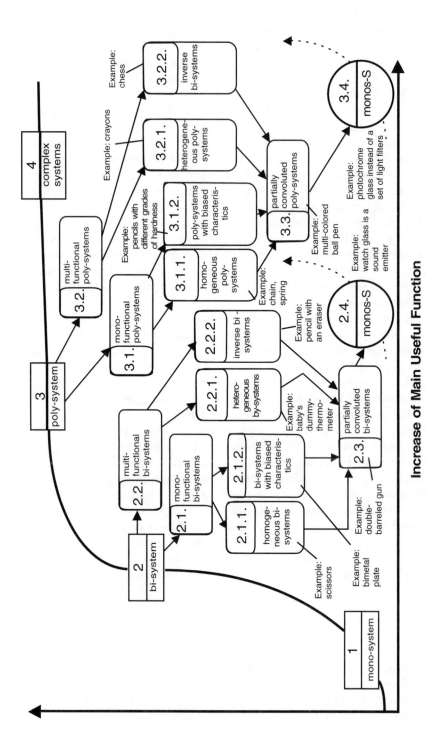

FIGURE 7.3 Expansion and convolution of technique. X-axis shows UF growth and Y-axis shows the increase of technique complexity [2 (b)].

Two kinds of vibrations are differentiated: intrinsic and forced. The intrinsic frequency of vibrations is an integral feature of any substance; it depends strongly on the substance's native characteristics (dimensions, mass, elasticity of parts in mechanical systems, induction characteristics in electrical systems, etc.). The intrinsic frequency of vibrations is a powerful indicator of many important characteristics of the substance. The subsystem's state can be determined from the changes in its intrinsic frequency of the monitored subsystem. Fields do not have a certain intrinsic frequency — any periodic change of the field parameters can be considered as native. Usually the corpuscular (substance-like) properties of a field become more important when its native frequency increases.

But, most interesting, an external force or field effect may coincide in frequency with the substance's native vibrations, giving rise to the well-known resonance effect. Resonance may be either useful or harmful. Therefore, to improve a subsystem's performance it is necessary to coordinate or de-coordinate the vibrations of its elements.

The rhythm trend provides a series of regulations:

- The field/energy flow should be coordinated/de-coordinated with natural frequency of the raw object or product and the tool in the techniques.
- The field frequencies used in the techniques should be coordinated/de-coordinated.
- If two effects are incompatible (e.g., transformation and measurement), one effect should be exerted when the other effect pauses. More generally, a pause in one effect should be filled by another effect.
- Use or prevention of resonance between subsystems should be through simple modification of its elements (dimensions, mass, frequency).

7.4   BASIC SUBSYSTEM TRENDS

The trends and paths, which are common for a whole technical system or techno-logical process, were described in the previous sections. As we know, in most techniques the six basic subsystems are working tool, energy source, engine, controls, transmission, and casing. These subsystems also move toward Ideality in accordance with some trends, including single- and bi-directional. Here we discuss the "individual" trends for the basic subsystems of a technique.

It was recognized that the *source of energy* for different kinds of technique uses different raw materials during the last couple of centuries along the direction

$$\textbf{solid} \rightarrow \textbf{liquid} \rightarrow \textbf{gas} \rightarrow \textbf{plasma} \rightarrow \textbf{field}$$

It can be illustrated by the example of the energy sources used by the technical systems for transportation: raw coal $\rightarrow$ petroleum $\rightarrow$ methane $\rightarrow$ solar energy were the raw materials for the steam engine $\rightarrow$ internal-combustion engine $\rightarrow$ gas motor $\rightarrow$ electrical accumulator $\rightarrow$ solar cells. These materials in these systems provide the primary function, motion. Of course, this trend (as many others) is not absolute. For example, in various kinds of techniques the first used sources of energy were liquid, e.g., the power of waterfalls.

It has been recognized that the casing (and some other inner subsystems) becomes more symmetric with time. This idea (proposed by French physicist Pierre Curie) has the following formulation:

> Casing of TS under a certain considerable effect of the environment in the form of substance, energy, or information flows has a certain type of symmetry caused by the combination and character of these flows.

If TS must have some symmetry, then it should be reflected in TS construction; otherwise the technical solution will be impaired. Hence, the analysis and evaluation of developed TS from the symmetrical point of view could be useful for a TS creator. In contrast, one TRIZ heuristic recommends in some cases to introduce an asymmetry in a TS or its subsystems (not necessary in the casing) in order to resolve the technical contradictions (see Chapter 13).

The *control subsystem* changes along two directions when it is necessary to measure some parameter of a technique. These directions are shown below.

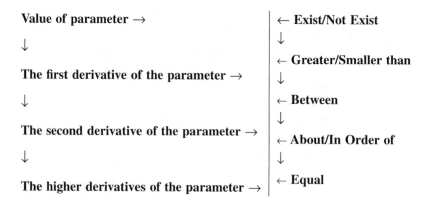

Usually the measurements for each entry in the left column start from a simple conclusion about the existence of a characteristic. Later, the accuracy of control moves vertically to the right column, then switches to the next entry in the left column, and so on. Note that (a) for many cases it is enough to recognize the existence of some value that plays the role of a switch for a control subsystem; and (b) some heuristics for control system development are presented in class 4 of Altshuller's Standards (see Appendix 5).

Any *uniform set of subsystems* (first of all a *working tool* or *transmission*), with the same or very similar useful functions, and the operating conditions possess partially coinciding sets of parameters Y that have a simple correlation with the parameter X characterizes the *working tool* or *transmission*. X determines the values of other parameters Y. For example, the bucket volume for an excavator (parameter X) determines the power of supporting energy systems. This trend was formulated by A. I. Polovinkin by analogy with the Vavilov's law of homological series in biology: "In genetically close species belonging to the same genus, the same characteristics vary similarly" [3]. When developing a uniform set of subsystems having qualitatively the same or very similar functions (primary, secondary, support, etc.) and operating conditions (from the viewpoint of interaction with a raw object and the environment) and different values of parameter X, this trend can be used for quick determination of other important constructive parameters Y. A study aimed at assigning the trend of TS uniform set allows sufficiently accurate prediction of appearance of new technical solutions due to simpler selection from the morphological table of variable parameters. This trend works during the cycle (α) only when contradictions between (or within) the selected and other subsystems or characteristics are absent.

Different classes of techniques (electronic devices, transport, petroleum refinement, telecommunication, etc.) have their own particular trends and paths of evolution that are, unfortunately, beyond the scope of this book.

7.5 CASE STUDY: ALTSHULLER'S "BALLAD ABOUT A BRICK"

Before beginning the next section and finishing the discussion of ET theoretical aspects, I cannot forego the pleasure of quoting Genrich S. Altshuller's discussion of [1(a)] the evolution of a very old but still viable technical system — the brick — which is one of the most important subsystems of many different static TS of the last several centuries. A brick's evolution reflects many trends and paths discussed above and can be considered a case study of ET. I have named this short story the "Ballad about a Brick," and I hope that the translation keeps the poetic and colorful language of the TRIZ creator, who believed that he discovered the laws of ET.

All laws of technique evolution are applicable to a brick (for example, transition to a bi-system: brick of "binary" substance). From the TRIZ positions, the technical contradiction is clearly seen here: a second substance must be introduced (it is the law!) and cannot be introduced (it complicates the technique). The solution is to use "nothing" as a substance: emptiness, air. Thus, we obtain a brick with inner cavities. As a result, it has less weight and better thermal-insulation properties. What's next? We increase the degree of cavity dispersion: from cavities to pores and capillaries. It is already nearly a mechanism. Porous brick saturated with nitrogenous material (according to USSR Patent # 283264) is put into molten cast iron; the brick slowly gets hot, and dosed feeding of gaseous nitrogen takes place. One more example: porous brick keeps a gas out but hinders open fire (USSR Patent # 737706) and water (USSR Patent # 657822). We again pass to a bi-system: capillaries can be partially filled (i.e., "emptiness" is again introduced), and then we can make a liquid flow over the brick (inner cover of thermal pipes).

Then the word "brick" should be put in quotation marks because the structure with capillaries filled with liquid can be anything. For example, it may be a ball bearing as in USSR Patent #777278: "Rolling-element bearing comprised of inner and outer races with hollow rolling elements arranged between them. Rolling elements are partially filled with heat carrier. Such a bearing differs from others by the fact that, in order to increase its life by ensuring the automatic balancing of rolling elements, inner surface of every rolling element has a capillary-porous structure."

The next invention (USSR Patent #1051026) proposes the brick with capillaries filled with magnetic liquid. Under exposure to a magnetic field, liquid goes up, thus creating the rarefaction in vacuum entrapment. Such a "brick" is nearly a machine. Generally, the level "a brick with capillaries filled with liquid" is very efficient. The number of inventory possibilities here is very large. Liquid can evaporate, thus creating a strong cooling effect. Liquid can be separated, filtered, moved.... Pores and capillaries may be of the same size or can vary in diameter, for example, over the "brick" length. In the latter case, they can pump liquid along the "brick" toward decreasing diameter (USSR Patent # 1082788).... However, porous brick is not yet a micro-level. We can deal with molecular groups — magnetic domains. Molecules, atoms, electrons.... Imagine the "brick" of Nitinol capable of changing the capillary diameter (and even the direction of its decrease!) under changing temperature. It is not "nearly a machine"; it is a machine.

An "ideal brick" features three main properties:

1. Useful work is done by all brick levels and by all substances of which it consists. A brick works at the level of stone, at the level of thermal-insulating cavities, at the level of pores and capillaries, at the level of crystal lattice, at the molecular level, etc.
2. The number of levels is comparatively small. However, at every level, dozens and even hundreds of effects and phenomena can be used. Finally, infinite possibilities to increase Ideality open up when using the interaction between levels.
3. Becoming more complicated, an ideal brick acquires properties of machines and mechanisms. The more complicated the ideal brick, the wider the set of its controllable functions and the more multipurpose the functions themselves.

Becoming more complicated, an ideal brick acquires properties of machines and mechanisms. The more complicated the ideal brick becomes, the wider is the set of its controllable functions and the more multipurpose become its functions themselves.

7.6   LIFE OF TECHNIQUE

Now that we know TRIZ ideas about trends and paths of TS and TP evolution, we can discuss the important peculiarities during a technique's "life." During the cycle (α), ET generally occurs in accordance with the S-curve that represents technique *performance* as the following function in time (see Figure 7.4). Much research distinguishes five major stages of technique evolution along the S-curve: (0) birth, (1) childhood, (2) growth, (3) maturity, and (4) decline.

0. **Birth** — The new technique appears. The step ($\beta 1$) occurs due to the recognition of a new need or requirement of society or due to a new scientific finding as a result of one or more high-Level breakthrough inventions or an outstanding achievement which produced the new technique. As a rule, the technique is primitive, inefficient, and unreliable, has many unsolved problems, and exists mostly in R&D laboratories as an operating prototype. Nevertheless, it provides some new mode of performing PF.

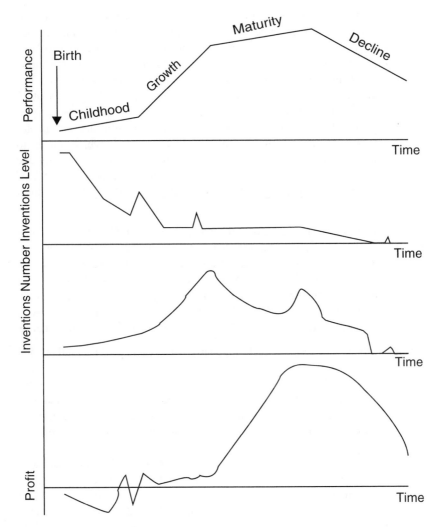

FIGURE 7.4 Evolution of technique. The upper S-curve is common for evolution of various objects and systems.

1. **Childhood** — The cycle ($\alpha 1$) begins. Development initially is typically very slow, due to the lack of profit from the new technique as well as limited human and financial resources. Most people do not know about or doubt the usefulness of this technique, but a small number of enthusiasts who believe in the technique's future continue to work toward its success and commercial introduction. Later, some support for the development of the new technique is found. Due to this support (and hence negative profit) the enthusiasts create many low- and middle-Level inventions, which turn the new technique from primitive to quite efficient.

2. **Growth** — This stage begins when society recognizes the value of the new technique. By this time, many problems have been overcome (usually due to a few high-Level inventions), efficiency and performance have improved, and a new market is created. As interest in the technique escalates, investment of money in R&D for the new TS or TP grows. The profitability of the new technique is positive, but not always. This situation accelerates the technique's development, improves PF performance, performs new UF, eliminates many HF, etc., and in turn, attracts greater investment. Thus, positive "feedback" is established, which serves to further accelerate the technique's evolution to higher "local" Ideality due to the growing number of middle-Level inventions in the framework of the new technique paradigm. Widespread adoption of the technique allows it to diffuse into other areas.

3. **Maturity** — The profitability of the new technique is high. It has high social and economic impact. The development slows as the initial concept upon which the technique was based nears exhaustion of its resources. Large amounts of money and labor may be expended; however, the results are inadequate. Improvements occur through technique optimization, tradeoffs, or low-Level inventions. When the cycle (α1) is exhausted, the S-curve is replaced by a saturation or even decreasing function of performance in time.

4. **Decline** — The limits of technique have been reached — no fundamental improvement is available. The profitability of the new technique goes down. The technique may no longer be needed due to a change in society's requirements or because it has been replaced by a newer technique with another paradigm for the mode of performing PF. The new technique stagnates, and no or occasionally only low-Level inventions support it.

At one of these stages (usually 3 or 4) the new step (β2) occurs and a fresh technique emerges and begins its own S-curve with the new cycle (α2). If we investigate a number of usage curves of successive models (from a profit point of view) of a technique family, we usually find that the wavelength of these curves is not constant — it reduces.

The cycle (α) or the usage life of techniques (from the viewpoint of the manufacturer) is becoming progressively shorter, and the steps (β) occur more often:

$$\tau^{\alpha i}_{E} > \tau^{\alpha j}_{E} \text{ and } 1/\tau^{\beta i}_{E} > 1/\tau^{\beta j}_{E} \text{ for any } \mathbf{j} > \mathbf{i}$$

Hence, the representative time τ_E of evolution is itself time-dependent; it decreases with time.

Each subsystem within a TS or TP has its own S-curve and path for development. These S-curves have different time scales according to the second postulate of ET. During its life, the evolution of a technique occurs because of one or a few of the paths discussed in the previous sections. The S-curve shown in Figure 7.4 represents a linear approximation of an ideal case. Of course, the real evolution of any technical system or technological process is very complicated because various influences modulate the actual evolution curves differently.

7.7   APPLICATION OF THE KNOWLEDGE ABOUT TECHNIQUE EVOLUTION

This section outlines the most important aspects of ET usage. Knowledge about the trends and paths of the evolution of a technique can be used for

- Qualitative technological forecasting,
- Technique creation, including directed technique synthesis and genesis,
- Problem solving, including selection of TRIZ heuristics,
- Marketing of innovations.

These interrelated branches of application of ET knowledge are discussed below within the framework of TRIZ.

Qualitative technological forecasting

Traditional technological or engineering forecasting attempts to find quantitative values for changing the parameter of a function that is performing in a selected technique after a certain amount of time. Very often the results of the traditional forecasting disagree with the facts because of wrong predictions about the values of the technique's parameters and the time interval when these parameters will achieve predicted values [7]. Such disadvantages are not surprising because of the following two factors.

1. It is possible to prove that TS and TP are nonergodic systems (see Appendix 2) that change not only due to technical solutions and competing technique, but also under the influence of economic, political, and social effects (such as embargo or crisis situations) as well as competing solutions with identical or related functions.*
2. The real evolution cannot be reduced to idealistic smooth and simple curves (such as the upper part of Figure 7.4) and cannot be described by theoretical nonsingular equations [7].

In addition to these disadvantages, traditional technological forecasting cannot describe *how* to achieve the desired performance of a technique. In fact, such questions are beyond the scope of such engineering forecasting [7].

In contrast, TRIZ's qualitative technological forecasting can answer *how*. ET postulates and some corollaries and paths serve as the foundation for qualitative technological forecasting. ET trends that have been recognized in the framework of TRIZ and information analysis (of patents, technical and economical reports, etc.)

---

* Unfortunately, it is impossible to calculate exactly the time τ_Q when a parameter of performing PF (or other UF) will achieve a certain value Q, because τ_E is not a constant.

enable forecasting "how can that be" with functions performing in particular classes of TS and TP. However, although the predictions are built on objective facts usually obtained from statistical and scientific investigations, the forecast predictions are not precise about the time-scale of possible changes and their numerical values.

Although TRIZ cannot answer *when* and *how much* questions, TRIZ-based technological forecasting is, nevertheless, very useful because it

- points to subsystems that should be improved,
- avoids development of subsystems (including whole TS or TP) that passed the Maturity (3) stage of evolution long ago or that are in the Decline stage,
- shows the set of possible paths for technique development (unfortunately, not almost exhausted yet),
- indicates the ways to build a patent fence (umbrella) around promising techniques at the initial stages of evolution (Childhood and Growth) and destroy the patent fences of competitors at the Growth and Maturity stages of evolution.

These advantages of TRIZ-based technological forecasting can be demonstrated through more detailed consideration of one of these useful aspects.

TRIZ allows one to evaluate how completely the current mode of performing PF is used for a technique based on investigation of the patents fund and other sources of technical information (see Chapter 8) and construction of the three bottom curves of Figure 7.4. If this mode has significant reserves (the beginning or middle of the cycle (α)), this observation can be used as a basis for formulating the real task for improvement. If the observation shows that the potentialities of the mode of performing PF are nearly exhausted (the end of the cycle (α)), then the justified conclusion is made about the necessity to move to a new mode. In this connection, one must seek and develop a more promising mode of performing PF, as well as avoid spending money in attempts to improve the well-matured technique.

Technique creation (I)

Research in guided evolution began only a few years ago [4], so this application is in its early childhood stage (see Figure 7.4). Therefore, it is too early to discuss this promising application. Note only that it combines many aspects of other applications presented in this section.

Technique creation (II)

The knowledge about evolution trends can also be used in analysis of functions of a developing technique to find a more efficient and more rational structure of the existing technique. This analysis includes the following major steps.

1. Evaluation of the functional value of every subsystem from the viewpoint of excluding the element and having other subsystems execute its functions. An appraisal sub-procedure asks the following questions:
   At what level of development is the considered subsystem?
   What restricts improvement of subsystem performance?

> Is the scientific, technical, and technological potential sufficient for transition to the more Ideal subsystem?
>
> Is the transition to the more Ideal subsystem worthwhile from social and economic points of view?

2. Evaluation of the efficiency of introducing new functional subsystems.
3. Evaluation of the efficiency of substance, energy, and information flows and, if needed, selection of a more rational sequence of functional links.
4. Separation of the neutral, auxiliary, and secondary functions as well as the functions executed by old and/or expensive subsystems or by a human being, followed by evaluation of the probability and efficiency of their execution by other subsystems.
5. Evaluation of the possibility of using subsystems of similar systems, anti-systems, and co-systems at higher stages of development to execute same or similar functions as the subsystems considered in the previous step.
6. Separation of a set of functions that could be executed by one self-sufficient subsystem.
7. Evaluation of the efficiency of function separation in subsystems executing more than one function.
8. Transfer the functions mentioned in Step 4 to the appropriate self-sufficient subsystems.

Notes:

1. Check technique completeness and energy, substance, and information transparency in accordance with the regularity of functional construction of a given class of TS or TP.
2. Use the appraisal sub-procedure creatively during various analysis steps.

This analysis (which should be performed as the first stage of new technique creation) helps start the directed evolution of the selected technique or subsystem.

Problem solving (I)

An important application of the knowledge about ET is the systematic hierarchical selection of the right problems and then of locally or globally Ideal technical solutions.

Initially for a given function (requirements and conditions), the most rational mode of performing PF and functional structure of a technique under development should be selected. At the second stage, the most efficient mode of operations (for set of UF) should be determined for the selected structure. Then the most rational technical solution that minimizes HF and NF for the chosen mode of operations should be sought. The last stage should be devoted to simulating the accepted technical solution and optimizing its parameters. Systematic hierarchical selection "prohibits" picking particular improved solutions as is generally practiced by engineers. Instead, it is oriented to the study and use of all possible ways of improvement.

An Ideal technical solution could be found in every case, if the problem is being solved with sufficiently complete information.

Particular emphasis should be placed on the necessity and utility of studies associated with comprehension and assignment of the postulate of progressive evolution to a specific class of TS, within which new generations are developed. The refined object-oriented formulation of the postulate will allow concrete definition and sufficiently complete separation of the criteria of development, main violations of the requirements, particular tendencies of changes in the technique, etc.

Problem solving (II)

Knowledge about trends, corollaries, and paths of ET in the methods of seeking more rational and efficient design and technological solutions allows us to filter out weak solutions that do not correspond to those trends, corollaries, and paths of evolution. In addition, other information can also be used to filter out weak solutions at the conceptual design stage, for example Nam Suh's Axiomatic Design [8] statements:

- The functional requirements of the design should be independent of each other.
- The design should have a minimum "information" content.

A solver should work carefully and remember that our knowledge is not complete. It means that a possible solution cannot yet be recognized as the strongest in the framework of TRIZ and other paradigms.

On the other hand, some corollaries are the base of TRIZ heuristics for creation of a new technique and improvement of existing TS and TP. Such studies allow us, first, to separate the most urgent criteria of development that are important for the next generation (quality, properties important for the aim and a user of technique), because changes in technique will occur only under the influence of these criteria. Second, they allow us to reveal particular trends in change of TS, which have decisive influence in the subsequent evolution. These trends, correlating with the separated criteria of development, give important information about most probable direction and character of changes in TS and TP in the next generation.

Marketing of innovations

Knowledge about ET (especially paths and corollaries) can be used in marketing of innovations, for example, as the input for Quality Function Deployment (QFD) [6]. Currently our understanding of the voice of a customer depends strongly on questionnaires prepared by a marketing department. These marketers, as well as specialists from R&D or engineering departments, often do not know many possible peculiarities of a future technique under investigation. Therefore, QFD questionnaires are subjective, incomplete, and outdated. In contrast, trends and paths of technical evolution extracted from patent information and from the history of a technique are objective, and many of them are domain-independent. Because there are numerous such paths and corollaries recognized in TRIZ, it is impossible to run a relatively inexpensive conceptual design in all these directions. It seems more reasonable from the economic point of view to ask the customer to choose the promising direction of technique development. First, QFD can identify conflicts between different performance measures. Second, QFD translates the voice of a customer into design requirements that rarely can be satisfied without problem resolution, i.e., the field where TRIZ is extremely powerful. TRIZ can reduce these conflicts to known contradictions and then overcome these problems by various heuristics and instruments (see Part 5). During the conceptual stage, an innovator familiar with TRIZ can forecast the whole picture of the socio-techno-economical tendencies. Moreover, TRIZ can be used for foreseeing the customers' competition evolution on the base of QFD data. Therefore, TRIZ can be used for creation of the patents fence for protection of the conceptual designs that are promising as an Ideal technique and then for marketing the technique.

7.8 CONCLUSION

A visit to any good technical museum or even long-term observation shows that all techniques, their subsystems, and super-systems have a history of development. There are also some finite nonzero probabilities of their further progressive development at any hierarchical level. TS and TP are not frozen entities; they are in the state of continuous change in order to satisfy everchanging needs of society and to survive in the very competitive global market. ET can be presented as a complex of processes that cover relatively large time spans of history by which a technique class originates, grows, and changes. In order to establish directions for development over longer time periods in various fields of engineering, we must start from the goals of society. Economic and market systems are thus important factors in operating the socio-technical system, and they affect, among others, those people (and their scope for action) who design and develop TP and TS, i.e., those who conduct and manage ET.

Direction and time postulates of ET, several corollaries, and a few trends and paths serve as part of the groundwork for other branches of TRIZ. The postulates cannot be proven in the framework of TRIZ, however, because they reflect the axiom of society (see page 95).

It should be kept in mind that the hierarchical exhaustion of a technique in the postulate of progressive evolution does not act formally.

1. Until the globally optimal parameters are achieved, a technique cannot transit to a new technical solution;
2. Until all potentialities of the Ideal technical solution (within a certain mode of operations) are exhausted, a technique cannot transit to a new mode of performing PF.

It should be kept in mind that the predicate about nonuniform evolution leads to unbalanced TS and TP; engineers often compensate for the imbalance with system or process complication by introducing various subsystems for auxiliary and neutral functions. The specialization is seen in design and manufacturing of each subsystem by an autonomous team and company. An expert in a tangible engineering field should recognize the hidden tradeoffs, while a TRIZnik should be able to discern contradictions that lead to the problem and then resolve them. At the next step, they together should choose the best solutions based on the ET trends and paths.

It should also be kept in mind that the essence of the most important corollaries and trends is that in a properly designed technique, every subsystem and every constructive parameter have a specific function in the TS or TP. Deprived of some subsystem, TS or TP either cease to operate (cease performing its function) or its performance worsens. In this connection, properly designed techniques have no "excess subsystems." This concept is the basis of all cognitive activity associated with analysis and study of existing technique, as well as the design activity for technique creation. Based on this corollary and ET trends, TRIZ experts have developed methodologies for new TS/TP design and construction of functional structures of a technique.*

Studies of ET and analyses of the status of a technique allow us to foresee future characteristics and parameters for the TS and TP, although without exact quantitative level. On the other hand, the application of past trends of ET allows us to direct technique developments because of TRIZ-aided design. In this case, trends can play the role of a filter that allows us to select promising solutions of a problem. Such a selection is inherent in some TRIZ heuristics and instruments; hence, a solver does not need to spend much time comparing and choosing solutions in contrast with what many other problem-solving methods entail. In each case of transition to a new generation of technique, particular trends are observed in changes of TS and TP. These trends determine, with high probability, the direction and character of TS or TP evolution in the next generation. ET trends and paths show, with sufficiently high probability, possible directions of change for the technical solution for eliminating revealed requirements violations or improvement of the corresponding criterion of development. Hence, ET understanding helps forecast the future of the technique and concentrate research and development in the most promising directions. Knowledge about ET also facilitates success in the engineering and marketing of innovations.

Unfortunately in reality, evolution trends are often ignored; that is why we encounter numerous techniques which are either poorly performing their goals or are far from the Ideal. Evolution of a technique is possible only by full use of all known natural and technical effects as fixed in numerous patents (see Chapters 8 and 9).

REFERENCES

1. (a) Altshuller, G. S., *To Find an Idea*, Nauka, Novosibirsk, 1986 (in Russian).
1. (b) Altshuller, G. S., Zlotin, B. L., Zusman A. V., and Filatov, V.I., *Search of New Ideas*, Kartya Moldovenyaske, Kishinev, 1989 (in Russian).

---

* Unfortunately, the limited volume of this book does not allow its presentation it here.

2. (a) Salamatov, Yu. P., "A System of Laws of Engineering Evolution," in *Chance for Adventure*, A.B. Selutsky, Ed., Petrozavodsk, Karelia, 1991, pp. 5–174 (in Russian).
2. (b) Salamatov, Yu. P., *TRIZ: The Right Solution at the Right Time,* Insytec, Hattem, 1999 (in English).
3. Polovinkin, A. I., *Laws of Organization and Evolution of Technique,* VPI, Volgograd, 1985 (in Russian).
4. Roza, V., Ed., *TRIZ in Progress,* III, Detroit, 1999.
5. Hubka, V., *Theorie technischer Systeme,* Springer-Verlag, Berlin, 1984 (in German). English translation *Theory of Technical Systems: A Total Concept Theory for Engineering Design*, by V. Hubka and W. E. Eder, Springer-Verlag, Berlin–New York, 1988.
6. Revelle, J. B., Moran, J. W., and Cox, C., *The QFD Handbook*, John Wiley and Sons, New York, 1997.
7. Martino, J. P., *Technological Forecasting for Decision Making,* North-Holland, New York, 1983.
8. Suh, N., *The Principles of Design*, Oxford University Press, Oxford, 1997.

Part 3

TRIZ Information

8 Inventions

8.1   INTRODUCTION

There are several known approaches to developing problem-solving methodologies for engineers (see references 1–6 and those to Chapter 1). Developers of problem-solving methods base their research on analysis of historical anecdotes and/or interviews with people who have successfully solved problems within the framework of the first approach. This approach represents an attempt to extract the techniques of problem solvers in order to create a generic set of problem-solving recommendations. Unfortunately, problem solvers, designers, doctors, engineers, and other white collar workers are often unable to articulate their process for resolving problems, hence the special (but not very successful) questionnaires and games developed through artificial intelligence studies to recreate the methods of design that experts employ.

The second approach to structured problem solving involves trying to define techniques based on the way the human brain is believed to work. Efforts by cognitive scientists have led to systems that collect and organize the knowledge of leading specialists in selected technical and nontechnical fields. These systems are quite useful in solving routine problems, especially problems that require numerous calculations and a great deal of information, but are not successful in solving creative or inventive problems. It is well known that reliable, well-organized knowledge helps solve problems. From ancient times, shallow knowledge has been distinguished from deep knowledge. From the point of view of a methodology for systematic resolution of technical problems, individuals, design teams, companies, and even industries have shallow knowledge with limited potential for resolving technical problems of a high difficulty **D**. Hence, the first and second approaches rely on shallow and limited knowledge of individuals instead of deep knowledge of the whole human society.

The third and final approach is the methodology developed by G. S. Altshuller and other TRIZ experts. It is based on

- documentation and following generalization of solutions of inventive problems stored in sources of deep knowledge, first of all in patents funds,
- abstraction of the current problem,
- analogy search for known successful solutions to unresolved inventive problems,
- creation of abstract solutions for the current inventive problem, and then
- transformation of the abstract solutions to specific solutions for the current problem.

All other approaches usually try to find a specific solution directly from the description of the current problem (see Figure 2.1).

Only the last problem-solving approach, TRIZ, relies on deep knowledge and provides possibilities for effective resolution of difficult technical problems. The strong advantage of TRIZ [6], in comparison with the other approaches [1-5], lies in Altshuller's choice of the right source for acquiring knowledge — the patent fund. Studies of the patent fund have enabled development of most of the modern TRIZ heuristics.

This chapter will

- describe the advantages of patents as compared with other sources of technical information,
- characterize a future of inventions and patents,
- outline TRIZ classification of patents,
- propose and discuss a procedure for patent study within the framework of TRIZ,
- describe the patent template used for TRIZ experiments,
- forecast and discuss the possibility of patent analysis by computer,
- outline the role of patent analysis stage in the TRIZ research cycle.

8.2   REASONS, GOALS, AND KINDS OF INVENTIONS

There are two principal reasons for new inventions:

- a problem arises *outside* the technique (e.g., demand from society or creativity of an inventor);
- a problem arises *inside* the technique (e.g., cost reduction and quality improvement efforts or new materials or technology available).

If an invention satisfies the criteria of novelty and few others, it may be registered as a legal document — a patent. Patents are the only legal type of monopoly allowed in many countries, which explains why they are so attractive. Nevertheless, even when a new TS or TP is patentable, the question "Will obtaining a patent have any advantage for an inventor or company?" should be asked because of economical drawbacks. Often a few independent inventors simultaneously but separately design very similar, new TS or TP, e.g., invention of the telephone, radio, semiconductor transistor, and integral circle. Patents establish the invention's legal priority.

Patents are usually classified by field of engineering. The majority of patents fall within primary areas: mechanical, electromagnetic, chemical, thermodynamics, electronics, optics, and software. Often a new device or technology belongs to more than one category. Unfortunately, such classification is not very useful because similar problems arise in different engineering fields and their solution often is the same. Nevertheless, this solution will be recognized as a patent even if the solution is known in a different field of engineering. Because narrow and intense specialization dominates in patents, other engineering fields which can benefit from an identical invention are often overlooked.

In spite of the fact that there are only two reasons for inventions, there are various motivations. Some of them have been summarized by Edward and Monika Lumsdaine [1] as the following:

- As a response to a threat → radar, weapons
- As a response to an existing need → can opener
- As a response to a future need → high-temperature ceramics
- For the fun of it — as an expression of creativity
- To satisfy intellectual curiosity
- As a response to an emergency → Band-Aid
- To increase one's chance of survival and security
- To increase comfort and luxury in lifestyle
- Better problem solving → hydraulic propulsion system
- Turning failure into success → Post-It notes
- To overcome flaws → "Magic" tape
- Accidentally, on the way to researching something else → polyethylene
- As a deliberate synthesis → carbon brakes for aircraft
- Through brainstorming with experts or outsiders → courseware
- From studying trends, demographic data, and customer surveys
- Through cost reduction and quality improvement efforts → float glass
- Through finding new uses for waste products → aluminum flakes in roofing
- Through continuous improvement of work done by others
- Through having new process technology → proteins from hydrocarbons
- By finding new applications for existing technology
- Through having new materials available
- To win a prize or recognition → human-powered aircraft
- Meeting tougher legal and legislated requirements → catalytic converter
- By having research funds available to solve a specific problem → superconductors
- By being a dissatisfied user of a product
- Responding to a challenge or assignment
- Because "it's my job" → "I do it for a living"
- To improve the organization's competitive position
- To get around someone else's patent

There are also numerous goals of inventions. It is interesting that in many cases strong solutions do not solve only the primary technical problem, but they also lead

TABLE 8.1
The Major Goals of Inventions

Technical	Economical	Functional
General	General	Automate/Mechanize
Control	Ergonomics (Human Engineering)	Eliminate/Simplify
Measurement	Ecology	Obtain/Use as
Application		Provide/Afford
		Reduce/Decrease
		Rise/Increase
		Expand/Limit

to additional unexpected, positive technical or economic benefit. Such phenomenon is called *extra-effect* in TRIZ. Quite often, an inventor concentrates too much on the resolution of the primary problem and does not recognize possible extra-effects in the solution.

It is often surprising to an engineer, when he looks back over a considerable number of his developments, to find that they are related to a relatively small number of different goals. Unfortunately, the possible goals of invention, other than those for which the invention was created, are often overlooked. The major goals of inventions, that can help in possible claims of future patents, are summarized in Table 8.1.

In order to avoid such losses by overlooking possible goals, one can prepare morphological boxes for each technical and economical category in Table 8.1; the primary and secondary functions of the system and subsystems should be written along one axis, and along the other axis should be the list of verbs from the "Functional" column and/or "manipulative" verbs from the Alex Osborn questionnaire (see Appendix 1 and Chapter 11), and/or verbs that are important for your technical field. Such boxes are a good aid in expanding possible goals and, consequently, patent claims.

It seems reasonable to present here some ideas about different kinds of inventions proposed by various American and European researchers.

John A. Kuecken [2] noted three circumstances for inventions:

1. **The Ingenious Solution** — A problem is observed, and then a suitable solution is found; e.g., a man *first* observed the problem of long-distance communication without wires and *then* radio was invented.
2. **The Ingenious Application** — A property is observed and then applied to solve a suitable problem; e.g., a primitive man *first* observed the property of fire and *then* utilized it for warming his cave.
3. **The Gap Filler** — A gap is observed by a specialist in the field and then creates something to fill it, e.g., chewing gum with apple taste.

Several years after Kuecken, inventor and writer Gilbert Kivenson [3] also separated inventions into a few categories as follow:

Single or multiple combination — A pen and pencil on opposite sides of the same holder create certain properties not possessed by either item alone. The user can conveniently change from pencil to ink without having to carry an extra writing utensil in his pocket.

Labor-saving concepts — Coupling an electric motor to a kitchen mixer and driving it at high speed permits the uniform addition of air to the food; this extends the use of the blender and imparts a pleasant modification of taste not possible when the mixer is operated by hand.

Whenever a new energy source is introduced, there is always a possibility for inventions that couple the new source with existing devices in order to save effort, produce more with the same effort, or achieve unattended operation.

Direct solution to a problem — The problem of reducing the size of vacuum tubes was solved by invention of the transistor.

Adaptation of an old principle to an old problem to achieve a new result — Pencil leads have been made for many years from powdered graphite fired with clay. It was not possible to produce these leads in diameters much less than one millimeter. The substitution of plastic (a well known material) for clay permitted sturdy pencil leads to be made in 0.6 and 0.3 millimeter sizes that are useful for draftsmen and artists.

Application of a new principle to an old problem — The application of transistors instead of vacuum tubes to hearing aid technology.

Application of a new principle to a new use — The deployment of communications satellites in stationary positions above the Earth and the use of these line-of-sight stations to relay programming and communications around the world to handle the increasingly heavy traffic and load on existing communications channels as well as the need for extra overland and submarine cables.

Serendipity, or the principle of exploiting lucky breaks — There are two types:

- a solver is stalled at a certain point in problem solving and a chance observation provides the answer: e.g., the famous case of rubber vulcanization by Charles Goodyear
- a solver suddenly discovers a new principle unrelated to the work in which he is engaged but related to some other area, and then successfully applies this discovery to that area: e.g., discovery of the explosive properties of a potassium nitrate, charcoal, and sulfur mixture.

Vladimir Hubka, well-known specialist in General Engineering and Design, proposed classifying TS by design originality or degree of novelty into the following classes [4]:

Re-used technical systems — A set of subsystems that exists or is available to fulfill the required functions, and the most appropriate ones can be selected and used without alterations. This group comprises standard parts such as screws, keys, valves, springs, etc., and standardized parts such as purchased components and assemblies. It also contains nonstandardized parts and groups that can be reused in a context other than that for which they were originally designed.

Adapted technical systems — Some technical systems may exist for required functions but do not comply with all requirements of the application. Some alterations must be made in terms of size, power, speed (rotational or other), connective or other dimensions, material, or manufacturing methods. New materials are generally used only to improve the quality of the technical system, reduce costs, modernize the system, or adapt it to other operating or environmental conditions (e. g., corrosion reduction). The main concepts, structures, and other important properties of the TS remain unaltered. The technical system is merely adapted to the particular conditions and requirements of the new task. Even this adapting may prove to be difficult under certain circumstances.

Redesigned technical systems — The existing technical systems do not completely fulfill the requirements with respect to their functions and other properties. In the conceptually redesigned version, only the functions, some of the parameters, and possibly the working principles remain unaltered. Parts may have their form, dimensions, material, or technology revised. In more complex technical systems, the alterations involve their construction and their structures, i.e., parts and constructional groups, their couplings, and spatial arrangements. Redesigning is probably the most frequent design activity.

Original technical systems — No existing technical system can fulfill the required functions, or the existing system has distinct deficiencies. A new working principle and additional technical properties are usually needed. As an example, a design task may be stated as "the effect 'heating,' which was accomplished by a technical system using the action principle 'burning a mineral oil derivative,' should be replaced by a newly designed system using 'electrical resistance heating.'"

All three categorizations are similar, although they are based on different assumptions. They are not, however, instrumental, which is why the ideas of G. S. Altshuller and his coworkers about quantitative classification for levels of inventions are so popular in TRIZ.

8.3   LEVELS OF INVENTIONS

Genrich S. Altshuller and Ralph B. Shapiro imported the idea of ranking, an old and common activity in many other fields (such as hotels, wines, etc.) into the world of inventions, ideas, and innovations. It was the beginning of TRIZ. Altshuller and Shapiro recognized five levels of inventiveness based on

- problem difficulty **D**,
- difference between an earlier known prototype and the new solution, and
- "distance" knowledge from the inventor's field used for the new solution.

Patents representing a simple modification to a design were assigned to the lowest level. Patents that changed the system in some way were considered more inventive, while patents introducing a new science were considered the most innovative. They showed that the level of inventiveness could be categorized in a range from Level 1 (personal knowledge) to Level 5 (universal knowledge, i.e., all known information). The Level is a kind of quantitative characteristic that cannot be calculated. It can usually be estimated by experts and is based on several criteria, such as the extent of technique change produced by the solution.

LEVEL 1: REGULAR

Level 1 includes routine design problems solved, after a few dozen attempts, by methods well known within the specialty or within a company. Approximately 32% of the solutions occurred at this Level.* Such solutions represent most recurrence and small changes of the earlier known prototype without its essential variations.

Example:
The ability to change the size of lead shoes for divers by adjusting their length was developed. (It is interesting that this development did not occur until the 1960s, some 70 years after the invention of divers' shoes; for 70 years all divers used uncomfortable shoes of the same size.)

Usually patents at the first Level are solved by trading off one subsystem (element, operation, etc.) for something else (as most engineers traditionally do). Level 1 patents are not inventions from the TRIZ point of view (Patent Offices of many countries have the opposite opinion). Instead, TRIZ considers them narrow extensions or improvements of an existing technique that ultimately is not substantially changed. Usually a particular feature is enhanced or strengthened. These solutions may represent good examples of "trade-off" engineering that usually is based on design principles where contradictions are not identified and resolved.

LEVEL 2: IMPROVEMENT

Development of an existing technique (approximately 45% of the solutions; a few hundred attempts). The earlier known prototype is changed qualitatively but not substantially, usually due to application of uncommon methods from the same engineering field as the technique with some additional knowledge from the inventor's specialization and/or some creative effort.

Example:
Welding two different metals together (such as copper and aluminum) can present a challenge. One useful technique is to use a spacer made of a metal that can be welded to each of the incompatible metals.

Level 2 solutions offer small improvements to an existing technique by reducing a contradiction inherent in the technique but requiring an obvious compromise; such solutions require knowledge of only a single engineering field. The existing technique is slightly changed, including new features that lead to definite improvements.

When problems are solved at Levels 2–5, the problem solver is considered an inventor from the TRIZ point of view.

---

* Between 1956 and 1969, patents under review from 14 different classes were evaluated to determine relative frequencies of solution levels. Percentages given here are from that period.

LEVEL 3: INVENTION INSIDE PARADIGM

Essential improvement and radical change of the earlier known prototype utilizing the methods or knowledge from other disciplines, sometimes far from the major engineering field or industry of the technique (18%; dozens of thousands of attempts). The changes are considerable and result in a new quality.

Example:
Cattle feed consists of various cut grasses which must be mixed by special equipment. Producing the grass mixture by sowing the various grasses together yields a crop that is difficult to till. Furthermore, one grass species may inhibit the others. The grasses can be sown in narrow parallel strips and then harvested across the strips. Thus, the grasses are mixed in the receiving bin of the mower.

Level 3 inventions significantly improve existing techniques, often through the introduction of an entirely new subsystem that usually is not widely known within the industry of the inventive problem. Novelty exists here from

Removal of false restrictions or resolution of contradictions,
Expansion of the sphere of application of the prototype,
Inclusion of the prototype as part of a whole — association of the prototype
    with similar or alternative systems.

The solution causes a paradigm shift within the engineering field. Level 3 innovation lies outside an industry's range of accepted ideas and principles.

LEVEL 4: BREAKTHROUGH OUTSIDE PARADIGM

Radical change of the prototype. A new idea that has practically nothing in common with the prototype. Creating a new technique generation, the solution usually cannot be obtained in engineering but rather can be found in science (4%; several hundred thousand attempts).

Example:
An electromechanical relay element has a finite number of switching cycles. Substituting a cheap semiconductor relay element increases the number of switching cycles and decreases the switching time and weight of the device.

Novelty exists here from replacement of a technique that carried out the primary function of the prototype. Level 4 solutions are breakthroughs, lie outside a normal paradigm of the engineering field, and involve use of a completely different principle for the primary function. In Level 4 solutions, the contradiction is eliminated because its existence is impossible within the new system. That is, Level 4 breakthroughs use physical or other effects that have previously been little known in engineering. Hence, the database of various effects is important for TRIZ (see Chapter 9 and Appendix 3).

LEVEL 5: DISCOVERY

Pioneer invention of a radically new technique is usually based on a major discovery in some basic or new science (less than 1%; millions of attempts) or recognition of

TABLE 8.2
101 Outstanding Human Achievements

Accumulator	Distillation of oil and	Lever	Sail
Agriculture	petroleum	Lift	Saw
Alphabet and ability	Domestication of wild	Map (geographical)	Screw (drill, groove)
to record	animals	Measuring devices	SCUBA
information	DNA	Medicine — drugs	Selection
(magnetic, optical,	Dyeing	Money	Semiconductor devices
writing)	Electrical	Musical instruments	Sewing needle, sewing
Artificial intelligence	communication	Newspapers, radio, TV,	machine
Artificial limbs	(telephone)	and internet network	Soap
Artificial rubber	Electric motor	Note (musical)	Spring
Assembly line	Electrical light	Paper	Steam engine
Bearing	Electronic computer	Pendulum	(locomotive,
Bicycle	Fabric weaver's	Photo	steamship)
Biotechnology	machine tool	Pipeline	String
Boat, vessel	Furnace	Piston	Superconductors
Book, typography,	Games (toys)	Plane	Synthesis of organic
library	Gear transfer	Plasma	substances
Brake	Gene engineering	Plastic	System of automatic
Bread	Glass	Pneumatics	control
Brick	Glue	Post (mail)	System of measures
Bridge	Greasing	Pottery wheel, turning	(mass, electricity)
Cement	Gunpowder	machine tool	Usage of water/wind
Cinema	Gyroscope	Preservation	power (mill)
Clocks (watches)	Hammer	Preventive inoculation	Usage of X-ray radiation
Clothes	Hinge	Pump	Vehicle (car, wagon)
Compass	House (dwelling)	Radar	Weapon (arrows, bow,
Controlled nuclear	Internal combustion	Railway	guns, etc.)
reaction	engine	Reception and transfer of	Wedge (axe, knife,
Cooking	Jet motor	electrical energy	shovel, plough, chisel,
Cosmic technique	Laser	Reception of a fire	cutter)
Dam	Lens (microscope,	Reception of metals and	Welding
Decimal notation	telescope)	alloys	Wheel

a new need. Principally a new idea arises because of a change of the primary function of the prototype and an occurrence of new subsystems for realization of the new function that include and/or substitute the old primary function.

Example:
Examples are given in Table 8.2, which lists the most important human discoveries and inventions.

Level 5 solutions exist outside the confines of contemporary scientific knowledge and usually stand between science and engineering. These discoveries require a lifetime of dedication. They occur when a new effect or phenomenon is discovered and afterward applied to an inventive problem.

Once a Level 5 discovery becomes well known, subsequent application or invention occurs at one of the four lower levels. For instance, the laser, technological wonder of the 1960s, is now used routinely as a lecturer's pointer and a land surveyor's measuring instrument. This evolution within the laser industry illustrates that a solution's level of inventiveness is time-dependent.

Usually a strong solution begins the chain of solutions at lower Levels. For simplicity, this chain can be represented by changes in the following attributes:

need $\rightarrow$ primary function $\rightarrow$ characteristic (property) $\rightarrow$ parameters $\rightarrow$ values

The evaluation of solutions by Level is, obviously, subjective to some extent. For example, some people think that the integrated circuit is the most important invention in the twentieth century, more important than a transistor. However, when a great number of solutions is classified, a small group of best inventions can be separated rather unanimously. All agree that the transistor and the integrated circuit are both very high Level inventions. Moreover, subjectiveness is not a problem* because the evaluators can get consensus with the help of the decision aids outlined in Chapter 1.

One open question is whether the distribution of inventions by Level changes with time. Altshuller studied the statistics of USSR patents from three patent classes again in 1982. The criteria were the same as in the study of patents from years 1956–1969. The results: 39% were Level 1, 55% were Level 2, and 6% were Level 3. Inventions of the fourth and the fifth Levels were not found at all. The change, perhaps, can be explained by the statistical variation between samples (note that the first collection was more representative).

In general, Level is related to the solution and to the problem, since a single problem can have solutions at different Levels and a single solution (that is usually close to the IFR) can resolve a few problems at various Levels.

For example, consider a problem of heavy machinery in a basement shop that produces excess vibration. The suggestion is made to place a rubber pad under the machinery to absorb the vibration. If this solution is adequate, the problem has a Level 1 solution. If, however, preventing transmission of the vibration is not effective enough, another solution should be sought, for example, attempting to cancel the vibration with anti-vibration; this solution might be estimated as a Level 2 solution. If this solution is satisfactory, the problem has a Level 2 solution. If not, a Level 3 solution might be suggested, for example, using an air or magnetic pillow.

The level of a problem can be estimated as **lg (D)**, where **lg** is the decimal logarithm and **D** is the difficulty of the problem (introduced in Chapter 1). Unfortunately, for many technical tasks it is hard to determine the difficulty **D** of the problem itself.

Level is not also a property of a new technique — it only mirrors the differences a new technical system or technological process has from the prototype.

---

* The only condition is that all evaluators should know the classification criteria of TRIZ.

8.3.1 REALIZATION OF INVENTIONS

Altshuller focused his investigation on the principles used in Level 2, 3, and 4 solutions. Level 1 solutions were ignored because they need not be innovative. Because Level 5 solutions require understanding of a new natural effect and/or recognition of a new humans' need, there were no recognizable parallels among these inventions. Altshuller believed he could help anyone develop Level 2, 3, and 4 inventions based on the extensive study of patents. Such studies (see the next section) allow the solver to identify heuristics for solving inventive problems. While searching the patent fund, Altshuller recognized that the same fundamental problems had been addressed by a number of inventions in different engineering fields. He also observed that the same fundamental solutions were repeatedly used but the implementations were often separated by several years. Level 1–3 solutions are usually transferable between different fields of technique, so approximately 95% of problems in any particular engineering field have already been solved in other fields. Access to the applications of these typical solutions would decrease the number of years between inventions. Consequently, problem solving and the innovation process become more efficient.

Unfortunately, there are many barriers to invention and innovation. The difference between the tempo of technological progress and social change complicates the success of upper Level inventions. In his book for children ...*And Suddenly the Inventor Appeared* [6] Altshuller* writes, "If one chooses to develop a completely new technical system when the old one is not exhausted in its development, the road to success and acceptance by society is very harsh and long. A task that is far ahead of its time is not easy to solve. And the most difficult task is to prove that a new system is possible and necessary." In other words, the inventor must be cautious because the public may not accept designs that are too advanced so support for further development will be limited. There are a few factors related to company practices and operations that also have to be considered in making the decision to proceed with new product or process development. These factors can be acquired by consulting with management, plant engineering, production, marketing, technical, and sales personnel. Introducing several incremental improvements to an existing system is often a good strategy. In the theory of innovation, this strategy is known as the diffusion process. For example, the initial reaction to radar when it was introduced during World War II illustrates a common response to new inventions. Radar radically increased a submarine crew's awareness of approaching aircraft. However, the submarine captains of one country refused to use the newly installed devices. Being newly aware of so many more aircraft, the captains thought the radar was attracting the planes. Today their reaction may seem strange, but it exemplifies people's initial resistance to technological innovation. For a more contemporary example, consider that computer programmers initially saw no reason to provide a CRT screen with the computer. There were two issues the designers did not understand: the anticipated needs of the customer and the usefulness of the monitor for

---

* He used the pseudonym H. Altov in some books.

themselves. The enigma of inventions' realization is discussed widely, so the reader can easily find many works about this topic. Although TRIZniks have accumulated some experience with inventions' realization, such discussion is beyond the scope of this book.

8.4 HOW TO STUDY PATENTS IN THE FRAMEWORK OF TRIZ

TRIZ heuristics and instruments, as well as TRIZ itself, are based on the selection and study of high Level patents. This section discusses the procedure of patents studies. Readers interested only in the application of TRIZ can skip this part.

8.4.1 SOURCES OF TECHNICAL KNOWLEDGE

Harold R. Buhl [5] wrote that "the only raw material available for solving problems is past knowledge." Insofar as this is true, three basic categories of facts are important for TRIZ:

- knowledge of basic heuristics used for technical problem solution,
- understanding of evolutionary trends in technical systems and technological processes,
- information about new scientific and technical effects and phenomena.

Such facts can be found in four major sources of technical information ordered here by *reliability* of information:

- encyclopedias, handbooks, and textbooks;
- monographs and reviews;
- patents;
- scientific papers and technical reports.

These sources are written documents or electronic databases. Brief summaries or digests of scientific books, papers, technical reports, and patents are often collected and published. These abridged documents help researchers find the original sources of useful information.

The list of information sources can be reordered from the perspective of *veracity* and from the degree of *novelty*. On the other hand, various sources of technical information play different roles in TRIZ (see Figures 8.1 and 8.2). Comparison of these figures indicates that patents can be the primary source of information for TRIZ. This conclusion is confirmed by a statement of the European Patent Office*: "Patents reveal solutions to technical problems, and they represent an inexhaustable source of information: more than 80 percent of man's technical knowledge is described in patent literature."

---

* See http://www.european-patent-office.org/espacenet/info/index.htm.

Serviceability of Information Sources

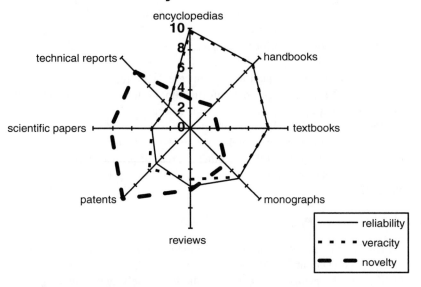

FIGURE 8.1   An estimation of the range of usefulness of different sources of information for TRIZ.

TRIZ Information Sources

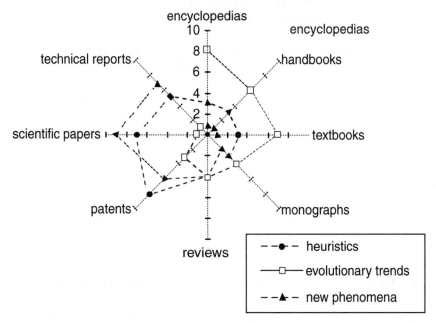

FIGURE 8.2   An estimation of the range of knowledge useful for TRIZ in different sources of information.

In contrast with all other sources, only patents

> have uniform structure for representation of information,
> report the "state of the art," and
> describe weak points in prior techniques.

Because of these unique qualities, patents are the preferred source of knowledge from which TRIZ experts can extract information. Fortunately, the patent databases of many countries are now accessible via the Internet, making the regular study of patents for experimental TRIZ more effective.*

Of course, not all engineering fields are represented equally in the patents fund; some techniques (e.g., software) have other sources of primary information. Nevertheless, such sources can be studied in order to gain deep knowledge in the engineering field.

8.4.2   PERUSAL OF PATENTS

More than 20 million patents have been issued worldwide, though there is some duplication among countries. Although approximately 85% of patents are not usable for TRIZ research because they belong to the low Levels, approximately 3 million patents warrant rigorous and detailed study. This study should be

> systematic,
> performed by a person (or computer, see below) that is familiar with TRIZ
>   and the field of engineering under the investigation, and
> well documented, containing reusable information.

In order to produce uniform patent study results that can be easily used by all TRIZniks, some procedures and terms were established and accepted:

1. Altshuller's scale for division of problem solutions into five Levels of inventiveness (described in the previous section);
2. the standard form for reports of experimental patent study shown in Figure 8.3;
3. the following terminology and rules:

Terms

TRIZ+trust — the registry of patents that represent strong solutions (Level 3 and above) that confirm TRIZ.
TRIZ-pool — the registry of patents that represent solutions that conflict with current TRIZ.
TRIZ Bank — the combined registries of **TRIZ+trust** and **TRIZ-pool**.

---

* The following URLs are popular for patent searches: http://patent.womplex.ibm.com/; http://www.uspto.gov/; http://patents.cnidr.org.

```
┌─────────────────────────────────────────────────────────────────────┐
│                      PATENT ABSTRACT FORM                             │
│                                                                       │
│ 00. Legal information                          Sign: +, -, 0 or X     │
│                                                                       │
│ 0.  Abstract of extracted knowledge that is new for TRIZ and keywords │
│                                                                       │
│ 1.  State of the art (before the patent)                              │
│                                                                       │
│ 2.  Elements and functions of the prototype system                    │
│                                                                       │
│ 3.  Resolved conflict                                                 │
│ a)  Type of contradiction                                             │
│ b)  Structure of problem                                              │
│                                                                       │
│ 4.  Invented system                                                   │
│ a)  Contradiction                                                     │
│ a1) Principle of solution                                             │
│ a2) Reduction rule                                                    │
│ b)  Effect (Phenomena)                                                │
│ b1) Natural                                                           │
│ b2) Technical                                                         │
│ c)  Heuristics                                                        │
│ c1) Engineering Principle                                             │
│ c2) Separation Principle                                              │
│ c3) Standard Transformation                                           │
│ c4) Prescript                                                         │
│ d)  Trend/path of evolution                                           │
│ e)  Special cases                                                     │
│                                                                       │
│ 5.  Notes (degree of efficiency, applicability, universality, etc.)   │
│                                                                       │
│ 6.  Information about this study (name of TRIZnik, date)               │
│                                                                       │
│ 7.  Changes of the entry classification                               │
└─────────────────────────────────────────────────────────────────────┘
```

FIGURE 8.3 Template for patent analysis.

InfoBank — the registry of patents that can be used as analogies and/or examples for new problems (usually solutions at Levels 3 and 2).

WasteDeposit — the registry of noninventive patents (Levels 1 and 2). For these patents, only legal information should be stored.

Rules

1. Completeness rule

$$\text{Patents fund} = \text{TRIZ Bank} \cup \text{InfoBank} \cup \text{WasteDeposit}$$

$$\text{TRIZ Bank} = \text{TRIZ+trust} \cup \text{TRIZ-pool}$$

2. Uniqueness rule

At any point in time, a patent can have only one entry in the patent fund.

Patent ∈ TRIZ-pool XOR TRIZ+trust XOR InfoBank XOR WasteDeposit

Entries are ordered in accordance with the level of curiosity of a particular patent search.

Note: TRIZ Experts use the signs "+" and "–" as shorthand references for TRIZ+trust and TRIZ-pool, respectively. InfoBank is auxiliary to TRIZ+trust and is designated by 0 (zero) while X designates the WasteDeposit.

One of most important goals of the uniform patent study procedure is the generation of reusable information. Extracted information can then be transferred to other manual or electronic databases and computer knowledge repositories.

8.4.3 BASIC FLOWCHART OF PATENT INVESTIGATION

Most patents are useless from the point of view of TRIZ experts. Nevertheless, all patents can be investigated in terms of traditional TRIZ (as shown in Figure 8.4), though Level 1 and 2 patents do not require a long time for analysis. These low Level patents are quickly scanned by TRIZniks and important legal information is noted in order to prevent other researchers from wasting their time. Conservation of researchers' time is also important when studying the higher Level patents. A small number of patents, Level 5, require meticulous analysis within the framework of techno-economical trends. A larger number of high Level patents, usually Levels 3 and 4, require focused, critical analysis to recognize a novelty in resolution of a problem that had contained a contradiction. Unfortunately, such tedious study of patents can be performed now only by human beings, not by computers. It seems that the first step for computers would be to separate the numerous low Level patents with already known TRIZ heuristics from the relatively small number of high level patents, which could be original from the TRIZ point of view (i.e., which could have new heuristics).

Altshuller and the author have shown that a single, new heuristic can be found in about 10,000 randomly chosen patents or in around 1,000 patents at Levels 3–5 (see Figure 8.5). Hence, patent studies are, unfortunately, not very productive. The increasing plethora of information in the worlds of science, business, technology, and government creates a need for tools and techniques that can analyze, summarize, and extract "knowledge" from raw data in huge and noisy databases of available facts. Most of the so-called "data mining" tools and techniques are based on statistics, machine learning, pattern recognition, semantics, and artificial neural networks. With some adaptation, it seems these tools could be used for knowledge discovery in the patent fund. Hopefully, computer-aided studies of patents will be possible in the future, as they should be faster and, perhaps, more effective than the manual studies conducted by TRIZniks now (see Figure 8.5).

Knowledge extracted from patents can gradually be included in TRIZ. It is well known that each science has its own experimental, theoretical, and application

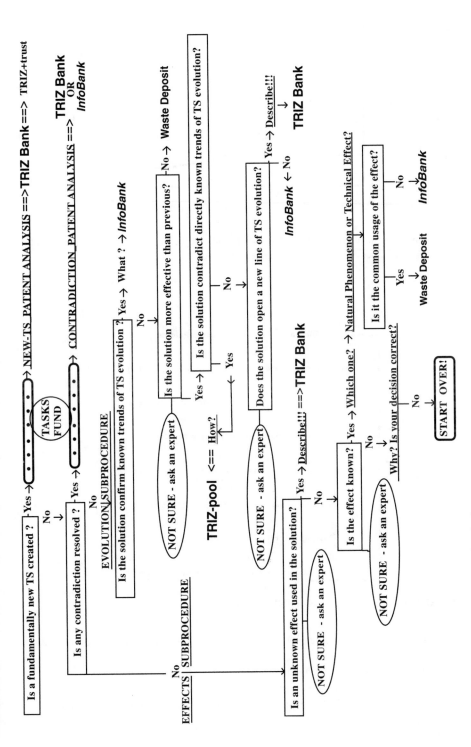

FIGURE 8.4 Schema for study of a single patent.

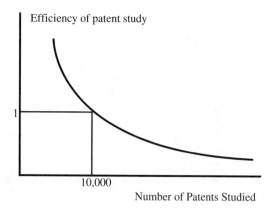

FIGURE 8.5   A sketch of the dependence of patent study efficiency for TRIZ (number of heuristics) on number of patents in an engineering field.

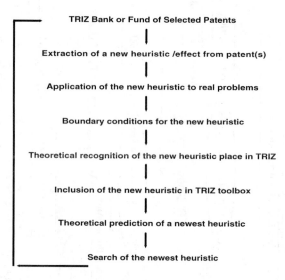

FIGURE 8.6   Part of flowchart for TRIZ research cycle.

components. The patent studies are the foundation of all these components for TRIZ (see Figure 8.6).

Using the concrete knowledge provided by information in patents, we can confidently apply TRIZ to new engineering fields. This concrete deep knowledge is distilled through our systematic procedure of patent studies within the framework of TRIZ.

8.5 CONCLUSION

The importance of the patent fund as the primary source of deep technical knowledge is confirmed by the fact that numerous attempts by psychologists, cognitive scientists, and specialists in artificial intelligence to design problem-solving methodologies equal to TRIZ have, as yet, failed. Fortunately, TRIZ has a source of deep knowledge — the patents fund. All patents can be classified into 5 Levels. The patents fund can be used for future development of TRIZ methodology. The main results of the studies of patents — TRIZ heuristics — are presented in the next chapters. In addition, TRIZ operates with a huge collection of scientific effects, which are presented in Chapter 9.

REFERENCES

1. Lumsdaine, E. and Lumsdaine, M., *Creative Problem Solving: Thinking Skills for a Changing World*, McGraw-Hill, New York, 1998.
2. Kuecken, J. A., *Creativity Invention and Progress,* H. W. Sams, Indianapolis, 1969.
3. Kivenson, G., *The Art and Science of Inventing,* Van Nostrand Reinhold, New York, 1982.
4. Hubka, V., *Theorie Technischer Systeme,* Springer-Verlag, Berlin, 1984 (in German). English translation: *Theory of Technical Systems: A Total Concept Theory for Engineering Design,* by V. Hubka and W. E. Eder, Springer-Verlag, Berlin–New York, 1988.
5. Buhl, H. R., *Creative Engineering Design,* Iowa State University Press, Ames, 1960.
6. Altov, H. (Altshuller, G. S.), ... *And Suddenly the Inventor Appeared.* Detskaya Literatura, Moscow, 1984, 1987, 1989 (in Russian); TIC, Worcester, 1996 (in English).

9 Effects

9.1   ROLE OF NATURAL SCIENCES IN TRIZ

Studies of numerous patents indicate that strong inventive solutions are frequently obtained by using natural effects that have rarely or never been used previously in a specific area of technology. Quite often knowledge of various physical, chemical, or geometric effects or their "uncelebrated" sides is necessary for solving a technical problem. An ordinary engineer usually knows about 100 effects and phenomena, while there are about 10,000 effects described in scientific literature. Each effect may be a key to solving a large group of problems. Since engineering students usually are not taught how to apply these effects to practical situations, engineers and designers often have problems using such well-known effects as thermal expansion or frequency resonance, let alone less recognized effects. On the other hand, scientists often do *not* know how to apply physical, chemical, biological, or geometric effects which they have discovered in various fields of engineering.

To assist in applying these effects to solving problems, a Database of Effects was proposed by Altshuller and developed later by Y. V. Gorin, S. A. Denisov, Y. P. Salamatov, V. A. Michajlov, A. Yu. Lichachev, I. E. Vikentiev, V. A. Vlasov, V. I. Efremov, M. F. Zaripov, V. N. Glazunov, V. Souchkov, and other TRIZniks. Their goal is to simplify the search of an effect, factor, or method capable of providing the required function or property.

The real effects in the Database are considered "a black box," i.e., they have no internal structures and cannot be divided into further objects, that provide a certain response to some entering influence. Such a way of classifying effects is quite convenient, although different from that used in traditional scientific encyclopedias. It allows the building of a hierarchical structure for a database and the possibility of quick searching of necessary effects or their combination.

From the point of view of simplicity of effects application, the Database of Effects is built on a functional principle: it contains a list of functions (applications) commonly encountered in practice, and a corresponding list of effects that may be employed to realize these functions. This structure helps engineers resolve technical problems. A short registry of frequently used physical effects is given in Table 9.1, while an expanded list of various physical, chemical, and geometric effects are presented in Appendix 3. (We do not include biological and materials science effects, being studied by A. Yu. Lichashev, V. I. Timochov, S. D. Savransky, S. E. Sofroniev, and others because these catalogs are not completed. We also do not provide a list of so-called mathematical effects begun by V. Tzurikov but not completed due to the evolution from analog to digital information and communication techniques.) A list of energies (Appendix 4) is supplied to the Database for convenience.

9.2   EXAMPLES OF EFFECTS APPLICATIONS

Let me illustrate by my own solutions of different problems how the Database works and how the problem solver can use the "Patent fund of Nature" with physical, chemical, and geometric effects.

PHYSICAL EFFECT

Electromechanical relays are not very reliable and can perform only 1–3 million switches only. Solid-state thyristors and opto-electronic devices are more reliable, but they are not very technological and require complicated, expensive, and precise technology for their production. For almost two decades numerous research laboratories were unable to resolve this problem.

Solution: Apply native electrical bi-stability of amorphous semiconductors, the resistance of which can be altered by several orders of magnitude under the influence of the applied electrical field (USSR Patents 997099, 1112925, 1136647, 1342277, 1453446, 1568017). The respective micro-electronic devices were produced. They have the extra-effect: ultra-high radiation stability that allows them to work in the cosmos.

CHEMICAL EFFECT

Flat panel displays use the physical effect of field electron emission from Silicon needle-like cathodes. It was necessary to obtain numerous needles with a tip diameter of about 500 Angstrom in order to build the prototype flat panel displays during research. Physical methods of needle sharpening do not work because of Silicon fragility, so chemical methods were chosen. Wet-chemical etching of Silicon wafers is one of the major steps of the semiconductor device process. The aim of this etch is to polish the wafer's surface. The widely used acid solution for this is the following:

65% Nitric acid HNO_3	19.5 parts by volume
98% Acetic acid CH_3COOH	19.5 parts by volume
40% Hydrofluoric acid HF	1 part by volume

The information about structural defects in Silicon wafers is often important for quality control of semiconductor devices. Such defects can be developed by the following etching acid solution:

65% Nitric acid HNO_3	3 parts by volume
98% Acetic acid CH_3COOH	11 parts by volume
40% Hydrofluoric acid HF	1 part by volume

TABLE 9.1
Short Registry of Physical Effects

1. Use of liquid and gas properties
   a. Pressure in liquids and gases transfers equally in different directions.
   b. Carrying capacity acts on an object immersed in liquid or gas.
   c. Volume of expelled liquid is equal to the volume of immersed part of a solid substance.

2. Use of thermal expansion
   a. Change of linear sizes of a body during thermal expansion may be due to considerable efforts.
   b. Change of body shape during thermal expansion occurs if the body consists of materials with different coefficients of thermal expansion.

3. Use of shape memory effect
   Bodies from special alloys deformed under mechanical forces may fully reconstruct their shape during heating and may produce large forces.

4. Use of phase transitions
   a. Phase transition of the first kind: the process of density and aggregate state change of the substance at the specified temperature, which is accompanied by heat detachment or absorption.
   b. Phase transition of the second kind: jump-like changes such as the change of a main substance's properties (heat conductivity, magnetic properties, superfluidity, plasticity, superconductivity, etc.) at the specified temperature and without energy exchange.

5. Use of capillary phenomena
   a. Liquids flow under the influence of capillary forces in capillaries and semi-open channels (micro-cracks and scratches).
   b. Liquid's rising inside a capillary depends on the capillary size.
   c. Existence of directed liquid flow inside the capillary and porous materials.
   d. Velocity growth and height of liquid rise inside a capillary under ultrasound influence.

6. Use of electrostatic fields
   Interaction between charged substances (attraction in the case of negative and positive charges and repulsion in the case of same charge).

7. Use of magnetic liquids
   a. It is possible to manage magnetic liquid migration with a magnetic field.
   b. Change of viscosity and pseudo-density of magnetic liquid in a magnetic field.
   c. Immediate solidification of magnetic liquid in strong magnetic fields.

8. Use of the piezoelectric effect
   a. Appearance of opposite electric charges on the diametric sides of some crystals under mechanical deformations, such as pressure, stretching due to direct piezoelectric effect.
   b. Opposite piezoelectric effect — the external electric field results in mechanical deformation of such crystals.

9. Electrokinetic phenomena
   a. Electrophoresis: The movement of discursive particles in a liquid or gas suspension under an external electric field.
   b. Electro-osmosis: The movement of liquid through capillaries or porous materials under an electric field.

TABLE 9.1 (continued)
Short Registry of Physical Effects

10. Use of electrolysis
The chemical reactions take place in electrolytes while a direct current runs through them. The electrolytes' positive ions move toward the cathode and negative ions toward the anode. The products of chemical reduction are located on the cathode while the products of oxidation are on the anode.

11. Use of corona discharge
a. Gas ionizes under the influence of the corona discharge.
b. Corona discharge parameters depend on the gas parameters (such as impurities, pressure, flow speed, etc.).
c. Corona discharge parameters depend on the electrode shape and size.

12. Use of ferromagnetism
a. Management of ferromagnetic particle movement by magnetic field.
b. Existence of the ferromagnetic self-magnetic field.
c. Screening of the magnetic field by ferromagnets.
d. The sharp change of magnetic properties of the ferromagnet near a specific temperature (Curie point). Above the Curie point, a ferromagnet transfers into a paramagnetic material.
e. Influence of the mechanical deformation on the ferromagnetic properties.

13. Use of phosphor
Appearance of luminescence under the influence of radiation (optical, ultraviolet, infrared) on a specific substance (phosphor).

14. Use of oscillations
a. Change of interaction type between substances when the oscillations are initialized (vibration, infrasound, sound, and ultrasound).
b. Dependence of the intrinsic frequency of the technique on its characteristics, such as mass, size, stiffness, and so on.
c. Resonance — sharp rise of the oscillation amplitude under the coincidence of the technique's intrinsic frequency with the frequency of the forced oscillations.

15. Use of foam
Change of various physical substances' properties (such as mass, size, volume at low density, thermo-isolation, sound absorption, and shock wave absorption) and chemical properties in the foamed condition.

16. Use of centrifugal forces
Centrifugal force arises in the rotating subsystem and acts on the elements of the technique. This force depends on the substance mass, its density, and its linear velocity of rotation.

Unfortunately, both solutions, as well as other known acid solutions, do not work for such small features as Silicon needles. A base solution cannot be used for field electron emission cathodes. A resolution of this problem was found: chemical etching plus electrolysis of Silicon needles with the following acid solution:

65% Nitric acid HNO_3	11 parts by volume
98% Acetic acid CH_3COOH	2 parts by volume
40% Hydrofluoric acid HF	7 parts by volume

(USSR Patent # 1384089, co-author V. G. Ivanov)

GEOMETRIC EFFECT

A delay device consists of a cylindrical glass tube filled with a special liquid. The time delay of the signal depends on its path in a liquid and on the liquid's temperature. Previously only one delay was needed in the machine. Three different delay times are required in the new machine. The routine solution is to install three different elements with various lengths of cylindrical glass tubes. However, it is not easy to keep the liquid in all these glass tubes at the same temperature. In order to resolve this problem, a bagel-shaped delay device with an asymmetrical "bridge" position (see Figure 9.1) was invented (USSR Patent # 1336107). Due to free thermal convection, the liquid in different parts of such a delay element has the same temperature. The active length of the liquid depends on the on-off positions of switches inside the delay device. This invention has an extra-effect use: instead of — the empty space (i.e., the resource) inside the "bagel" was used for location of the switching subsystem that controls the delay element.

Of course the boundaries between natural sciences are artificial, hence a single invention can include effects from different disciplines. For example, the author's development of a photo-relay (USSR Patent # 1342277) based on the physical effect of photoconductivity would be impossible without the special chemical composition of active semiconductor material and works appropriately due to the uncommon geometry of its elements (electrodes, active semiconductor, optical condenser, etc.)

In conclusion, any individual, design team, company, or industry has shallow knowledge with limited potential for resolving technical problems of a high difficulty **D**. The extracted knowledge about different natural effects increases the power of TRIZ for solving various technical problems.

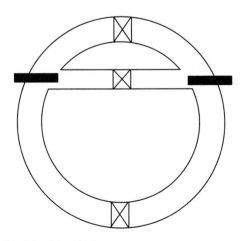

FIGURE 9.1   Sketch of the delay device.

Part 4

Preparations to Problem Solving

10 Before Starting

10.1 INTRODUCTION

There are two important elements for solving any problem — the problem itself and the experience and creativity of the solver. In Chapter 1, we discussed some ideas about problems. Now, before we begin to study the TRIZ heuristics and instruments for resolution of problems, it seems suitable to discuss this topic in more detail.

In order to be prepared to solve a problem, a person should

have some information (specifically discussed in this chapter),
eliminate his or her own psychological inertia (discussed in this and the following chapter), and
have a model of the object under investigation (the simplest one for a technique is presented in Chapter 12).

Any application of TRIZ starts with the understanding of the technical system or technological process and the situation in which the problem appears. Then a set of problems is recognized, correct statements of the problems are formulated, and at the end the appropriate heuristics are applied to find the best possible solution of the problems. Therefore, it is appropriate to discuss some important issues of technical problems and activities of their solving before we begin to study problem-solving heuristics and the instruments themselves. Although the human factors of creativity in problem solving are discussed in detail in the next chapter, some aspects of coping with psychological inertia to allow a solver to better handle technical problems are covered in this chapter.

10.2 CORRECT STATEMENT OF THE PROBLEM

The two most important steps in problem solving are defining the problem to be solved and finding a potential solution. A good, clear, and simple definition can prevent a false start in the search of a solution. Engineers and scientists often repeat John Dewey's idea that a problem properly defined is virtually solved [1]. That is why the correct statement of a problem (CSP) plays such a vital role in TRIZ.

Creativity begins when a person is able to initiate a question or recognize a problem and then devote himself to seeking an answer. At the beginning of the problem-solving process, one usually faces an *initial situation* associated with some system's disadvantages that should be eliminated or the requirements to improve a technique. These disadvantages can occur because of new requirements of society for the technique or some drawbacks inside the technique. In real practice, a solver deals with the initial situation, which invariably has many uncertainties and unclear points. It is important to understand the initial situation as well as possible and to transfer it into the CSP. Various correct statements of the problem can be formulated within the same initial situation for different goals of the problem solution. Hence, to formulate a problem statement more accurately means to give more important information needed for problem solution. The possible solutions of these technical problems can be achieved by changing a given technique, altering one of its sub-systems, modifying some super-systems, or creating a conceptually totally new technique. It is possible to distinguish the category of problem that should be resolved during the developing or improving of a technique based on the degree of execution of the PF:

1. PF is not executed at all (technique is unavailable; it is to be designed);
2. PF is executed only partially (technique is to be improved);
3. PF is executed, but a contradiction is observed between positive and undesired effects (technique is to be transformed).

In the first case, the so-called maxi-problem (genesis of a new subsystem) should be resolved; in the last case, the so-called mini-problem (change of the existing subsystem) is usually enough. In the second case, the category choice depends on qualitative value of the main parameter corresponding to PF. In an efficient technique this value must be no less than some minimal level, named by Boris I. Goldovsky as "parameter threshold" [4]. Provision for lift power exceeding aircraft weight by 10 to 20% was a threshold. This condition was necessary for reliable flight of an aircraft. Another threshold was connected with the distance a steamer could travel without refueling. This threshold alone has determined the transition from a steam-boat to a steamship and then to an ocean liner. A necessity to overcome the parameter threshold of the currently limited technological capabilities of a society determines the mode of performance of a technique to be invented and categorizes the problems to be solved in the second case.

Usually a creative solver wants to resolve the maxi-problem (because it offers a more radical solution, more fun, and potentially bigger benefit) even for the third case, although difficulty involved in implementing a solution is usually higher as well.

Modern techniques are a subsystem of economy; therefore, various governing factors can influence problem formulation, the selection of problems themselves, and their solutions with the best cost/benefit ratio for the company. A number of factors related to company practices and operations must be considered by technical personnel, management, plant engineering and production, marketing, and sales personnel when making the decision to proceed with new technique development with the help of TRIZ. Because a radical solution may prove unacceptable, depending on an organization's culture and psychological inertia, it is useful to discuss the following questions before starting the problem-solving process:

1. Is the company's top management really interested in this new technique development and will they support it fully?
2. Is the company able to manufacture the new technique?
3. Does the company have personnel with the required skills for the technical effort and will they be available for this development?
4. Can the existing marketing and sales departments handle the new technique?
5. How will the entry of the new technique compete with present company products?

If there are doubts in the answers to the previous questions, the next set should be asked:

1. Will special equipment be needed?
2. Will personnel have to be hired?
3. Will a new marketing and sales group have to be established?
4. How will the marketing and sales effort be handled?

If the company is not ready to use the effective solutions of a maxi-problem because of fear of new unevaluated concepts or other restrictions, it is often better to solve a few mini-problems.

10.2.1 PROBLEM CLARIFICATION

B. Zlotin, A. Zusman, and A. Zakharov noted that sometimes uncertainties with technical problems occur only because nobody tried to clarify the situation previously or because the formulation of the technical problems was given incorrectly [2]. The purpose of Table 10.1 is to *define the problem* to be solved.

An efficient solution of many problems can be obtained only when the source of a harmful function effect is determined. Here, an effect of a harmful function is any deviation from expected values in any technique's characteristics or parameters. The following questions often help to find the source of a harmful function:

Who — the degree of direct human participation in creation of the HF effect
Where — the place where the HF effect manifests
When — the time when the HF effect occurs at the above place
What — the essence of HF effect, what parameters are abnormal
Why — the reason of the HF effect's appearance or HF effect's cause
How — under what conditions the HF effect occurs

These questions can be memorized easily with Rudyard Kipling's short poem:

I keep six honest serving men
They taught me all I knew:
Their names are What and Why and When
And How and Where and Who. [5]

Let me stress again that *correct statement of a problem itself can contain the elements necessary for a solution* before discussing these elements in the next sections and chapters.

TABLE 10.1
**Typical Mistakes in Technical Problem Formulation and Methods
to Eliminate Them**

Mistake	Explanation	Method
Excessively global **or** overly concrete statement	The statement of a problem is too general **or** too narrow (in the last case the problem is understandable only for those who formulated it).	Concretize the problem, having attached it to the concrete situation. Or Explain the problem in simple words using TRIZ slang (see below).
False problems	1. The problems' solutions do not give any positive effect. Or 2. An attempt is made to solve a much more complicated problem instead of the one that really needs to be solved. Or 3. The short-sighted statement of a problem does not take into account the variations that might take place during the time of solution of the problem and the reduction thereof to practice. Or 4. A tangle of interacting problems is mistaken for one problem.	1. Understand the result of solving this problem. Or 2. Reconstruct a problem situation of invention and choose another problem, which is capable of providing the result needed. Or 3. Elicit the prospects of production and consider the time needed for implementation of the solution to practice. Or 4. Study the structure of the problem, pick out all elementary problems, and solve each of them individually, assuming that the other problems in this group have already been solved.
Deadlock and Secondary explanation	An old problem statement directs a search in a way having no prospects, and the specialists explain some effects or peculiarities in a structure not by the true reasons but by some false opinion that has become traditional and legitimate due to its long use.	Reconstruct the original situation of invention and choose another problem, the solving of which is capable of providing the effect needed. and Investigate the physical nature of the process under consideration, not relying upon the explanations offered by specialists.
Nonsystematic approach	The statement of only the most obvious problem is given. When the solution of this problem is obtained it becomes clear that the problem is only a link in a chain of problems which impede the development of a TS or a TP.	Elicit the entire chain of problems and find the key one among them.

10.3   INFORMATION, CONSTRAINTS, AND PRESUMPTIONS

INFORMATION

Nobody knows *a priori* what information will be needed for a solution and what is excessive for a problem. The label "available scientific and technical level," which is often used in the presentation and evaluation of information, incorporates current technologies, energy sources, materials and substances, information about techniques used in the past, as well as new (yet unimplemented) techniques that will be used in the future, and information about natural and technical effects (see the previous chapter) which are currently used or can be used, and so on.

TRIZ offers assistance by pointing out the significant ingredients in any technical problem:

1. systems and/or processes
2. functions
3. contradiction
4. scientific and technical knowledge
5. actions to cross the gap between the initial and desired situations

The reader is already familiar with the first four ingredients, so let us briefly consider now the informational aspects of problem solving. The next part of this book provides detailed descriptions of TRIZ heuristics and instruments, which should help perform those actions mentioned in the last ingredient.

One of the most important pieces of information about a technical problem is a list of constraints for future solutions. These constraints can be natural (limitations from nature), economical, technical, legal, etc. It is interesting that the degree of difficulty of a problem **D** depends on the number of constraints. Such dependence, confirmed by numerous simulations, leads to an unusual situation: Sometimes it is necessary to increase the number of constraints for resolution of a problem [1].

Of course, different classes of problems that can be resolved with the aid of TRIZ require various types and amounts of information. Therefore, a solver should divide all information for such problems into the important and unimportant.

Important information is that which characterizes the problem to be solved: it makes clear the problem, contradictions, and subsystems needed to define that problem. It is assumed that an analysis of functions and constraints has been done, so that the root causes of problems in the system have been identified. It is always important to have information about a good solution itself and know how to achieve this solution for technical problems.

Unimportant information is not needed to characterize the problem, i.e., to understand the core of the problem to be solved. Such information tends to introduce extraneous objects, attributes, and their functions — too many facts in the wrong places. Often data about numerical parameters, materials specifications, etc. are not beneficial for conceptual solutions of a problem. TRIZ recommends suspending such engineering considerations until later stages of technique design. Usually in the problem statement, numbers, parameters, and detailed specifications are irrelevant.

The process of simplification of a problem is more effective if engineering details can be postponed until the conceptual solution has been found.

Neither information redundancy nor lack of information is good for a solver. In the first case a solver stating a problem in an attempt to facilitate its solution obtains so much information that the data that is really needed becomes lost in it. To prevent this situation, the essence of the problem should be isolated and all unimportant information should be discarded. The important information can be presented in mind-maps (see Chapters 1 and 6). In the second case a solver stating a problem has omitted some important data or is trying to find a new approach without having studies of known solutions, among which might be a solution for this problem. To prevent this situation, a solver should communicate with specialists in the corresponding field or familiarize himself or herself with existing patents and other technical information on the problem.

PRESUMPTIONS

There are various constraints that the problem solver uses to select information. Some, such as constraints by the laws of nature, are useful, while some, such as *presumptions*, can be useless. If the solution to a problem requires violating the laws of nature and using nonexistent materials or technologies, then the problem should be reformulated for creation of a principally new system capable of solving the original problem. To overcome such constraints, one should choose another problem formulation, providing the needed effect but without violating laws.

Usual presumptions such as "That will never work because ..." can destroy the search of solutions. Such killer phrases collected by a few American psychologists are presented in Table 10.2. Do not pay too much attention to such statements; usually there is no foundation to them. You need to remember that presumptions can be temporary. Today's conceptual solution that is rejected by presumptions can be tomorrow's real solution after (or even without) small modifications. Examples of some engineering, business, and administrative presumptions collected by Edward N. Sickafus [3] are given in Table 10.3.

Often excessive limitations or presumptions are presented, such as:

- the requirement "not to change anything,"
- the requirement to solve the problem in a strictly specified way,
- only acceptable proposals are for modifying a part of a system or process designed to rectify the faults that arose due to a previous operation imperfection.

They can be bypassed by

- questioning the prohibitions and their validity,
- formulating a new CSP,
- postponing the prohibitions, limitations, and presumptions during problem solving.

TABLE 10.2
Innovation Killer Statements

No way.

Yes, but...

That's irrelevant.

It is not in the budget.

We tried that before.

Don't rock the boat!

We haven't got the manpower.

The boss will eat you alive.

It will be more trouble than it's worth.

Obviously, you misread my request.

That's not in your job description.

You can't teach an old dog new tricks.

Great idea, but not for us.

Get a committee to look into that.

If it ain't broke, don't fix it.

You have got to be kidding.

It isn't your responsibility.

We've always done it this way.

Put it in writing.

Don't waste time thinking.

What will people say?

It will never fly.

Don't be ridiculous.

People don't want change.

Let's stick with what works.

We've done all right so far.

The boss will never go for it.

It's too far ahead of the times.

Do you realize the paperwork it will create?

I'm the one who gets paid to think.

It's OK in theory...but.

Be practical!

Don't fight city hall!

Because I said so.

I'll get back to you.

That will never work!

TABLE 10.3
Examples of Presumptions

Engineering	Administrative	Business
Blueprints	Approvals	Competition
Dimensions	Budget	Cost
Materials	Deadlines	Infrastructure
Constraints	Personnel	Market niche
Specifications	Records	Patentability
Tolerances	Timing	Suppliers

Hence, a solver should

- recognize *presumptions* about the system and its possible improvements,
- eliminate special technical *slang* and replace it with simple generic names, and
- express the system via elementary *sketches* that contain only necessary subsystems and links (energy/signal flows) presented as simply as possible.

These points depend strongly on the solver's inventiveness (see Chapter 11).

10.4 CONCLUSION

While analyzing a technical problem, it is important to know when (or after what events) and where (or under what conditions) the problem has grown in order to select a right direction for its solution.

One of the strongest advances of TRIZ compared with all other problem-solving methods, design approaches, and creativity aids is its systematic ability to provide information about

- the result of a problem-solving process (concept of Ideal Final Result),
- steps during a problem-solving process (TRIZ heuristics and instruments),
- clarification of the initial situation of a problem (concept of contradictions),
- simplification of technique and problem (Su-Field Models, discussed in Chapter 12, and generic problem structures).

This set allows us to solve technical problems very effectively. Nevertheless, often psychological inertia hinders the effectiveness. That is why various methods to decrease psychological inertia have been developed by TRIZniks. Those methods are reviewed in Chapter 11.

REFERENCES

1. Dewey, J., *How We Think*, D.C. Heath and Co., Boston-New York, 1910.
2. Altshuller, G. S., Zlotin, B. L., Zusman, A. V., and Filatov, V.I., *Search of New Ideas,* Kartya Moldovenyaske, Kishinev, 1989.
3. Sickafus, E. N., *Unified Structured Inventive Thinking — How to Invent*, Ile, Grosse, 1998.
4. Goldovsky, B. I. and Vaynerman, M. I., *Rational Creativity,* Moscow, Rechnoj Transport, 1990.
5. Kipling, R., *The Elephant's Child and Other Just So Stories,* Dover, Mineola, NY, 1993.

11 Inventiveness

11.1 INTRODUCTION

Genrich S. Altshuller and recently Larry Smith identified the necessity of inventiveness when they recognized that engineers often come to weak, obvious solutions. Using the definition of problem difficulty D, one can conclude that any strong solution can be found as easily as any weak solution. Let me show that this conclusion can be incorrect. The probability of finding a strong solution is $Ps = Cs/(Cs+Cw)$ and the probability of finding a weak solution is $Pw = Cw/(Cs+Cw)$. Often Pw is larger than Ps because concentration Cw of weak solutions is usually higher than the concentration Cs of strong solutions. On the other hand, because of the psychological inertia, the barrier Bs that separates a new strong solution from the known solution is higher than the barrier Bw that separates a new weak solution from the known solution. Hence, we obtain $Ps/Pw \sim \exp(Bw - Bs/Ef)$, where $\exp \sim 2.72$ and Ef is the effort to find a new solution (it is assumed that B and E have the same dimension). As a result, we will have the simple equation $Ps/Pw = Cs/Cw * \exp(Bw - Bs/Ef)$ that explains the mentioned observation.

That is why Altshuller decided that the Creative Imagination Development Course should be useful for engineers and TRIZniks. Such a course must increase the problem solver's inventiveness, the ability to create new real and imaginary objects (systems, processes, concepts). Psychologists have counted only 16 main characters of people and there are many more intermediate characters. Such great variety means it is impossible to propose one common course for stimulating inventiveness and creativity for all people. Different methods and exercises developed in Western countries for stimulating creativity have already been reviewed (see references for Chapters 1 and 8).

This chapter presents a short review of some methods and games* that are helpful for learning TRIZ. Many TRIZniks, notably P. R. Amnuel, G. S. Altshuller, N. N.

---

* Many such tools are based on word games and, unfortunately, cannot be proposed for people who do not know the Russian language.

Khomenko, E. Yu. Sviridenko, A. A. Gin, Yu. G. Tamberg, A. L. Sorkina, I. M. Vertkin, S. S. Litvin, L. I. Shragina, M. S. Gafitulin, L. A. Kogevnikova, I. N. Murashkovska, A. G. Rokakh, A. A. Nesterenko, A. L. Shtul, M. N. Shusterman, and others have created or adopted various exercises and elaborated on them during their TRIZ lessons [1-10].

The discernible peculiarity of the methods and games presented in this chapter is their compatibility with TRIZ concepts and heuristics. These methods and games can decrease psychological terminological inertia, image inertia, inertia for specific functional orientation, and classification inertia, and as a result improve the inventiveness of a problem solver. Inventive thinking has five distinctive features:

- the ability to present the world as a system with links between phenomena and objects;
- the ability to consider various resources;
- the ability to formulate contradictions, that is, to discern the core of the problem;
- the ability to consider each object in evolution (and trace its past, present, and future) to Ideality;
- the ability to classify objects and to understand the relativity of any classification.

A flexibility of imagination and rigid dialectics are two important attributes of inventiveness. TRIZ cultivates these two contrasting sites of inventive thinking as will be shown in the remainder of this chapter.

11.2 TRIZ SLANG GRAPHICS APPROACH

TRIZ differs from the majority of problem-solving methods by its recommendation to start from an "exact big picture" of the problem definition by using simple slang, sketches, or symbols about the most important details of technique and problem.

Specialized terms related to a tool, product, raw object, environment, etc. should be replaced by simple words in order to discard terminological psychological inertia.

Terms

impose old ideas about the element or operation;
veil features of substances and fields taking part in the problem; and
influence ideas about possible states of the technique.

Possible practical applications of a technique beyond its PF and secondary UF diminish terminological inertia. The function of the same technique can be formulated with a different degree of functional generalization. For example, let us show the hierarchy of the primary function for a technical system that you probably have in your kitchen — a meat chopper:

break meat
↓
break food stuffs

↓

break materials (any)

↓

produce the material of a required dispersed composition (not only by breaking).

TRIZniks prefer to operate with plain words with a high degree of functional generalization that can be understandable to a child in order to handle broad but simple information. It is possible to replace special terms by a simple word in the following steps:

1. Proceed from special terminology to common engineering lexicon.
2. Replace common engineering lexicon with functional, action-oriented words if the lexicon at step 1 does not emphasize the action which is necessary to be carried out.
3. If the functional simple terms continue to focus on a certain predetermined means, make a list of synonyms for each term or jump directly from technical slang to simple terms.

The following word chains illustrate such a replacement:

humidity reduction → evaporation → water removing → dry,

OR

ion implantation → doping → particles insertion →

→ change of material composition → mixture preparation

It is suggested in TRIZ to keep these accessible, simplest, and appropriate names during all problem solving and to return to the initial special terminology only after the solution is obtained.

In addition, for many problems it is useful to:

- formulate an operative model of the solution by a phrase that reflects Ideality as much as possible (e.g., the phrase for a detection model can be "It is visible only enough to find it");
- reflect briefly and symbolically possible forms of connection of the incompatible requirements in the definition of the physical contradiction.

The goal in the last case is to stress the physical contradiction even for a nonspecialist in the field of the problem (e.g., instead of the physical contradiction formulation "the part of the object should be liquid to cool, and solid to polish," one can use statements such as "a solid liquid," "a solid coolant," "a liquid polisher," "liquid hardness").

The plain words of TRIZ *slang* can often bring wider analogies for solution of a problem and speculations on the object of properties.

Instead of a detailed technical drawing common in engineering, TRIZ operates with almost primitive *sketches* that represent the root cause or problem situation in simple, understandable graphics of a technique or subsystem. Usually these hand drawn color or black and white sketches do not have all dimensions, tolerances, materials specifications, etc. in contrast with detailed technical blueprints and play the role of an aid in problem solving. The most important feature of such sketches is their emphasis on the

- root cause of the problem and/or
- operative zone and period and/or
- functions of the technique.

These sketches are presented and discussed in Chapter 17.

Another very important idea of TRIZ is a possibility to present any technique in terms of a model or to describe it by some symbolic "language." One of the ways of cognitive human activity is to reduce information about a complex and difficult, multidimensional object or phenomenon to an abstract, single-mode model, and then to study it. This way spreads into the methodology of modeling — researching objects of study through their material or mathematical models: circuits, equations, descriptions, and images. The symbolic "language" of TRIZ, called Su-Fields models, is more fully discussed in Chapter 12.

11.3 MULTI-SCREEN APPROACH

The multi-screen approach, also known as the system thinking operator, was developed by Altshuller in order to combine the idea about technique evolution from past through present to future with the ideas about TS/TP functions and their organizational hierarchy from elements to super-systems.

Engineers usually think concretely but nonsystemically. After statement of a problem, an engineer often concentrates his attention on a particular object that should be improved. For example, if the problem describes a tree, an engineer considers only a tree. In system thinking it is necessary to imagine not only the tree itself (investigated system), but simultaneously a forest (super-system) and separate branches and leaves (subsystems); moreover, it is probably useful to include in the analysis the climate (super-system of a forest), wood (another sub-system of a tree), and cells in a leaf (tree elements).

When trying to solve a problem, we usually develop in our minds an image of the subsystem in the technique being improved, or an image of the subsystem at present only. However, in many instances it is much more productive to see a wider presentation of the subsystem. In order to handle a dynamic picture, the multi-screen is very helpful because it allows the solver to remember that any division of a technique into subsystems is arbitrary and therefore to perform a gradual transition between different subsystems and states of the technique.

The system-thinking operator is a mental exercise with representation of the technique under development. The template for the most common "nine-screen" system operator is shown in Figure 11.1. Note that the diamond-like structure for

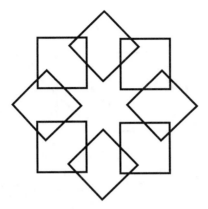

FIGURE 11.1 Multi-screen approach template.

representing the system thinking operator allows you to build as many screens as you need (much more easily than the rectangle-based Altshuller's picture). Some people feel that just looking at the diamond-like structure of 8 empty squares organized in two sets "clicks on" their visual imagination.

As we know, a technique consists of some subsystems and it, itself, exists as a subsystem of a more general super-system. Therefore, in the template, the middle horizontal line corresponds to the technique under consideration; the bottom horizontal line shows subsystems and elements formed by the technique; and the top horizontal line represents the super-system, including the technique under consideration as one of the elements. As we also know, each technique has its own individual history (similar to ontogenesis, the individual development of an organism from a germ to old age) and participates in a history of group development of the super-system (similar to phylogenesis of biological groups — from elementary virus to multi-cell organism, from amoeba to animal). Therefore, the left vertical line of screens reflects the previous conditions of the technique and its sub- and super-systems; the middle vertical line shows the present condition; and the right vertical line forecasts the future of the technique and its sub- and super-systems. Moreover, because expansion and convolution trends always exist in technique evolution (see Chapter 7), the multi-screen approach reflects them with two diagonal arrows. The filled template, known as the "nine-screen" approach for a generic technical system, is shown in Figure 11.2 (without concrete information about the technical system since this information is problem dependent).

Sometimes one has to see not only the technical system but also an *anti-system* (a system with opposite primary functions or properties; e.g., instead of to heat up, the anti-system would cool, instead of mixing it would divide, and instead of being heavy it would be light), a *concurrent or alternative system*, often called a *co-system* (a system with complementary functions or properties, such as a device for resistivity measurements and a checker for the type of charge of carriers in semiconductor wafers production), and a *nonsystem* (the system with the same primary functions or properties, such as a car and a bicycle, both built for ground transportation and

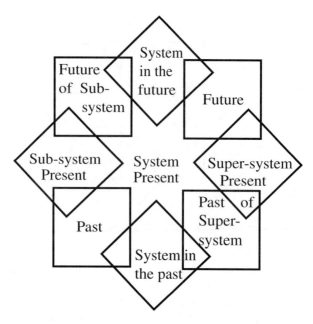

FIGURE 11.2 Simple nine-screen model of an evolving technical system.

having the same friction peculiarities). It is possible to expand the number of screens accordingly:

Screen I (system + function) — there is only one unique technical system (the only system in the world). It is sure to perform its primary function

Screen II (Super-system + function) — super-system incorporates a multitude of identical systems *or* super-system performs the useful functions of the system

Screen III (nonsystem + function) — no system, the primary function is delivered by other systems

Screen IV (anti-system + function) — the same function, an opposite system

Screen V (subsystem + function) — the primary function is delivered by a subsystem of the initial system

Screen VI (system + super-function) — one system, many useful functions (in addition to the primary function)

Screen VII (system + nonfunction) — the system is here, but its function is redundant (or there is no need to achieve the goal)

Screen VIII (system + anti-function) — the same system with an opposite function

Screen IX (system + subfunction): incomplete (partial) function.

For the reflection of each type of development new screens are necessary. The popular "eighteen-screen" approach means analyzing the past, present, and future of six systems (system, subsystem, super-system, anti-system, co-system and non-system), or analyzing the two "nine-screen" diagrams of two different systems, or analyzing the single system in time with six functions, etc. It is possible to imagine that the screens of the anti-, co- or nonsystem are placed on an opposite wall of a room, in which our screens for the initial system are established. A system and its anti-, co- or nonsystems can have different properties (object, phenomena, process, etc.) and complexity.

The whole multidimensional multi-screen picture (or the complex multi-screen approach) should include the following elements:

Primary Function	Antifunction	Secondary Functions
Support Functions	Auxiliary Functions	Nonfunction
Subfunction	Super-function	HF and NF
System	Super-system	Subsystem
Nonsystem	Anti-system	Co-system
System in the past	System in the present	System in the future
Subsystem in the past	Subsystem in the present	Subsystem in the future
Super-system in the past	Super-system in the present	

The multi-screen approach to a technique provides the ability to see, perceive, and represent the TS/TP in its complexity as one integrity, with all the links, its alterations combining different yet mutually complementary approaches: component approach studying the technique's composition (availability of subsystems in it, its super-systems) and structural approach (subsystems arrangement in relation to each other in space and time, its interrelations) together with the functional converge (primary function, useful functions, sub- and super-functions, and their interrelations). Due to this multi-screen approach, a solver can not only see the technique as it is but also should try to understand how it grows and changes from the past into the future.

The skill to visualize a system as hierarchical and time structures and to employ a whole system family (i.e., to apply the multi-screen approach) should be developed, since it is very important for solving inventive technical problems. To make an invention, a problem solver should be able to

- see a potential system in a group of separate, uncoordinated elements that represent subsystem;
- develop certain links between the subsystems that can create a new and useful property;
- understand the super-system(s) of the system under consideration;
- know limits on changes in the system and super-systems;
- have knowledge about the anti-system, co-system and nonsystems;
- be familiar with previous developments of the system (as well as super- and subsystems);
- have knowledge about the anti-function, super-function, and nonfunction;
- be informed about the next future forecast of the system.

The questionnaire below is useful for training in the multi-screen approach.

Object	Question
Technique	Is it an object (system, process)?
Primary function	What is the technique doing?
	What is the main behavior of the technique?
	What is this technique for?
Secondary functions	What is this technique doing in addition to the primary function?
	How can the technique be used?
Anti-function	What is the opposite action to the major activity (primary function)?
Super-system	What is a more general technique that includes this technique?
Subsystem	What does the technique consist of?
	What subsystems does it consist of?
Technique in the past	What was the technique under consideration in the past ("childhood" of the technique)? What other technique performed the function of this technique in the past, when this technique did *not* yet exist (the technique's "parents")?
Technique in the future	What will the technique be in the future ("senility," "death" of the technique)? What technique will replace this technique (the technique's "kids")?
Anti-system	What other technique performs the anti-function?
	What other technique acts with the opposite result?
Nonsystem	What other technique performs the same primary function as this technique but acts differently?
	What other technique has a similar function?
Co-system	What other technique almost always and everywhere accompanies, exists, or is used with the technique?
	What other kind of technique is included with the technique in the super-technique, which is the same and common for both joint techniques, at the next hierarchy level?

The multi-screen approach is illustrated by the following three examples

1. **Super-system**

In order to eradicate the weevil, a crop-damaging pest, it was necessary to measure its temperature. Due to the small size of the bug, just a few millimeters long, conventional thermometers could not be used. Complex electronic devices were designed for measuring the temperature of the weevil. However, the best technique was proposed by an independent inventor who suggested using a conventional thermometer to measure the temperature of a multitude of weevils.

Consider this problem from a systemic viewpoint. A weevil can be considered a physical system that has a temperature to be measured. The conventional thermometer cannot be used for one system (weevil). However, the problem changes if there are many such systems (i.e., a super-system of weevils). It was suggested to put the weevils in a container with a conventional thermometer inside.

2. **System in the future**
   To reduce gravitational forces on electronic devices in a rocket during liftoff, the devices can be submerged in plastic that then evaporates in open space (U.S. Patent 3,160,950).

3. **Nonfunction**
   A laboratory balance weighs small masses (micrograms) against a set of weights. The scale's accuracy is higher with more weight in the set. However, precise selection of the weights slows the process. Thus, one should resolve a conflict: Accurate weighing requires numerous weights (ideally, infinitely many), but there should be fewer weights (none, ideally) to shorten the process. It was suggested to use the pressure of light to resolve the conflict (U.S. Patent 3,590,932). A flow of photons (in a way, "micro-weights"), generated by a laser, exerts a mechanical force onto the scale pan, thus bringing the balance to equilibrium. Usually light pressure, discovered about a century ago, is not considered in engineering.

As a result of careful analysis of the technique's structure and major links between its subsystems, as well as between the technique and the super-system, the multi-screen approach helps

- find the place of the technique in the super-system,
- see restrictions inherent to the given technique,
- reveal the basic conflicts, technical contradictions, problems for future decisions.

11.4 FANTASY

Fantasy is necessary for any solver of inventive problems. Three levels of fantasy are distinguished:

- creating a new (modified) object experienced earlier (such as a tree with several different fruits);
- creating a new (developed) object that the person has not encountered earlier but has some initial information about (such as magnetic corn, glass car);
- creating (synthesizing) a new nonexistent object of which no initial information is given (extraterrestrials, gaseous plants, solid echo).

It appears that TRIZ beginners have a rather moderate level of inventiveness and usually their fantasy is confined to two forms:

1. simple amalgamation of known (often homogeneous) systems (such as a watch with a mobile phone);
2. wild, uncontrolled fantasy with enigmatic properties and unclear functions (pencil with mini-saw).

Many researchers have admitted the importance of science fiction for fantasy. The reason is, in Altshuller's opinion, that science fiction is the source of promising technical ideas, which sparks human inventiveness and diminishes psychological inertia. That is why it is the usual practice in TRIZ courses to encourage students to read and then evaluate fantastic stories. But in order to do this effectively, one has to analyze the science fiction literature in the way that the world patent fund has been analyzed (see Chapter 8). The registry of fantastic ideas by Altshuller and Amnuel is a classified collection of fantastic ideas. It shows regularities and common roots in fantastic ideas. The following three tools were developed during the creation of this registry:

Science Fiction Evaluation Scale

Any science-fiction novel, story, or movie can be evaluated according to the following criteria:

- Novelty of the idea;
- Convincing presentation of the idea;
- Additional knowledge about technique, human nature, and society.

The highest personal evaluation score is given to those works which the evaluator will remember throughout his or her life because of the high scores on these criteria and the good artistic value.

Fantastic Ideas' Generation Scheme

The above named scheme has four levels:

1. Ideas connected with the usage of only one fantastic object in order to gain some fantastic results
2. Ideas connected with the usage of many fantastic objects in order to receive another fantastic result (the system effect)
3. Fantastic ideas connected with gaining fantastic result(s) without use of any objects
4. Fantastic ideas connected with no need of previous fantastic result(s)

The fantastic ideas on every level can be of a high or low level. The fourth level isn't better than the third or first; it just should have a description of conditions under which the need for results disappears. It is the different and simply internal logic of the fantastic ideas' development that is considered. The four-level building can be built for every fantastic topic.

Fantogram

The Fantogram is one of the finest tools for creative idea generation. It is a morphological box with the possible ways an object can change (the left column of the following table) listed horizontally and with the levels of change (the right column of the table) listed vertically. Each cell of this box is connected with new idea generation — nearly every fantastic idea from existing fantastic literature can be placed into a specific cell.

The Possible Changes	Level of Change
Increase	Physical state
Decrease	Chemical consistency
Join	The object itself
Separate	Elements of micro-structure of the object
Decompose	Super-system for the object
Substitute property with anti-property	Direction of the object's development
Accelerate	Reproduction, self re-creation, regeneration
Decelerate	Energy feeding
Move back or forward in time	The method of movement
Make a property changeable in time or make it constant	Goal (what this object is intended for), sense of existence
Separate function from object	Field of distribution
Change connection with environment or the environment itself	Control

The list of possible changes can be expanded based on analysis of science fiction.

The usual exercise in the Creative Imagination Development Course is to challenge the student to invent a fantastic animal or a new natural phenomenon (rain, snow, wind) using the Fantogram. The students are asked also to invent a new fantastic idea or to write a science fiction tale using the tools. When students receive a science fiction story to read, they evaluate it using the Fantasy Scale. Students are also invited to improve some low-level-idea stories. These tools allow a solver to develop *system thinking* and to receive very strong fantastic ideas. Moreover, they are supported by a few methods and games that are discussed in the next two subsections. All these tools, methods, and games have a common goal — to help one learn and apply TRIZ.

11.4.1 GAMES

"The Clouded Planet" Game

John Arnold, from Stanford University, proposed the "Clouded Planet" game. In the game, you are to imagine that your spaceship comes near an unknown planet. Strong clouds enclose the planet. Unmanned stations can go through the clouds but any wire or wireless connection is impossible. On the planet there are the same conditions, laws, and factors as on Earth, but there one "X factor" is changed. By sending programs to the automatic stations, you must find the X factor with a minimum of attempts. The teacher plays on the "planet" team and students play on the spaceship team.

For example, during this game a student loses station after station trying to discover the X factor. The mystery factor could, for example, be that the speed of light is 1 millimeter per hour. When students send the next station and it does not return, they decide to change strategy and send the station step by step. For example, they program for the station to dive a couple of meters under the clouds, to take probes, and to return to the ship, or to return after 1 second. In such a way the students are able to find the right answer.

This game is used for breaking psychological inertia in search of a problem's causes. It can also be used for training the planning and design of experiments.

Good-Bad Game
According to this game, students try to find in every bad phenomenon good ones and vice versa. For example, the bad phenomenon is that I got ill and cannot go to work. The good phenomenon is that I don't go to work, so I can be home with my family. It is bad to be home with the family because they will make me crazy with the stories that they will tell me. It is good to go crazy, because then I can kill this idiot from the flat above and not go to prison ... and so on.

This game requires sharp classification abilities and stresses the relativity of any classification (e.g., separation of a technique's functions into UF, NF, and HF).

11.4.2 Methods

The Snow Ball Method
This method is aimed at developing any fantastic idea. A fantastic idea is always connected with a changing system (artificial or natural object) which is connected with other systems. Thus, those connected systems also are changed, and so on. For example, if we imagine that cars use apple cider instead of gasoline, this change will affect not only farms but also the pipe industry and metallurgy. Feedback also is taken into account and increases the quantity of changes. This quantity snowballs, or grows rapidly and exponentially. Perceiving the world as a system appears to be the chief merit of dialectic and system thinking. The clear and simple statement that "everything is interrelated" bears deep significance and enables anyone who follows it to view every object as multifaceted and to conceive its past and future. Psychological inertia, on the other hand, pushes us backward, encouraging us to consider only a single familiar and evident property of the object.

The Value's Changing Method
This method is for generating new fantastic ideas. In order to do this we assume that something that normally has high value (gold, for example) accidentally has zero value, and conversely that an object that normally has nearly zero value (sand, for example) accidentally has high value. This idea is then developed with the help of the Snow Ball Method.

To this same group belongs also the Natural Law Removal game, in which we remove a natural phenomenon or object (e.g., weight). We then imagine how our lives and techniques would be changed without that phenomenon or object.

These methods and games allow us to understand various links and resources in the world.

The Trend Extrapolation Method
This method has four major steps:

1. Choose two real but visibly unrelated trends of evolution of an object, such as technique, science, culture.
2. Extrapolate each trend separately in the future until it acquires a leading position.

3. Reveal the contradiction between the two extrapolated tendencies.
4. Suggest a new idea to eliminate the contradiction using one of the known methods.

According to this method, we are extrapolating, or increasing, one or more trends until it creates a contradiction with other aspects of human life. Resolving this contradiction we gain a new high-level fantastic idea or group of ideas, which is then developed with the help of the Snow Ball Method.

The TRIZ tendency method differs from similar methods used in other sciences because it is geared to the regularities of dialectics (see Chapter 7). For example, forecasting uses the tendency method by merely projecting today's tendencies into the future and ignoring the possibility of conflicts of tendencies and of emerging contradictions.

Ideal-Real Transition Method

One of the main characteristics of inventive thinking is the ability to see the unusual within the usual and vice versa. Every fantasy or inventive situation consists of two parts: real things and a fantastic grain. The aim of the Ideal-Real Transition Method (often called "Golden Fish Method" in honor of the famous tale) is to extract this fantastic grain. In order to do this, a fantastic situation is divided, step by step, into two parts — real and fantastic — until it cannot be divided any more. This indivisible part is called the "fantastic grain." Altshuller gave a recurrent formula for resolving every fantastic situation

$$F0 = R1 + F1,$$

$$F1 = R2 + F2 \ (F2 < F1 < F0),$$

$$F2 = R3 + F3 \ (F3 < F2 < F1)$$

Here R is a real part, and F is a fantastic part. The equation recurs until Fi will be so small that we may not consider it an unbelievable fantasy.

Let's see how this method works on the example of the "tale of the golden fish."

The old man came to the sea and began to call the golden fish. The fish got to him and asked by human voice...

Let's analyze this situation:
Could an old man go to the sea? Yes, he could. So that part is real.
We remove that part and are left to consider

The old man began to call the golden fish. The fish got to him and asked by human voice...

Could an old man call the golden fish? Yes, he could. So that is also real. We are now left with

The fish got to him and asked by human voice...

Could some golden fish (we know that there are such fish) get near the old man? Yes, they could. So this bit is also real.

The fish asked by human voice...

Could the old man hear a voice from the fish? Yes, he could! We know that some fish make sounds. So that part is also real!

human ...

Could this voice be human? No, this could not. That is it! The fantastic grain of the situation is that the voice of the fish was human.

But if we take even this fantastic thing of the golden fish story, we cannot consider it, because it can have a real explanation: Could it seem to an old man who does not hear well because of age that the golden fish voice is human?

Note that if the situation were a technical one, we would come to the physical contradiction determining the fantastic grain of the situation. For example, take the problem of creating pressure by a liquid, with the help of centrifugal forces, on a cylinder that is placed on the axis of the centrifugal rotation. The fantastic grain of the situation is that the direction of the centrifugal force is opposite to the direction of the needed pressure. One can easily formulate the physical contradiction now and then to find its solution in the list of physical effects. (Try to do it yourself!)

This method builds mastery of skills in the backward search of a problem's solutions that is important for some TRIZ instruments.

Real-Ideal Transition Method

As we know any real object is a system, i.e., a set of constituent elements and/or operations, and at the same time it is also a subsystem (part of a larger super-system). Each artificial system has its primary useful function, for the purpose of which the system exists. Every artificial system consumes energy, substance, and information, occupies space, and produces harm and wastes (cost of a system). The less the mass, size, and energy consumption are, the cheaper the system. An utterly economical or Ideal artificial system performs its function at a zero mass-size-energy consumption.

The Real-Ideal Transition method has four major steps:

1. Select the object to be changed.
2. Determine its primary useful function.
3. Increase the Ideality of the object
   - by transferring its function to other objects (Mass-Size-Energy consumption is zero);
   - by transferring one, some, or many functions of other objects to the chosen object (other objects disappear).
4. Describe the changes in the real situation and in the environment. What changes can be traced in the life of the person, society, and nature?

This method can lead to good command of one of the most important TRIZ concepts — Ideality.

11.5 YES-NO TRIALS

Yes-No Trials, or "Conversation Game with a Computer," is well known and quite popular inside and outside the TRIZ community. The principle of the Yes-No Trial is revealing, bounding, and successive narrowing of the area of solution. The essence of this game is as follows.

The teacher provides an interesting situation and asks students to explain it by asking a minimum number of questions. The questions have to be asked in such a form that the teacher can answer only "yes," "no," or "no information." The teacher has to be ready to explain situations brought by a student in order to show the method of asking the "right" questions. The optimal way is to ask more general questions first, in order to eliminate "empty" fields, and concrete questions at the end. Good textbooks, encyclopedias, etc. can be used as the source for such yes-no problems. Some yes-no problems and their solutions are presented on the Internet at http://www.kulichki.com/puzzles/selected/danet-g.html.

N. N. Khomenko [8] proposed to graduate this game into levels of various search difficulty as follows.

Find a Value — It is most convenient to introduce the Yes-No game to students when a numerical value of a parameter (which can be ordered along the linear axis) should be determined (for example, the dates of an event). One can find this value sufficiently quickly by separating the set of values (ordered along the numerical axis) into two parts: "Greater or Smaller than…" This lesson demonstrates the feasibility of finding a solution without applying the trial-and-error method to all possible versions.

Find an object or its attribute — The next step in mastering classification skills is the search for objects or their attributes (properties, parameters) which cannot be ordered along a numerical axis. For example, "Guess what city I am thinking of." In this case, the attribute is a name of a city, and the set of attribute values is the list of all cities.

For each such object (usually well known) we can construct several types of classification of the set of its attributes. It is sufficient to ask simply: "What object am I thinking of?". These problems are a little bit more complicated; they serve as a bridge to the next class of problems in Yes–No Trials.

Describe an Object — Problems of this type are to describe the object (usually not well known) by finding its attributes. Answering questions, students must reveal as many attributes (functions, properties, parameters, and their values) describing the object to be found as possible. The aim here is to describe the object in detail. The goal of such manipulations is to form the necessary skills of classification of objects: to reveal the names of characteristics, parameters by which these characteristics are classified, and the values of these parameters by which the objects to be classified are grouped (or joined). This can lead to understanding better the meaning and relations among the terms object, attribute (functions, characteristic), parameter, and value. Several additional goals are also achieved, namely, the ability to

- ask questions which significantly narrow the area of solution at once;
- use morphological analysis (see Chapter 1);
- use the multi-screen approach.

Questions cannot be formulated correctly without consideration of a situation by the multi-screen scheme. Considering a question, a student has to think at different levels of abstraction, as well as to see the whole and parts. This process involves development of the elementary skills of abstracting and concretizing, separating the operative zone and period, and other attributes, through which the problem's objects are described. These skills are important for solution of inventive problems.

Problems that involve the use of physical or chemical effects are especially well-suited for the Yes–No game. These problems, being formulated in a form seemingly not related to physics or technique, induce the students to repeatedly formulate contradictions, refine the initial situation, reveal subproblems, and analyze resources (for their capability to perform the needed actions). Such problems eventually become so transformed that they have nothing in common with the initial statement.

Problems-situations — The previous levels of the Yes–No game are considered preparatory in TRIZ. The creative problem-situation is based on systems of contradictions; it forms the basic skills for analysis and solution of complicated problems. The problem-situation can be separated into deadlock situations (for which a creative solution should be found) and final ones (which need an explanation of why they arose). These groups are reminiscent of inventive and diagnostic problems* well known in TRIZ.

The first class of such problems should be presented during the Yes-No game in the form of a single contradiction, which can be resolved usually in only one way. The problem to be solved is successively transformed until all subproblems are removed. Such problems are useful for engineers (especially in quality assurance) who must reveal causes of defects in production.

Problems can be solved far more efficiently if a trainer strictly instructs students to search for contradictions through analysis of available resources and then to resolve these contradictions. It is then feasible to demonstrate to students how productive problem resolution is through the major concepts of TRIZ. A trainer should emphasis working with contradictions and resources, as well as on formulating inconsistent requirements to an operative zone and period. A trainer can demonstrate how to use the most general principles of work by abstracting and concretizing the contradiction. Even just this exercise is the starting point for mastering TRIZ skills in separating a contradiction into elementary components, creating an image of the solution and its concretizing on the basis of physical, chemical, and other effects.

The second type of problem-situation is introduced, little by little, as the students master the complex of mechanisms needed for solution of problems of the first type. These problems are more complex; they may have several solutions, so, unlike with problems of the first type, a solution found by one student might differ from that found by another. A trainer should reward students for atypical solutions; then the

---

* Unfortunately, diagnostic problems cannot be considered in this book. The reader can find a review of Diagnostic Problem Solving, written by Gregory Frenklach and the author, at the TRIZ web site: http://www.jps.net/triz/triz0000.htm or http://come.to/triz.

search for the solution-prototype should be continued. To solve a difficult creative problem-situation, one needs to resolve a system of contradictions. Thus, the skills for manipulating TRIZ mechanisms are developed. These skills are important for solving real problems, and they simplify further learning, which includes detailed study of these mechanisms.

Several aims of TRIZ training are achieved with the problem-situation:

1. Some basic ideas of TRIZ (resources, functions, super-system, subsystem, anti-system, etc.) are successively introduced against this background without their strict definition.
2. The main habits of working with inventive technical problems are consolidated:
   - search, sharpening, and separation of a contradiction into parts (based on analysis of attributes of available resources);
   - resolution of contradiction and creation of an abstract image of the solution;
   - filling of this abstract object with a specific content for that type of resource which has properties to serve as a basis for the solution;
   - synthesis of a real solution — TS or TP.
3. The intuitive understanding of the contradictory character of a situation under consideration is cultivated in the process. This understanding relieves the fear of a sharp and contradictory problem. The students learn not to go from one extreme to another. A new solution is sought far beyond stereotypes. The students acquire the information about most general principles of resolving contradictions formulated in a somewhat different form.

To do the Yes-No Trials, it is necessary to ask a clear question and to analyze the obtained information. This game is about the planning of experiments which eliminate "empty" probes as the answer to our questions. The Yes-No Trials sharpen skills needed to operate with the main TRIZ concepts and a technique's attributes.

11.6   PARAMETERS OPERATOR

In order to avoid *presumptions* about the technique under development, a trick named "manipulative verbs" was proposed by Alex Osborn [4]. Based on this idea, the "Parameters Operator," or "Size-Time-Cost Operator," was introduced by G. S. Altshuller. He proposed that the inventor should consider three questions:

1. What will happen if the *size* of the system is decreased or increased?
2. What will happen if performance *time* of the system increases or decreases?
3. What will happen if the *cost* of the system is zero or very high?

The answers to these questions should be placed in a simple morphological box:

PARAMETER	ZERO	INFINITE
SIZE		
TIME		
COST		

This box can easily be expanded for parameters of other characteristics of a technique, e.g., shape, material, temperature, pressure, speed, color, mutual and time arrangement of subsystems (for TS and TP correspondingly), electrical conductivity, and necessary power.

Moreover, a morphological box with parameters of other characteristics of a technique and the manipulative verbs can be constructed. The most popular manipulative verbs collected by American and Russian psychologists are listed below.

Multiply	Unify	Freeze	Lighten	Widen
Divide	Magnify	Soften	Combine	Protect
Rotate	Harden	Minimize	Repeat	Segregate
Eliminate	Distort	Adapt	Thicken	Integrate
Subdue	Flatten	Fluff-up	Stretch	Modify
Invert	Squeeze	Bypass	Rearrange	Symbolize
Alter	Complement	Substitute	Accelerate	Abstract
Separate	Reverse	Add	Extrude	Dissect
Transpose	Submerge	Subtract	Repel	Compare

These verbs and the morphological box are not intended to solve the problem. The task of the box is only to overcome a solver's presumptions and psychological inertia, which block the thinking process, and to push a solver to clarify the Ideal Final Result (IFR).

Often problems contain contradictions such as "many objects and not many objects." If the contradiction of the type "not many objects" is strengthened, it should be reduced to the one type "no objects" (or "absent object"). If a contradiction of the type "many objects" is strengthened, it should be reduced to the one type "infinite number of objects" (or "too many objects," "myriad objects").

An "action" of this Parameters Operator can be expanded for any of the so-called opposite parameters that are discussed in Chapter 14.

11.7 CONCLUSION

Psychological inertia is based on the strong hard connection between concrete objects and their images in the mind of a specific person. The above methods and games are aimed to break this harmful connection. They achieve this in the following object- and solver-focused ways:

by changing the *object* and/or its functioning (Parameters and System Operators, Fantogram, Yes-No Trials, etc.);

by correcting the *human* behavior in the process of problem solving (Yes-No Trials, Golden Fish Method, Silver Clouded Planet, etc.).

The exercises, methods, and games of the second group (prejudiced) have a clear trend to transfer to the first one (unprejudiced) with an increase in our understanding of the regularities these methods are based on.

During these exercises, the skills in clarifying a contradiction as well as intuitive skills in realizing Ideality are developed. They remove nonrealized fear for sharp and inconsistent problems and enable development of skills for working with such problems without tradeoffs, both of which are important for solving real problems. Initial skills are fulfilled in application of abstraction-concretion, allocation of an operative zone and time, and other attributes of techniques through which objects of a problem are described. Some TRIZ concepts can be introduced without their precise definition during these exercises. Nevertheless, such an approach is also valuable because it allows the construction of preliminary definitions during the games.

The described tools, methods, and games serve to teach inventive thinking, to master TRIZ skills in working with real technical problems, and to develop competence in transferring these skills to nontraditional technical and nontechnical problems. A reader should decide what methods and exercises work best for him or her to develop inventiveness. The described methods and games are also a good starting point for developing creative people and groups. Actually, modern TRIZ has two primary goals — to be a powerful theory of nonroutine problem solving and to be a methodology for development of human creativity. The second goal, outlined in a few Russian books [1, 6, 7], will be described elsewhere in English.

REFERENCES

1. Amnuel, P. R. and Mikhailov, V. A., *Development of Creative Imagination,* ChuvSU, Cheboksary, 1980 (in Russian).
2. Altshuller, G. S., ... *And Suddenly the Inventor Appeared,* Detskaya Literatura, Moscow, 1984, 1987,1989 (in Russian); TIC, Worcester, 1996 (in English).
4. Litvin, S. S., *Course of Creative Imagination Development,* Samizdat, Leningrad, 1981 (in Russian).
5. Shragina, L. I., *The Logic of Imagination,* Chernomor'e Publishing House, Odessa, 1995 (in English and Russian).
6. Altshuller, G. S. and Vertkin, I. M., *How to Become a Genius: Life Strategy of a Creative Person,* Belarus, Minsk, 1994 (in Russian).
7. Tamberg, Yu. G., *How to Teach a Child to Think,* Tersziya, St. Petersburg, 1999 (in Russian).
8. Khomenko, N. N., *Usage of Yes-No Games during TRIZ Studies,* 1994. Available in Russian from Web sites: http://www.triz.minsk.by/e/yes-no.htm and http://www.jps.net/ triz/Xomenko1paper.htm
9. Rokakh, A.G., *Logic and Heuristics of Scientific and Technical Solutions,* SarGU, Saratov, 1991 (in Russian).
10. Gin, A. A., *Principles of Educational Technology,* Vita-Press, Minsk, 1999 (in Russian).

12 Su-Fields

12.1 INTRODUCTION

Dividing a large problem into smaller pieces is a common process in science and engineering. Often scientists and engineers work with numerical or material models of real objects. In this chapter we describe briefly the TRIZ approach for simplifying and modeling a technique.

A graphical model of a minimal working technique in TRIZ is called Substance-Field, or Su-Field. Su-Field Analysis is an instrument for modeling the most important parts of TS and TP for the particular problem and identifying the core of a problem related to this technique. Su-Field models and Su-Field Analysis, created by G. S. Altshuller [1], provide a fast, simple description of subsystems and their interactions in an operation zone and period via a well-formulated model of the technique in which all subsystems, inputs, and outputs are known or can be quite easily determined. Any technique can be presented as the ordered set of Su-Fields.

12.2 SU-FIELD TERMS AND SYMBOLS

As any model of nature, society, or technique, a Su-Field model has some simplifications and conditional agreements, which are described in this section.

The term *substance* (S) has been used in TRIZ to refer to a material object of any level of complexity. S can be a single element (bolt, pin, cup) or complex system (car, spacecraft, or mainframe computer).

The states of substances include not only the typical physical states (i.e., vacuum, plasma, gas, liquid, and solid) but also a large number of in-between and compound states (such as aerosol, foam, powder, gel, porous, or zeolit) as well as those states having special thermal, electrical, magnetic, optical, and other characteristics (thermoinsulators, semiconductors, ferromagnets, luminophor, etc.). Substance is itself a hierarchical system. With accuracy sufficient for practical use, the hierarchy can be represented as follows:

recognizable substance (e.g., a shirt);
minimum treated (simplest) substance (e.g., fiber);

"super-molecules": crystal lattice, polymers (e.g., nylon), association of
    molecules;
complex molecules;
molecules (e.g., NaCl);
parts of molecules, groups of atoms (e.g., –OH);
atoms (e.g., C, H, O);
atom's parts (nucleus);
fundamental particles (electrons, neutrons);
subparticles (quarks, gluons).

The term *field* (F) has been used in TRIZ in a very broad sense, including the
fields of physics (that is, electromagnetism, gravity, strong and weak nuclear interactions). Other fields can be olfactory, chemical, acoustic fields, etc. A TRIZ field
provides some flow of energy, information, force, interaction, or reaction to perform
an effect. The presence of a field always assumes presence of a substance, as it is
a source of the field.

Fields that often act upon substances in techniques are given in Table 12.1 and
arranged by frequency of occurrence and their importance in various techniques.

Note that the boundaries between different fields are not strong, as often happens
in TRIZ. For example, the interference belongs to one of the A, O, or R fields in
dependence to the nature of waves, while the piezoelectric effects always belong to
both M and E fields. Since the field is a form of interaction between substances,
strictly speaking, energy and field analysis are equal. TRIZ uses the terms as synonyms, although representation in terms of energy is more common for technique
designers while representation in terms of fields is more common for problem

TABLE 12.1
TRIZ Fields

Symbol	Name	Examples
G	Gravitational	Gravity
ME	Mechanical	Pressure, inertia, centrifugal force
P	Pneumatic	Hydrostatic, hydrodynamic
H	Hydraulic	Aerostatic, aerodynamic
A	Acoustic	Sound, ultrasound
T	Thermal	Heat storage, conduction, insulation and transference, thermal expansion, bimetal effect
C	Chemical	Combustion, oxidation, reduction, solution, bonding, converting, electrolysis, exothermic and endothermic reaction
E	Electrical	Electrostatic, inductive, capacitive
M	Magnetic	Magnetostatic, ferromagnetic
O	Optical	Light (infrared, visible, ultraviolet), reflection, refraction, diffraction, interference, polarization
R	Radiation	Xrays, nonvisible electromagnetic waves
B	Biological	Fermentation, decay, disintegration
N	Nuclear	Beams of α-, β-, γ-particles, neutrons, electrons; isotopes

solvers. More detailed information about different fields and energies is given in Appendix 4. TRIZ neglects the particle-field dualism known in physics because it is not yet important for most TS and TP.

The letters associated with the applied field are used in the Su-Field model of the different systems; for example, Su-A_Field means Substance–Acoustic Field model and Su_C_Field means Substance–Chemical Field model. Traditionally information is represented through fields in Su-Field models.

Su-Field is the model of any subsystem of a technical system comprising, as a rule, three components: two substances and a field usually, although a transformer can be modeled by two fields and a substance. The identification of substances (S1 and S2) depends upon the application. Often S1 is a product or a raw object and S2 is a tool.

In the case of analysis and modeling of technological processes (that can be represented as technical systems expanded in time), most effective are the models comprising one substantial component and two field components, where one field represents input and the other output. Often F1 is an input and F2 is an output.

Su-Field symbols mirror relationships between the subsystems of a technique. They are shown in Figure 12.1 by different connecting lines. It is common practice to place the F symbol above substances for an *input* field, and below the symbols of substances for an *output* field. Usually a harmful substance or field is marked with a tilde (~) above its symbol. If spatial-temporal organization is important for a technique, it is shown as S(x,t) and F(x,t) in the Su-Field model.

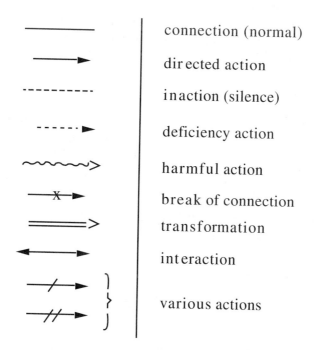

FIGURE 12.1 Connections in Su-Field analysis.

The complete simple Su-Fields can be graphically represented as

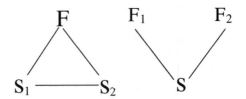

The model of two substances and one field (the left part of the figure above) presents, as a rule, a subsystem of TS, while the model of two fields and one substance (the right part of the figure above) presents, as a rule, a subsystem of TP.

The incomplete Su-Fields with only one or two components can be graphically represented as

$$S_1, \ F_1, \ S_1 \text{——} S_2, \ S_1 \text{——} F_1$$

Sometimes, a complete Su-Field is shown as a simple triangle if its details are not important. It is interesting that a triangle is the smallest building block for trigonometry as well as for a technique. Complex systems and processes can be modeled by multiple, connected Su-Field triangles.

Transformations of technical systems and technological processes can be shown using graphical formulas. For technological processes, a time arrow or time notation should be designated.

12.3 SU-FIELD PROPERTIES

Because Su-Field is a model of a technique, usually of some important part of a TS or a TP, when we speak about properties and actions of a Su-Field we mean the properties and actions of a subsystem of the technique. The five most important properties of Su-Fields are the following.

1. If you consider a subsystem as an incomplete Su-Field component, any of its characteristics can be changed, and the subsystem can be a component of a complete Su-Field:

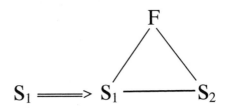

By the term "change" we mean here any transformations or modifications of a subsystem, other than "measurement" or "detection," such as transfer, regulation, shape change, separation, etc. To modify a substance it is often necessary to use a field; and for transformation of a field it is often necessary to use a substance.

2. Differentiated action upon one Su-Field component causes differentiated transformation of other components. The main practical importance of this property is the possibility to apply the control action or several actions to the most controllable Su-Field component in order to obtain one or several corresponding changes in other Su-Field components that are uncontrollable from outside.

3. If one Su-Field component has a specific spatial-temporal structure, then a similar structure can be created in other Su-Field components. This means that to create a specific spatial-temporal structure, it is often worthwhile to act not upon a subsystem but on any component making up a complete Su-Field in combination with the subsystem to be changed.

4. The number of fields or types of interaction between Su-Field substantial components is not limited. This number is determined by their physical properties and character of interaction.

5. Any component of any Su-Field can be a component of another Su-Field at the same time:

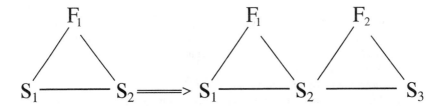

12.3.1 BASIC RULES OF SU-FIELD TRANSFORMATION

Su-Field Analysis can be used at the macro- as well as at the micro-level to transform a problem to a Standard Solution (presented below in this chapter). Such transformation rules help avoid psychological inertia during problem solving.

A. To solve a problem, the missing component is introduced to the incomplete Su-Field to make it complete.

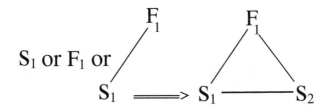

If there is a problem with an existing subsystem and any of the three components is missing, Su-Field Analysis indicates where the model requires completion and the Standards show directions for the completion. The Agents Method and Structure-And-Energy Synthesis (discussed below) can usually help identify exactly the missing components. Note that this rule is equivalent to the Standard 1.1.1 (discussed below).

B. To increase the efficiency of an existing Su-Field, its substantial component, since it is a tool, can be expanded into an independent Su-Field, connected to the initial one (the obtained Su-Field is referred to as a chain one):

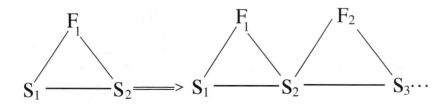

Notes:

1. If a Su-Field has three required components, but the TS or TP is ineffective, Analysis helps find the "weak" component and replace it with another component of the Su-Field or adds Su-Fields in order to increase effectiveness of the system and/or to modify the system for better performance (if radical changes in the design are possible).
2. Often the transformation into Su-M_Field, i.e., introducing ferromagnetic substantial component and magnetic field, is useful for such tasks.
3. A chain Su-Field is a representative of complex Su-Fields which have more than two substances and more than two fields.

C. Problems in detection and measurement can be expanded into a Su-Field having fields at both input and output — called a "measuring" Su-Field. The output field (F_2) carries out the information about a system:

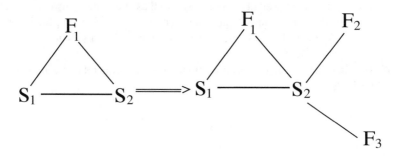

D. The most efficient way to destroy harmful, unwanted, or unneeded Su-Fields is to introduce a third substantial component that is a modification of one or both substantial components composing the given Su-Field.

Note:
Problems of the complete Su-Field system with improper connections between components functioning inefficiently are usually solved by increasing efficiency via forcing the technique's dynamism or flexibility, its transformation into Su-Fields with well-controllable fields, magnetic (Su-M_Fields) or electric (Su-E_Fields). In the case when substantial components cannot be replaced by the problem conditions, additions are introduced into the system (complex Su-Fields) or into the environment (external complex Su-Fields). When additions cannot be used, modifications of initial substance components are used: other aggregate states of the matter, mixtures with unlimited available materials from the environment (air, water, soil, wastes of super-system, emptiness). Usage of physical and chemical effects greatly increases the efficiency of Su-Field systems.

E. If the field F2 is needed at the system output (with field F1 at input), the Su-Field should be transformed using the F1–F2 physical transformation:

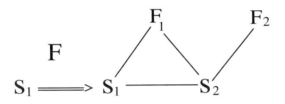

Chain-like transformations of Su-Fields using 2–3 sequential physical effects are sometimes required by a problem.

12.4 CONCLUSION

Su-Fields serve quite well when we need to model a part of the technique that is involved in the technical problem. A few attempts to improve Su-Fields, by V. A. Korolyev, J. Kowalick, and Z. Royzen, have not been completely successful. Su-Field Models and Analysis are still the most popular and simple TRIZ language. The Su-Field is not a unique model for TS and TP; other models have been developed, in different countries, such as SADT, IDEF, CASE, etc. [2]. Special research is needed to compare these models, including Su-Fields, for the description of technical problems.

REFERENCES

1. Altshuller, G. S., *Creativity as an Exact Science: The Theory of the Solution of Inventive Problems,* Gordon and Breach Science Publishing, New York, 1984.
2. Marca, D. A. and McGowan, C. L., *SADT: Structured Analysis and Design Technique,* McGraw-Hill, New York, 1988.

Part 5

TRIZ Heuristics and Instruments

13 Resolution of Technical (Pair) Contradictions

13.1   INTRODUCTION

Attempts to compile *rules of thumb* that inventors use were made by many researchers during the twentieth century based on interviews with several well-known inventors. The selection of rules of thumb was carried out subjectively by both the compiler and the inventor and reflected only their professional experience. Therefore, these rules are not very useful for any engineer. Nevertheless, it can be useful to collect such rules in your engineering field and to study those found by other TRIZniks. The common recommendation is to use such rosters as checklists during design, although no clear instruction is supplied.

    The patents fund analysis shows that a relatively small number of generic rules of thumb or heuristics is used over and over in originating, developing, and applying for improvement of parameters of various techniques. Each heuristic is usually not confined to a single scientific rule but incorporates several laws of physics, chemistry, or engineering. Consideration of these heuristics is valuable to the inventor because of the generalized and concise survey of present day science and engineering they afford. TRIZ provides detailed recommendations and instructions which allow inventors to use these heuristics effectively. This section provides detailed discussion of various heuristics recognized in TRIZ and instruments built on them. It consists of six chapters, the first of which deals with heuristics for pair contradictions. The heuristics for point contradictions and the Standards are discussed in Chapters 14 and 15. As Appendix 6 shows, these heuristics are not independent of each other. Nevertheless, the "artificial" division of heuristics used here is useful during studies of TRIZ.

13.2   ALTSHULLER'S ENGINEERING PARAMETERS

Most heuristics for pair or technical contradictions were recognized and identified by G.S. Altshuller between the late 1940s and 1970s [1, 2], but they are still relevant for many inventive problems today. Altshuller selected about 40,000 patents at

Level 2 and above, from the several hundred thousand patents issued in the USSR, US, Germany, and other countries. He extracted from them 40 Inventive Principles and 39 Engineering Parameters that are generic for many engineering fields. These Engineering Parameters are presented in the following table.

Generic Engineering Parameters

1. Weight of moving object	2. Weight of binding object	3. Length of moving object	4. Length of binding object
5. Area of moving object	6. Area of binding object	7. Volume of moving object	8. Volume of binding object
9. Speed	10. Force	11. Tension, pressure	12. Shape
13. Stability of object	14. Strength	15. Durability of moving object	16. Durability of binding object
17. Temperature	18. Brightness	19. Energy spent by moving object	20. Energy spent by binding object
21. Power	22. Waste of energy	23. Waste of substance	24. Loss of information
25. Waste of time	26. Amount of substance	27. Reliability	28. Accuracy of measurement
29. Accuracy of manufacturing	30. Harmful factors acting on object	31. Harmful side effects	32. Manufacturability
33. Convenience of use	34. Repairability	35. Adaptability	36. Complexity of a system
37. Complexity of control	38. Level of automation	39. Productivity	40. Savransky's list*

* The author decided not to leave this cell empty and filled it with parameters from the extended matrix that has been developed from the 1980s.

Representing technical contradiction as a combination of two of these parameters requires a broad interpretation of any Engineering Parameter, so they are quite generic. As can be seen, many of these parameters distinguish moving and binding objects:

Moving objects can easily change position in space, either on their own or as a result of external forces, e.g., a boat.

Binding objects do not change position in space, either on their own or as a result of external forces, so they are stationary, e.g., a building.

All Engineering Parameters belong to the following clusters:

1. common physical and geometric parameters (mass, size, energy, ...);
2. technique-independent negative parameters (waste of substance or time, loss of information, harm, etc.);
3. technique-independent positive parameters (productivity, manufacturability, etc.).

It seems that other parameter clusters could be found in the patents fund, but such research has not yet been done.

Explanation of the Engineering Parameters terminology, where an object can be a technique, its subsystem, or a single element, is presented briefly below.

Cluster 1. Common Physical and Geometric Parameters

1 and 2 — Weight: The mass of the subsystem, element, or technique in a gravitational field. The force that the body exerts on its support or suspension, or on the surface on which it rests.

3 and 4 — Length: A geometric characteristic described by the part of a line (straight or curved and not necessarily the longest) that can be measured by any unit of linear dimension, such as meter, inch, etc.

5 and 6 — Area: A geometric characteristic described by the part of a plane enclosed by a finite continuous line that can be measured in a square unit of dimension. The part of a surface occupied by the subsystem.

7 and 8 — Volume: A geometric characteristic described by the part of a space that can be measured in a cubic unit of dimension. The part of a space, either internal or external, occupied by the subsystem.

9 — Speed: The velocity of the subsystem. The rate of a process or action in time that can be measured by any linear unit of length divided by a time unit.

10 — Force: Any interaction that can change the subsystem's condition due to the interaction between subsystems.

11 — Stress or pressure: Tension on or inside the subsystem.

12 — Shape: The external contours, boundaries, that separate the subsystem from the environment or other subsystems. The appearance of the subsystem in the space.

17 — Temperature: The thermal condition of the subsystem. Liberally includes other thermal parameters, such as heat capacity, that affect the rate of temperature change.

18 — Brightness: Light flux per unit area. Also any other illumination characteristics of the subsystem, such as light intensity, degree of illumination.

21 — Power: The time rate of energy usage due to which the subsystem's functions are performed.

See also parameters 13–16 and 26. Of course, the variety of physical, chemical, biological, geometric, and other parameters can be expanded for different fields of technique. Therefore, the Principles for resolution of technical contradictions between such Parameters should include miscellaneous special materials phenomena known in the natural sciences (see Subsection 13.3.1).

Cluster 2. Technique-Independent Negative Parameters

15 and 16 — Duration of action: The time during which the subsystem can perform useful and/or neutral functions (durability). It can be estimated as the average period between failures, the service life.

19 and 20 — Energy spent by the subsystem: The subsystem's requirement (such as electricity or rotation) to perform a particular function. Often energy is provided by the technique or super-system.

22 — Waste of energy: Use of energy (such as heat) that does not contribute to the job being done (compare with 19 and 20). Reducing energy loss sometimes requires heuristics that are different from the heuristics for improving energy usage. Consequently, energy waste is a separate Parameter.

23 — Waste of substance: Partial or complete, permanent or temporary loss of some of the subsystem's materials or elements.

24 — Loss of information: Partial or complete, permanent or temporary loss of data or access to data in or by the subsystem. Frequently includes sensory data such as aroma, texture, etc.

25 — Waste of time: Time is the duration of an activity. Improving the loss of time means reducing the time taken out of the activity. "Cycle time reduction" is a common term.

26 — Amount of substance: The number of the subsystem's materials or elements that might be changed fully or partially, permanently or temporarily.

30 — Harmful factors acting on subsystem: Susceptibility of the subsystem to externally generated harmful effects.

31 — Harmful side effects: A harmful effect that is generated by the subsystem as part of its operation within the technique, and that reduces the efficiency or quality of the functioning of the subsystem or whole technique.

See also Parameters 14, 36, 37.

CLUSTER 3. TECHNIQUE-INDEPENDENT POSITIVE PARAMETERS

13 — Stability of the subsystem: The ability of the subsystem to keep its integrity (wholeness). Steadiness of the subsystem's elements in time. Wear, chemical decomposition, disassembly, and growth of entropy are all decreases in stability.

14 — Strength: The ability of the subsystem to resist a change in response to force. Resistance to breaking.

27 — Reliability: The subsystem's ability to perform its intended functions in predictable ways and conditions.

28 — Accuracy of measurement: The closeness of the measured value to the actual value of the subsystem parameter.

29 — Accuracy of manufacturing: The closeness of the actual characteristics of the subsystem to the specified or required characteristics that can be achieved during the subsystem production. (Note that manufacturing precision is often connected with quality of the subsystem.)

32 — Manufacturability: The degree of facility, comfort, ease, or effortlessness in manufacturing or fabricating of the subsystem.

33 — Convenience of use: Simplicity and ease of operation. The technique is not convenient if it requires many steps to operate or needs special tools, many highly skilled workers, etc. Often a convenient process has high yield due to the possibility to do it right.

34 — Repairability: Quality characteristics such as convenience, comfort, simplicity, and time to repair faults, failures, or defects in the subsystem.

35 — Adaptability: The ability of the subsystem to respond positively to external changes, and the versatility of the subsystem that can be used in multiple ways under a variety of circumstances.

36 — Complexity: The number and diversity of elements and element interrelationships within the subsystem. The user may be an element of the subsystem that increases the complexity. The difficulty of mastering the subsystem is a measure of its complexity.

37 — Complexity of control: Measuring or monitoring the subsystems that are difficult, costly, and require much time and labor to set up and use, that have fuzzy relationships between components, or that have components that interfere with each other, demonstrating "difficult to detect and measure."

38 — Level of automation: The ability of the subsystem to perform its functions without human interface. The lowest level of automation is the use of a manually operated tool. For intermediate levels, humans program the tool, observe its operation, and interrupt or reprogram as needed. For the highest level, the machine senses the operation needed, programs itself, and monitors its own operations.

39 — Productivity: The number of functions or operations performed by the subsystem or whole technique per unit of time. The time for a unit function or operation. The output per unit of time or the cost per unit of output.

See also Parameters 19–21.

As we can see, the boundaries of the clusters are not sharp, and the division of the Engineering Parameters into positive and negative is somewhat arbitrary.

Although many TRIZniks think that Altshuller's collection of Engineering Parameters is comprehensive and exhaustive, the author's studies include the following additional technique-independent Parameters:

Safety
Stability of parameters
Accuracy of operation
Information
Tolerances
Susceptibility
Ergonomics
Aesthetics, etc.

Common physical Parameters include

Electrical impedance
Optical transparency
Viscosity
Friction
Corrosion resistance
Noise
Transient processes in condensed matter, etc.

Moreover, I believe this so-called Savransky's List can be expanded. These Parameters can be included in the Matrix for removal of typical technical contradictions; the slightly expanded Matrix is presented later in this chapter. In order to be included in the Matrix, each new Parameter should be confirmed by several high-Level inventions from various engineering fields. Some of the Parameters from Savransky's List have already passed this test (described later in this chapter).

13.3 ALTSHULLER'S INVENTIVE PRINCIPLES

Let us now list the 40 Inventive Principles selected by Altshuller and generalized by the author. Some illustrative examples have been collected by TRIZniks from different countries and students of the Virtual TRIZ College (T. A. Kengerli, G. Mazur, E. Domb, M. G. Zapol'sky, Tz-Chin Wei, M. Schlueter, M. A. de Carvalho, and others). Note that some Principles are lengthened to compare with the original Altshuller description [1, 2].

Principle 1. Segmentation (Fragmentation)
  A. Divide an object into independent parts.
    Examples
    Multiwire cables
    Replace mainframe computer with personal computers and network.
    A cargo ship is divided into identical sections. If necessary, the ship can be made longer or shorter.
    Garden hoses can be joined together to form any length needed.
    A large air duct has a 90-degree elbow that is segmented to avoid strong turbulence and to improve air flow.
  B. Make an object modular.
    Examples
    Modular furniture, modular computer components, folding wooden ruler.
    The pole of a temporary street light consists of a few elements linked by flexible joints for easy transportation and installation.
    Quick-disconnect joints in plumbing.
  C. Increase the degree of fragmentation or segmentation.
    Examples
    Replace solid shades with Venetian blinds.
    Use powdered welding metal instead of foil or rod to get better penetration of the joint.
    Roller conveyor.

Principle 2. Removal/Extraction
  A. Separate (extract) an interfering part or property from an object, or single out the only necessary part (or property) of an object.
    Examples
    Place a noisy compressor outside the building where compressed air is used.
    Use fiber optics or a light pipe to separate a hot light source from where the light is needed.

Frighten birds away from the airport by using a tape recorder to reproduce a sound known to scare birds. (The sound is thus separated from the birds).

Principle 3. Local quality

A. **Change an object's structure from uniform (homogeneous) to non-uniform, change an external environment (or external influence) from uniform to nonuniform.**

Example

Doping silicon wafers in special windows prepared in a photolithography process* allows for making semiconductor devices from this material.

B. **Make each part of an object function in conditions most suitable for its operation.**

Example

Lunch box with special compartments for hot and cold solid foods and liquids.

C. **Make each part of an object fulfill a different and useful function.**

Examples

Pencil with eraser, hammer with nail puller.

Multifunction tool that scales fish and acts as pliers, wire stripper, flathead screwdriver, Phillips screwdriver, manicure set, etc.

Ultrasonic drill consists of heat conductive head and heat resistant body.

Principle 4. Asymmetry

A. **Change the shape of an object from symmetrical to asymmetrical.**

Examples

Asymmetrical mixing vessels or asymmetrical vanes in symmetrical vessels improve mixing (cement trucks, cake mixers, blenders).

The outer side of a car's tire has higher strength in order to improve resistance to impact with a curb.

Put a flat spot on a cylindrical shaft to attach a knob securely.

Dust filter membranes have different porosities.

Change from circular O-rings, to oval cross-section, to specialized shapes to improve sealing.

B. **If an object is asymmetrical, increase its degree of asymmetry.**

Examples

Use astigmatic optics to merge colors.

Make one side of a tire stronger than the other to withstand impact with the curb.

Principle 5. Merging (Joining/Combining)

A. **Merge identical or similar objects, assemble identical or similar parts to perform parallel operations.**

Examples

Personal computers in a network.

Millions of transistors in a single microprocessor chip.

Catamaran.

---

* See case study in Chapter 17.

B. Make operations contiguous or parallel; bring them together in time.
Examples

Link slats together in Venetian or vertical blinds.

Medical diagnostic instruments that simultaneously analyze multiple blood parameters.

The working element of a rotary excavator has special steam nozzles to defrost and soften the frozen ground.

Mulching lawnmower (demonstrates also Principle 6).

Principle 6. Universality

A. Make a part or object perform multiple functions; eliminate the need for other parts.
Examples

Handle of a toothbrush contains toothpaste.

Child's car safety seat converts to a stroller.

Sofa converts into a bed.

Minivan's seat adjusts to accommodate seating, sleeping, or cargo carrying.

Charge-coupled device with microlenses on the surface.

PC in a library functions as a reference, instructional aid, news source, etc.

Principle 7. Nested structures

A. Place one object into another; place each object, in turn, inside the other.
Examples

Measuring cups or spoons

Russian nesting doll (*Matrioshka*) (often this name itself is used for Principle 7).

Portable audio system (microphone fits inside transmitter, which fits inside amplifier case).

B. Make one part pass through a cavity of the other.
Examples

Telescoping radio antenna.

Extending pointer.

Zoom lens.

Mechanical pencil with lead stored inside.

Chairs that stack on top of each other for storage.

Seat belt retraction mechanism.

Retractable aircraft landing gear stow inside the fuselage (also demonstrates Principle 15, Dynamism).

Principle 8. Anti-weight (Counterweight)

A. To counter the weight of an object, merge it with other objects that provide lift.
Examples

Inject foaming agent into a bundle of logs, to make it float better.

Boat with hydrofoils.

A rear wing in racing cars that increases pressure from the car to the ground.
Use of helium balloon to support advertising signs or a cable above a river.
Paintbrush with lightweight handle that floats.

B. To compensate for the weight of an object, make it interact with the environment (e.g., use aerodynamic, hydrodynamic, buoyancy, and other forces).

Examples

Shape of aircraft wing reduces air density above the wing, increases density below the wing, to create lift. (This also demonstrates Principle 4, Asymmetry.)

Vortex strips improve lift of aircraft wings.

Hydrofoils lift ship out of the water to reduce drag.

Principle 9. Preliminary anti-action (counter-action)

A. If it is necessary to do an action with both harmful and useful effects, this action should be replaced with anti-actions to control harmful effects.

Examples

Buffer a solution to prevent harm from extremes of pH.

A cutting method using a dish cutter rotating on its geometric axis during the cutting process. In order to prevent vibrations, the dish cutter is charged in advance with forces close in size and direction and directly contrary to the forces arising in cutting process.

B. Create actions in an object that will later oppose known undesirable working actions.

Examples

Prestress rebar before pouring concrete.

Reinforced concrete column.

Reinforced shaft made from several pipes which have been previously twisted to a specified angle.

Corrugated and covered paper for cartons is bent in opposite directions. Carton becomes flat when the glue between the papers dries.

Masking before harmful exposure: lead apron covers parts of the body being exposed to Xrays; masking tape to protect the part of an object not being painted.

Principle 10. Preliminary action

A. Perform, before necessary, a required change of an object (either fully or partially). Carry out all or part of the required action in advance.

Examples

Pre-pasted wallpaper.

Self-adhesive stamps.

Rubber cement in a bottle is difficult to apply neatly and uniformly. However, if formed into a tape, the proper amount can be applied more easily.

Sterilize all instruments needed for a surgical procedure on a single sealed tray.

World Wide Web search engines (Lycos, AltaVista) browse through all possible links and create a fast-access index of keywords to Internet locations. The query engine then looks up answers in the local index database instead of directly retrieving web pages, which would take much more time.

B. Pre-arrange objects so that they can act from the most convenient place and without losing time for their delivery.

Examples

Predeposited blade in surgery cast. The blade works during cast removal.

Utility knife blade made with a groove allowing the dull part of the blade to be broken off, restoring sharpness.

Principle 11. Beforehand cushioning (Cushion in advance)

A. Prepare emergency means beforehand to compensate the relatively low reliability of an object.

Examples

A strip on photographic film that directs the developer to compensate for poor exposure.

Backup parachute. Tape backup of critical data (unreliable computers or power systems).

Alternate air system for aircraft instruments.

Merchandise is magnetized to deter shoplifting.

Usage of old tires at sharp road turns for safety.

A cover of steel by a material that resists oxidation.

Principle 12. Equipotentiality

A. In a potential field, limit position changes (e.g., change operating conditions to eliminate the need to raise or lower objects in a gravity field).

Examples

Spring-loaded parts delivery system in a factory.

Engine oil in a car is changed by workers in a pit to avoid using expensive lifting equipment.

Locks in a channel between two bodies of water (such as the Panama Canal).

A device for raising and lowering heavy presses that takes the form of an attachment with a roll-gang fastened to the press table.

"Skillets" in an automobile plant that bring all tools to the right position.

Principle 13. Reverse ("The other way around")

A. Invert the actions used to solve a problem (e.g., instead of cooling an object, heat it).

Example

To loosen stuck parts, cool the inner part instead of heating the outer part.

B. Instead of an action dictated by the requirements, one implements the opposite action.

Example

Rotate the part instead of the tool.

C. **Make movable parts or the external environment fixed, and fixed parts movable.**

Examples

Flow water in short pool moving against a swimmer.

Moving sidewalk with standing people. Treadmill (for walking or running in place)

Abrasively cleaning parts by vibrating the parts instead of the abrasive.

D. **Turn the object or process "upside down."**

Examples

Turn an assembly upside down to insert fasteners (especially screws).

Empty grain from containers (ship or railroad) by inverting them.

Principle 14. Spheroidality — Curved

A. **Instead of using rectilinear parts, surfaces, or forms, use curvilinear ones; move from flat surfaces to spherical, from parts shaped as a cube (parallelepiped) to ball-shaped structures.**

Examples

Use arches and domes for strength in architecture.

A circular landing way in airports with "unlimited" length.

B. **Use rollers, balls, spirals, domes.**

Examples

Spiral gear (Nautilus) produces continuous resistance for weightlifting.

A device for welding pipes into a lattice has electrodes in the form of rotating balls.

Computer mouse uses ball construction to transfer linear two-axis motion into vector motion.

Ballpoint and rollerball pens for smooth ink distribution.

C. **Go from linear to rotary motion, use centrifugal forces.**

Examples

Linearly move the cursor on the computer screen using a mouse or a trackball.

Spinning clothes instead of wringing in a washing machine to remove water.

Use spherical casters instead of cylindrical wheels to move furniture.

Principle 15. Dynamism

A. **Allow or design the characteristics of an object, external environment, or process to change to be optimal or to find an optimal operating condition.**

Example

Adjustable car steering wheel (or seat, back support, mirror position...).

B. **Divide an object into parts capable of movement relative to each other.**

Examples

The "butterfly" computer keyboard. (Also demonstrates Principle 7, Nested structures.)

Scissors instead of knife.

A flashlight with a flexible gooseneck.

A transport vessel with a cylindrical body. To reduce the draft of the vessel under full load, the body is composed of two hinged, half-cylindrical parts which can be opened.

C. If an object (or process) is rigid or inflexible, make it movable or adaptive.
Examples
The flexible boroscope for examining engines.
The flexible endoscope for medical examination.
A strip electrode in an automatic arc welding pre-bent at different angles along its length that allows control of the shape and dimensions of the weld bath during welding.

Principle 16. Partial, satiated, or excessive actions

A. If 100 percent of an object is hard to achieve using a given solution method, the problem may be considerably easier to solve by using "slightly less" or "slightly more" of the same method.
Examples
Deposit excess chemical for photolithography in semiconductor production, then remove excess by spinning.
Fill, then "top off" when filling the gas tank of a car.
Software: various image encoding algorithms such as JPEG, GIF, etc.
To obtain uniform discharge of a metallic powder from a bin, the hopper has a special internal funnel, which is continually overfilled to provide nearly constant pressure.

Principle 17. Another dimension

A. Difficulties involved in moving or relocating an object along a line are removed if the object acquires the ability to move in two dimensions (along a plane). Accordingly, problems connected with movement or relocation of an object on one plane are removed by switching to a three-dimensional space.
Examples
Infrared computer mouse moves in space, instead of on a surface, for presentations.
Five-axis cutting tool can be positioned where needed.

B. Use a multi-story arrangement of objects instead of a single-story arrangement. Use a multilayered assembly of objects instead of a single layer.
Examples
Cassette with several CDs to increase music time and variety.
Greenhouse that has a concave reflector on the northern part of the house to improve illumination of that part of the house by reflecting sunlight during the day.
Fourier transform-based software, in which digitized signals are transformed from the time domain to the frequency domain for processing.

C. Tilt or re-orient the object, lay it on its side.
Example
Dump truck

D. **Use another side of a given area.**
Examples
All devices with Mobius belt.
Electronic chips on both sides of a printed circuit board.

E. **Use optical lines falling onto neighboring areas or onto the reverse side of the area available.**

Principle 18. Mechanical vibration

A. **Oscillate or vibrate an object.**
Examples
Electric carving knife with vibrating blades.
Vibrate a casting mold while it is being filled to improve flow and structural properties.

B. **If oscillation exists, increase its frequency.**
Example
Distribute powder with vibration.

C. **Use an object's resonant frequency.**
Examples
Destroy gall stones or kidney stones by ultrasonic resonance.
An instrument for cutting timber without a saw whose pulse frequency is close to the inherent frequency of vibration of the timber.

D. **Use piezoelectric vibrators instead of mechanical ones.**
Example
Quartz crystal oscillations drive highly accurate clocks.

E. **Use combined ultrasonic and electromagnetic field oscillations.**
Example
Mix alloys in an induction furnace.

Principle 19. Periodic action

A. **Instead of continuous action, use periodic or pulsating actions.**
Examples
Hitting something repeatedly with a hammer.
Replace continuous siren with pulsed sound.
A flashing warning lamp is more noticeable than one that is continuously lit.

B. **If an action is already periodic, change the periodic magnitude or frequency.**
Examples
Use Frequency Modulation to convey information instead of Morse code.
Replace a continuous siren with sound that changes amplitude and frequency.

Principle 20. Continuity of useful action (uninterrupted useful effect)

A. **Continue on actions; make all parts of an object perform UF and/or NF at full load, all the time.**
Examples
Flywheel (or hydraulic system) stores energy when a vehicle stops, so the motor can keep running at optimum power.
A drill with cutting edges, which permits cutting in forward and reverse directions.

B. Eliminate all idle or intermittent actions.

Examples

Print during the return of a printer carriage — dot matrix printer, daisy wheel printers, inkjet printers.

Crystal growing machine with permanent raw material supply mechanism.

Principle 21. Skipping (Rushing through)

A. Conduct a process or certain stages (e.g., destructible, harmful, or hazardous operations) at high speed.

Examples

Use a high speed dentist's drill to avoid heating tissue.

Cut plastic faster than heat can propagate in the material to avoid deforming the shape.

Principle 22. Convert harm into benefit

A. Use harmful factors (particularly harmful effects of the environment or surroundings) to achieve a positive effect.

Examples

Use waste heat to generate electric power.

Sand or gravel freezes solid when transported through cold climates. Over-freezing (using liquid nitrogen) makes the ice brittle, permitting pouring.

Recycle waste material from one process as raw materials for another.

B. Eliminate the primary harmful action by adding it to another harmful action to resolve the problem.

Examples

Add a buffering material to a corrosive solution.

Use a helium-oxygen mix for diving to eliminate both nitrogen narcosis and oxygen poisoning that are a danger with air and other nitrox mixes.

C. Amplify a harmful factor to such a degree that it is no longer harmful.

Examples

Setting a backfire to eliminate fuel from a forest fire.

When using high frequency current to heat metal, only the outer layer became hot. This negative effect was later used for surface heat-treating.

Application of extremely low temperatures to frozen aggregate materials to speed the process of restoring their flow capability.

In TRIZ folklore, this principle is also known as "Blessing in disguise," "Turn Lemons into Lemonade," or "Spin harm to gold."

Principle 23. Feedback

A. Introduce feedback (referring back, cross-checking) to improve a process or action.

Examples

Automatic volume control in audio circuits.

The level of liquid is self-adjusted by a floating valve inside a tank.

Feedback inside the software program, analogous to feedback in mechanical or electrical systems, is commonly used to control the operation of various elements.

Signal from gyrocompass is used to control simple aircraft autopilots.

B. If feedback is already used, change its magnitude or influence.

Examples

Change sensitivity of an autopilot when within 5 miles of an airport.

Change sensitivity of a thermostat when cooling vs. heating, since it uses energy less efficiently when cooling.

Principle 24. Intermediary

A. Use an intermediary carrier article or intermediary process.

Examples

Carpenter's nail set, used between the hammer and the nail.

Word processors and spreadsheets include conversion filters to read and write files in competitive product formats.

B. Merge one object temporarily with another (which can be easily removed).

Examples

Pot holder to carry hot dishes to the table.

To reduce energy loss when applying current to a liquid metal, cooled electrodes and intermediate liquid metal with a lower melting temperature are used.

Suspensions (adhesive parts can be dissolved or burned out).

In TRIZ folklore this principle is also known as "Go between" or "Mediator."

Principle 25. Self-service and self-organization

A. Make an object serve itself by performing auxiliary helpful functions.

Examples

A soda fountain pump that runs on the pressure of the carbon dioxide that carbonates the drinks. This assures that drinks will not be flat and also eliminates the need for sensors.

To weld steel to aluminum, create an interface from alternating thin strips of the two materials. Cold–weld the surface into a single unit with steel on one face and copper on the other, then use normal welding techniques to attach the steel object to the interface, and the interface to the aluminum. (This concept also has elements of Principle 24, Intermediary, and Principle 4, Asymmetry).

B. The object should service/organize itself and carry out supplementary and repair operations.

Examples

Halogen lamps regenerate the filament during use — evaporated material is redeposited.

Software programs employ some form of self-checking to verify their integrity.

C. Use waste resources, energy, or substances.

Examples

Use heat from a process to generate electricity, "Cogeneration."

Use animal waste as fertilizer.

Use food and lawn waste to create compost.

A cone-shaped concrete dam sink in sand on a river's bottom will seal itself in the event of an earthquake.

Principle 26. Copying

A. Instead of an unavailable, expensive, fragile object, use simpler and inexpensive copies.

Examples

Virtual reality via computer instead of an expensive vacation.

Use the sound of a barking dog, without the dog, as a burglar alarm.

Modeling stage in design.

Listen to an audio tape instead of attending a seminar.

B. Replace an object or process with optical copies.

Examples

Do surveying from space photographs instead of on the ground.

Photolithography in semiconductor production.

Measure an object by making measurements in the photograph.

The height of tall objects can be determined by measuring their shadows.

Use sonograms to evaluate the health of a fetus, instead of risking damage by direct testing.

C. If visible optical copies are already used, move to infrared or ultraviolet copies.

Example

Make images in infrared to detect heat sources, such as diseases in crops or intruders in a security system.

Principle 27. Inexpensive short-lived objects

A. Replace an expensive object with multiple inexpensive objects, comprising certain qualities (such as service life, for instance).

Examples

Use disposable supplies to avoid the cost of cleaning and storing durable objects.

Plastic cups in motels, disposable diapers, many kinds of medical supplies.

Single-roll disposable camera for tourists.

Principle 28. Mechanics substitution

A. Replace a mechanical means with a sensory (optical, acoustic, taste, or olfactory) means.

Examples

Replace a physical fence to confine a dog or cat with an acoustic (electronic) "fence" (signal audible to the animal).

Use a bad smelling compound in natural gas to alert users to leakage, instead of a mechanical or electrical sensor.

B. **Use electric, magnetic, and electromagnetic fields to interact with the object.**

Examples

To mix 2 powders, electrostatically charge one positive and the other negative. Either use fields to direct them, or mix them mechanically and let their acquired fields cause the grains of powder to pair.

To increase the bond between metal coating and a thermoplastic material, the process is carried out inside an electromagnetic field which applies force to the metal.

Application of magnetic fields for reducing oxygen effects of performance of semiconductor wafers.

C. **Change from static to movable fields, from unstructured fields to those having structure.**

Example

Early communications used omnidirectional broadcasting. We now use antennas with a very detailed structure of the pattern of radiation.

D. **Use fields in conjunction with field-activated particles (e.g., ferromagnetic).**

Example

Heat a substance containing ferromagnetic material by using a varying magnetic field. When the temperature exceeds the Curie point, the material becomes paramagnetic and no longer absorbs heat.

Principle 29. Pneumatics and hydraulics

A. **Use gas and liquid parts of an object instead of solid parts (e.g., inflatable, filled with liquids, air cushion, hydrostatic, hydroreactive).**

Examples

Comfortable shoe sole inserts filled with gel.

Store energy from decelerating a vehicle in a hydraulic system, then use the stored energy to accelerate later.

Using reducing air pressure in "vacuum holders."

B. **Use the Archimedes forces to reduce the weight of an object.**

Example

To produce an all-metal shell dirigible without high-priced adjustments, the montage is realized by floating on pontoons in water.

C. **Use negative or atmosphere pressure.**

Example

In order to prevent displacement of friable cargo in a ship, at time of transportation the free surface of cargo is covered by a hermetic rubber-band layer, creating a vacuum which ensures shrinking of the layer by atmospheric pressure.

D. **A spume or foam can be used as a combination of liquid and gas properties with a light weight.**

Examples

Spume lubricant is used for a stamp instrument.

Cavities of electrical soldering pen are filled by spume mass.

Principle 30. Flexible shells and thin films

  A. **Use flexible shells and thin films instead of three-dimensional structures.**

    Examples

    Use inflatable (thin film) structures as winter covers on tennis courts.

    Coating glass windows with sapphire for high-temperature applications.

  B. **Isolate the object from the external environment using flexible shells and thin films.**

    Examples

    Float a film of bipolar material (one end hydrophilic, one end hydrophobic) on a reservoir to limit evaporation.

    For shipping fragile products, air bubble envelopes or foam-like materials are used.

Principle 31. Porous materials and membranes

  A. **Make an object porous or add porous elements (inserts, coatings, etc.).**

    Examples

    Drill holes in a structure to reduce the weight.

    To avoid pumping coolant to a machine, some of its parts are filled with a porous material soaked in coolant liquid. The coolant evaporates when the machine is working, providing short-term uniform cooling.

    Porous wafers for GaAs devices.

  B. **If an object is already porous, use the pores to introduce a useful substance or function.**

    Examples

    Use a porous metal mesh to wick excess solder away from a joint.

    Store hydrogen in the pores of a palladium sponge. (Fuel "tank" for the hydrogen car — much safer than storing hydrogen gas).

Principle 32. Color changes

  A. **Change the color of an object or its external environment.**

    Example

    Use safe lights in a photographic darkroom.

  B. **Change the transparency of an object or its external environment.**

    Example

    Use photolithography to change transparent material to a solid mask for semiconductor processing. Similarly, change mask material from transparent to opaque for silk screen processing.

  C. **In order to observe objects or processes that are difficult to see, use colored additives. If such additives are already used, employ luminescence traces.**

    Examples

    Fluorescent additives during UV spectroscopy.

    A water curtain used to protect steel mill workers from overheating blocked infrared rays but not the bright light from the melted steel. A coloring was added to the water to create a filter effect while preserving the transparency of the water.

Principle 33. Homogeneity
A. **Make objects interacting with a given object of the same material (or material with identical properties).**
Examples
Make the container out of the same material as the contents, to reduce chemical reactions.
The surface of a feeder for abrasive grain is made of the same material that runs through the feeder, allowing a continuous restoration of the surface.
Make parts of semiconductor equipment from silicon.

Principle 34. Discarding and recovering
A. **Discard (by dissolving, evaporating, etc.) portions of an object that have fulfilled their functions or modify these directly during operation.**
Examples
Use a dissolving capsule for medicine.
Biodegradable plastics.
Rocket boosters separate after serving their function.
Sprinkle water on cornstarch-based packaging; it reduces its volume by more than 1000X!
B. **Conversely, restore consumable parts of an object directly in operation.**
Examples
Self-sharpening lawn mower blades.
Automobile engines that give themselves a "tune-up" while running.

Principle 35. Parameters and properties changes
A. **Change an object's physical aggregate state (e.g., to a gas, liquid, or solid).**
Examples
Freeze the liquid centers of filled candies, then dip in melted chocolate, instead of handling the messy, gooey, hot liquid.
Transport oxygen, nitrogen, or petroleum gas as a liquid, instead of a gas, to reduce volume.
B. **Change the concentration or consistency (see also Principle F below).**
Examples
Liquid hand soap is concentrated and more viscous than bar soap at the point of use, making it easier to dispense in the correct amount and more sanitary when shared by several people.
As a rule, the resistivity of semiconductors changes by concentration of impurities.
C. **Change the degree of flexibility (see also Principle 15).**
Examples
Use adjustable dampers to reduce the noise of parts falling into a container by restricting the motion of the container's walls.
Vulcanize rubber to change its flexibility and durability.

D. Change the temperature.

Examples

Raise the temperature above the Curie point to change a ferromagnetic substance to a paramagnetic substance.

Raise the temperature of food to cook it. (Changes taste, aroma, texture, chemical properties, etc.).

Lower the temperature of medical specimens to preserve them for later analysis.

Dry objects by hot air.

Thermocouple.

E. Change other characteristics of a technique.

Example

Sphingometer.

Principle 36. Phase transitions

A. Use phenomena that occur during phase transitions (e.g., volume changes, loss or absorption of heat, etc.).*

Examples

Heat pumps use the heat of vaporization and heat of condensation of a closed thermodynamic cycle to do useful work.

Crystallization.

Superconductivity.

Principle 37. Thermal expansion

A. Use thermal expansion (or contraction) of materials.

Examples

Fit a tight joint together by cooling the inner part to contract, heating the outer part to expand, putting the joint together, and returning to equilibrium.

To control the expansion of ribbed pipes, they are filled with water and cooled to a freezing temperature.

B. If thermal expansion is being used, use multiple materials with different coefficients of thermal expansion.

Examples

The basic leaf spring thermostat: 2 metals with different coefficients of expansion are linked so that the object bends one way when warmer than normal and the opposite way when cooler. To control the opening of roof windows in a greenhouse, bimetallic plates are connected to the windows, so temperature change bends the plates, causing the window to open or close.

A lid of a hothouse is controlled by hinged hollow pipes with spreading water inside. The center of gravity of the pipes shifts with a change in temperature and therefore the pipes raise and lower the lid.

---

* The meaning of phase transition in TRIZ and in physics is the same. Note that thermal expansion (Principle 37) can be the result of a phase transition.

Principle 38. Strong oxidants

A. Replace common air with oxygen-enriched air.
Example
Scuba diving with non-air mixtures have high concentration of oxygen for extended endurance.

B. Replace enriched air with pure oxygen.
Examples
Cut at a higher temperature using an oxy-acetylene torch.
Treat wounds in a high pressure oxygen environment to kill anaerobic bacteria and aid healing.

C. Expose air or oxygen to ionizing radiation.
Example
Ionize air to trap pollutants in an air cleaner for ultra-cleanrooms (class 100 and better in the semiconductor industry).

D. Use ionized oxygen.
Example
Ionize oxygen to increase the speed of semiconductors' surface oxidation.

E. Replace ozonized (or ionized) oxygen with ozone.
Example
Speed chemical reactions by ozone.

Principle 39. Inert atmosphere

A. Replace a normal environment with an inert one.
Examples
Prevent degradation of a hot metal filament by using an argon atmosphere.
Prevent cotton from catching fire in a warehouse by treating it with inert gas while being transported to the storage area.
Foam is used to isolate a fire from oxygen in air.

B. Add neutral parts or inert additives to an object.
Example
Increase the volume of powdered detergent by adding inert ingredients. This makes it easier to measure with conventional tools.

Principle 40. Composite materials

A. Change from uniform to composite (multiple) materials.
Examples
Composite (Copper–Ceramic–Copper) substrates for power semiconductors have high thermal conductivity and high electrical insulation.
Military aircraft wings are made of composites of plastics and carbon fibers for high strength and low weight.
Composite ferro-electrics for high-frequency applications are smaller, lighter, and more reliable than traditional parts.

The Principles are formulated in a general way with the effort to demonstrate the solution concept. If, for example, few Principles recommend the "flexibility" heuristic, it means that the solution of the problem relates somehow to changing the degree of flexibility or adaptability of a technique being modified. Some of these

Principles have a remarkably broad range of application, including management, advertising, marketing, etc.

Some of these Principles are dual (e.g., 4 and 14) or can be inverted (e.g., 24) as was shown by Tat'yana A. Kengerli, Genady Filkovsky, Irina Flikshtein, and Sergey I. Perniszky. Some of these Principles are complementary in space and time (e.g., 17 vs. 11) as shown by Yevgeny Karasik. Altshuller's initial 40 Principles are not independent. It is possible to show that all principles can be divided according to those that are general, or can work in several engineering fields, and those that are particular or work in a specific engineering field. Such ongoing research is very important for creating a system of Principles and other TRIZ heuristics. On the other hand, the common principles can be used during problem solving and/or design only by specific interpretation. Such separation of principles into common, specific, and particular, and the investigations of patents greatly increases the number of principles. Continuous search for new Inventive Principles and sub–Principles is one of the current activities of TRIZ experts. This research is illustrated by the following new Inventive Principles and sub–Principles.

A. Multistep principle

Efficiency of action increases due to consequent usage of a group of uniform objects instead of the single object.

Examples

A multicylinder combustion engine.

Thousands of cells in a power MOSFET transistor.

B. Dissociation-association (sub–principle to Principle 34).

This principle allows division and coalition on the molecular level.

Examples

Use some alloys prepared by a molecular beam epitaxy in opto-electronic devices.

Use some materials as a working body for sidebars of binary cycle of energy equipment. These materials can dissociate during the warming semicycle with absorbing heat and associate back during the cooling semicycle with the extraction of same portions of energy.

C. Use of pauses (sub–principle to Principle 19)

Use pauses between actions to perform similar or different actions; i.e., one action is active during pauses of other action.

Examples

The auto-control of a personal computer that runs when the computer is not running other applications.

In cardiopulmonary respiration, breathe for victim once every 5 seconds.

Squeezing of channels in telephone communications.

D. Use of epenthetic (insert) parts

Overcome problems in fabricating an object by using some temporary insert during the manufacturing process, deleting the insert when finished.

Examples

During fabrication of semiconductor devices, the surface at some operations is protected by photoresistors or silicon dioxide.

Bullet casings are ejected after the gun fires.

E. Match of impedances

During design, determine the input impedance level and set the system internal impedance to that input signal. If an exact match is impossible, minimize losses by amplifying or attenuating the input signal or dispersing the input signal via a few channels, each of whose impedance can be matched with the system impedance.

Example

System impedance (complex resistance) matches input impedance to provide a maximum transfer of energy across systems, such as electrical, fiber optics, hydraulic, gas, information and transmission lines, measurement apparatus, and devices with distributed parameters.

F. Concentration-dispersion principle

Concentration/Dispersion is a systems transformation method consisting of a set of objects (at least two), managing their mutual arrangement and relative quantity.*

a. Concentrate essential resources, elements, actions in a key place and moment of time in order to achieve the purpose (e.g., increase technique effectiveness).

Example

Sublimated food products, encyclopedias, notebook computers vs. desktop.

b. If concentrated objects or actions cause undesirable effects, they should be disseminated, dispersed in space and/or time.

Example

Preparing food from concentrates, thermal treatment of semiconductors after ion implantation, harmful effects of medical drugs at high concentrations, activation of diffusion.

Typical ways of realizing this Principle:

Changing distance between objects or interval of time between actions;

Changing quantity of objects in given area or distribution density of objects in space;

Changing amount of objects — dilution;

Expanding/convoluting a flow in different or same directions;

Moving from or to the center.

TRIZ experts apply stringent tests to any new Principles in order to confirm their importance. For a new Principle to become a TRIZ heuristic and to enter a specific cell of the Contradiction Matrix, it must recur in many high-Level patents in various engineering fields. There is no consensus on the number of such inventions, but the lowest is 10 and the highest is 500, according to different TRIZniks'

---

* A concentration can be absolute (expressed by number of objects in unit of volume or in the given spatial area) and relative (expressed by the relation of number of the given objects to total number of all objects in the given area). S. I. Perniszky, the author of this Principle, noted that the essence of the Dispersion sub–Principle is partially crossed with essence of Splitting and Removal Principles, as well as of the Concentration sub–Principle with the Association and Local Quality Principles; however, they are not reduced to each other. On the other hand, a part of this Principle is included in Principle 35.

opinions. Altshuller estimated a threshold of several dozen patents, but he was not consistent and sometimes incorporated a Principle that he had discerned in only a few inventions in the Matrix; sometimes he did not include another Principle that he had discerned in many patents. Of course, this threshold is also connected with the overall number **S** of Inventive Principles that can be found. We do *not* know how many various TRIZ heuristics (including Principles) exist, but almost all TRIZniks believe that their quantity is a finite number. Some estimation can help find an answer to this question.

Assume for each Principle that we can find only N patents at Level 4 for each Inventive Principle, 10N inventions at Level 3, and 100N patents at Level 2. We can write the total number of patents P as

$$P = P_1 + P_2 + P_3 + P_4 + P_5 \approx 20,000,000,$$

where P_i is the number of patents at Level **i**. It is simple to calculate each P_i due to known statistics of inventions that belong to each Level (see Chapter 8).

Assume that S is one to three orders of magnitude smaller than $P_5 \approx 200,000$, which means that finding a new Principle is as or more difficult as discovering a new phenomenon. This assumption is taken from the history of science that shows that scientists more often discovered a new phenomenon than a new approach to conduct research. This assumption leads to S values from 200 to 20,000. Because we are interested in the Principles that would work for various engineering fields only, set S = 2,000 or 0.01% of all patents, i.e., $S = P/10^4$. Therefore, each Principle on average has been applied in 5000 different technical solutions that are incorporated into existing patents. Note that each Inventive Principle ("old" or "new") assists in finding many new patentable solutions of different problems.* Now it is easy to calculate that N is approximately 200 because approximately only 10% of all patents have been studied so far in the framework of TRIZ. This fact leads to the estimation that the threshold is approximately a couple of dozen patents. In other words, a new Inventive Principle can be incorporated into the Matrix if it can be confirmed by 20 or more patents from different engineering fields studied by TRIZniks.

Another good sign to include a new Inventive Principle into the Matrix is the applicability of this new Principle for resolution of contradictions between a considerable number of different inventive Parameters within various clusters. A TRIZnik can study approximately 10,000 patents during a year. (Unfortunately, TRIZ has not become the object of regular academic research yet, so no TRIZnik can dedicate all his time to studying patents.) Let us assume that a TRIZnik's goal is to find *one* new Inventive Principle in a year, concordant to our previous estimation and also reasonable for newcomers in TRIZ who have thorough knowledge of their engineering field. Note that when a new Inventive Principle is discovered** in

---

* By the way, this is why some patent attorneys think that TRIZ can destroy the current patent system, because now each of about 5000 different patents is actually based on the same Inventive Principle.

** I believe that the word *discovered* is absolutely correct here because TRIZniks as well as physicists search for what already exists but has not been recognized. Moreover, the recognition of a single new Inventive Principle is really the result of TRIZ research as well as a discovery in the natural sciences.

so-called theoretical TRIZ, it is easier to figure out this particular Inventive Principle in different technical solutions conducting experimental TRIZ research. Right now several Inventive Principles are pending approval. However, because the updated version of the Matrix, as well as new Parameters and Inventive Principles, is proprietary, most TRIZniks still use the original Altshuller's Matrix, presented in a slightly expanded form later in this chapter.

13.3.1 A CLASSIFICATION OF THE INVENTIVE PRINCIPLES

The Inventive Principles, as published by Altshuller, do not represent a system. The numbering of the Principles reflects only the order in which they were included in the TRIZ fund. K.A. Sklobovsky [3] proposed a more accurate approach, presented here in a slightly expanded form, which classifies the Principles by functional attributes.

1. **Principles Relating Mainly to Substrates of the Technical System**
   1.1. Hierarchy of the subsystems
       1.1.1.   Segmentation
             1.1.1.1. Discarding and recovering of parts
             1.1.1.2. Inexpensive short-lived objects
             1.1.1.3. Taking out
             1.1.1.4. Dissociation-Association
             1.1.1.5. Usage of centrifugal forces
       1.1.2.   Unification
             1.1.2.1. Bi-principle (unification in a bi/poly-system)
             1.1.2.2. Homogeneity
             1.1.2.3. Merging (joining/combining)
       1.1.3.   Replacement, change of subsystem
             1.1.3.1. Usage of intermediary
             1.1.3.2. Usage of copies
             1.1.3.3. Taking out
             1.1.3.4. Assembly of large-sized designs in/on water
             1.1.3.5. Usage of phase transitions
   1.2. Qualitative characteristics of subsystems
       1.2.1.   Universality
       1.2.2.   Uniformity
       1.2.3.   Local quality
   1.3. Spatial (dimensional) relations between subsystems
       1.3.1.   Symmetry
       1.3.2.   Asymmetry
       1.3.3.   Spheroidality-Curvature
       1.3.4.   Equipotentiality
       1.3.5.   Nested structures
       1.3.6.   Another dimension
       1.3.7.   Structures with unusual topological features
       ...
       1.3.999. See list of geometric effects (Appendix 3).

1.4. Special materials and phenomena
  1.4.1. Strong oxidants
  1.4.2. Inert environment, Vacuum
  1.4.3. Explosives and Flammable Materials
  1.4.4. Self-regenerating materials
  1.4.5. Composite materials
  1.4.6. Usage of Epenthetic Parts
  1.4.7. "Smart materials"
  1.4.8. Use of phosphors
  1.4.9. Use of the ferromagnets
  1.4.10. Electrokinetic phenomena (electrolysis, electro-osmosis, electrophoresis)
  1.4.11. Magnetic liquids
  1.4.12. Piezoelectric effect
  1.4.13. Shape memory effect
  1.4.14. Porous and capillary materials
  1.4.15. Flexible shells and thin films
  1.4.16. Spume/foam
  1.4.17. Pneumatics and hydraulics
          1.4.17.1. Liquid and gas properties
  1.4.18. Corona discharge
  …
  1.4.999. See lists of chemical, biological, materials science effects (Appendix 3).

2. Field Relations Between the Subsystems of TS
2.1. A direction of interaction
  2.1.1. Reverse ("The other way around")
  2.1.2. Convert harm into benefit
  2.1.3. "Futurism"
          2.1.3.1. Preliminary action
          2.1.3.2. Preliminary antiaction
          2.1.3.3. Beforehand cushioning
  2.1.4. Antiweight
2.2. Rate and rhythm characteristics of interactions
  2.2.1. Skipping
          2.2.2.1. Continuous action
          2.2.2.2. Periodic action
          2.2.2.3. Multi-step, repeated action
  2.2.3. Pauses
  2.2.4. Oscillations
2.3. Qualitative characteristics of interactions
  2.3.1. Dynamism
  2.3.2. Partial, superfluous, or excessive action
  2.3.3. Vibration or oscillation (mechanical, electromagnetic)

Note: 1.4 above contains an index of some materials and effects that are usually not mentioned in Altshuller's Principles. Nevertheless, we include in this index materials and phenomena that are used quite often in modern techniques. Let us illustrate one of them:

1.4.3 Use of Explosives and Flammable Materials

Execute the operations of mixing and/or warming of objects with pure access by means of explosives and termites.

An anchor-like device, with an explosive for electro-transmission lines, radio-masts, etc., locates the necessary depth with help of a drill machine and then blasts the explosive, creating a cavity in the soil and firmly bolting expanding links of the anchor-like device within the cavity [*Patent USA 3 281 153*].

An even more expanded index belongs to the lists of physical, chemical, and other effects created after Altshuller had developed the Contradiction Matrix.

13.4 MATRIX FOR REMOVAL OF TYPICAL TECHNICAL CONTRADICTIONS

As we already know, a technical contradiction is when one characteristic or sub-system A improves, another, B, gets worse; the opposite is also true: if B improves, then A worsens (see Chapter 4). Because A and B are different, the technical contradiction represents a *pair* problem, reflecting the fact that two statements can be constructed for such contradictions. In other words, at least two technical contradictions can be formulated for each single technical problem. For some problems, both contradictions are valid because both Parameters need improvement, and it is difficult to decide which improvement is more important. Fortunately, TRIZ makes it possible to solve both contradictions.

The Matrix for removing typical technical contradictions, the Contradiction Matrix, is the first original TRIZ instrument proposed by Altshuller for general technical systems that can be in conflict (have technical or physical contradictions). In many cases, the individual Principles themselves, as a selection of good heuristics, give excellent results in solving problems, but in the Matrix their benefit exceeds the arithmetical sum of the values composing the total heuristics.

The horizontal (row) elements of the Contradiction Matrix are the Engineering Parameters to be improved, and the vertical columns contain the Engineering Parameters that can be adversely affected and/or degraded as a result of improving the parameters. The numbers at the intersection cells guide a solver to a number (1–4 in the original Altshuller Matrix and 1–6 in the modern Matrix) of Inventive Principles that might be of help in resolving the technical contradiction. Not all cells are filled, but even so the Matrix indicates the Principles for more than 1200 types of technical contradiction. Note that the Matrix is *not* symmetrical; i.e., if the Parameters are reversed, often other Principles are suggested to compare with the initial set. The diagonal cells indicate where a physical contradiction can occur (strength has to be great and small, for example). Some nondiagonal cells are empty, indicating either where only a few or no patents were found illustrating resolution of that particular contradiction or where such contradictions do not occur in modern TS or TP.

It is very simple to use Altshuller's Contradiction Matrix in the sequence of the following algorithm:

1. Name of Technique:
2. Define the main function of the Technique (formulate the goal of the Technique).
   The ... is designed to ...
3. List the main subsystems of the Technique and their primary and secondary functions:

   Subsystem   Primary Function   Secondary Functions

   A.

   B.

   ...

4. In plain language describe operation of the Technique.
   ... operates as follows:
5. Determine characteristics that should be improved. Determine characteristics that should be eliminated.
   Note: The origin (root cause) of the problem is identified in this preliminary step in the framework of TRIZ, and characteristics of the technique to be changed are formulated. Improve UF and NF or exclude HF and NF of the technique (UF, NF, and HF are useful, neutral, and harmful functions).
6. Reformulate the found characteristics in terms of the Engineering Parameters.
   Note: Ask "Can I improve/eliminate this characteristic using the conventional methods without deterioration of another subsystem or characteristic?" If your answer is YES, you face a routine or low **D** problem and

do *not* need to use the Contradiction Matrix or other TRIZ heuristics. If your answer is NO (i.e., "I know how, but it makes things worse and creates conflict between performance considerations"), you need to convert your problem statement to a pair contradiction that is a conflict between two different characteristics (subsystems) within the technique.

TRIZniks require persistent and disciplined thinking in order to find the Engineering Parameter that best describes the contradiction. It will occur automatically after practice with the Matrix. On the other hand, consider the time wasted trying to reach a solution through trial and error. Clearly, the benefits of this precise analysis of your problem speak for themselves.

7. Formulate the Technical Contradiction as follows: If the positive Engineering Parameter is improved by (state how) then the other characteristic (state which one) will get worse.

8. Formulate the Technical Contradiction as follows: If the negative Engineering Parameter is reduced by implementing (state how) (2b), then the positive Engineering Parameter (state which one) will get worse or another negative characteristic will be intensified.

9. Find the cells in the Matrix that correspond to the two Technical Contradictions, which were found in steps 7 and 8.

10. Using the Matrix, look for the solutions of the conflict of these two parameters.

Notes: The Engineering Parameters have numbers associated with them. Look at the corresponding row and column numbers' cell that will have a list of numbers in the cell. These numbers in the cell are solution Principle numbers. All numbers are given in the original Altshuller notation.

The Principles in the Matrix cells are presented in order of decreasing frequency of their use in patents selected by Altshuller. The statistics on usage of Principles extracted from the initial Altshuller's Matrix is given in Figure 13.1.

11. Apply the Principles from these cells to your problem(s).

Note: There is usually more than one Principle that has been used in the past to solve any particular contradiction. These Principles show *only* the promising direction of solutions. Nevertheless, try to apply each suggested Principle to your technical system. Do not reject any idea at this stage. If all suggestions are completely unacceptable, reformulate the technical contradiction and try again.

12. Find, evaluate, and implement conceptual solutions for your problem.

Note: Sometimes the introduction of a Principle creates a secondary problem. Do not automatically reject it. Find a way to solve the secondary problem — and, if necessary, even a third problem. This method is often used to solve linear, star, and other technical problems. If the secondary problem is easier to solve than the original primary problem, progress has been made. This line of reasoning is used for most of the TRIZ heuristics and instruments presented in this book.

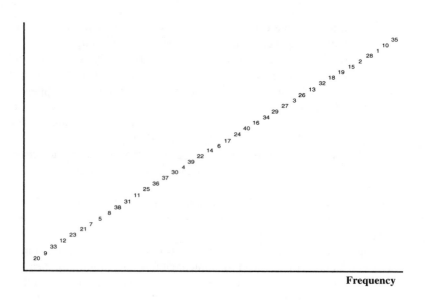

FIGURE 13.1   Statistics on the usage of the Inventive Principles in Altshuller's Matrix.

There are more rows and columns in the modern Contradiction Matrix because of consideration of new Parameters, enrichment of the Matrix cells. For example, a new cell, Safety (as the Undesired Effect) vs. Productivity (as the Feature to Improve), contains Principles 10,14, 25, 35, and 38. The idea of specific and particular Principles was extended to the set of engineering field-dependent Tensors. It is worth mentioning that the Matrix works only with two parameters, while a Contradiction Tensor can operate with the same or higher number of engineering Parameters, so it is multidimensional. Design solutions can be obtained in a more timely manner, although Contradiction Tensors cannot be used as widely as the Matrix. It is clear that the Matrix/Tensors cannot exist in some permanent forms because new fields of technique and engineering will enrich the contradiction matrix/tensor. Consequently, we are convinced that TRIZ needs development. The possibility of creating the universal Tensor seems reasonable and should be researched.

The statistics of Inventive Principles, in order of frequency of their presence in Altshuller's Matrix, were extracted by Solomon D. Tetelbaum and Stan Kaplan (Figure 8.1). We will illustrate the usage of the Contradiction Matrix with a popular TRIZ problem solution based on one of the most frequently used Principles.

13.5   A CASE STUDY

Consider an example of plastic piping necessary to pneumatically transport metal shots. The existing system was originally designed to move plastic pellets (Figure 13.2), but it was found that the metal shots or pellets were more useful for production.

Unfortunately, the metal shots destroy the pipe's elbow; the pair elbow and shots conflict. What should we do? Should we replace the pipe or can we find a cheaper solution?

Before using the Contradiction Matrix, investigate ordinary solutions (as it is suggested in the notes to step 1) as well as the primary function and the goals of the TS:

> *Technique Subsystems:* Straight section of pipe, pipe elbow
> *Technique Goal:* To move the metal shots as fast as possible
> *Conflicting Subsystem:* Metal shots and pipe elbow
> *Conflict:* Metal shots destroy pipe elbow
> *Strength Conflict:* Consider increasing the speed of the metal shots
> *Primary Function of Conflicting Subsystem:* Change direction of shot movement
> *Possible Ordinary Solutions:* Reinforce elbows, use quick-change elbows,
>   redesign shape of elbow, select another material for elbow

Once this information is recorded, several contradictions emerge between the ordinary solutions and the system's goals. Two desired improvements include increasing speed and decreasing energy requirements.

There are several degrading Parameters associated with each improvement, e.g., the object stability (#13). There are also several Parameters in this system that can be improved. The Parameters that degrade, when the speed of the shots is increased, are presented along with the Principles used to reduce the contradiction. This information was taken from the intersection of the relevant parameters on the Contradiction Matrix. A record of the Principles used in all of the contradictions suggests considering those that occur most frequently (see Figure 13.1). One of them is Principle 28, Mechanics substitution. It suggests using electric, magnetic, and electromagnetic fields to interact with the object (elbow and shots are mechanical objects in our problem). In our case, it suggests placing a magnet at the elbow to bind metal shots to a plastic material, providing a blanket of shots that then absorb the energy.

This suggestion leads to the solution of the problem (Figure 13.3). It is also possible to obtain other solutions, as is demonstrated in Chapter 15.

13.6   CONCLUSION

In this chapter we learn TRIZ heuristics for resolving pair or technical contradictions. There are two ways to handle these instruments:

1. Use the Contradiction Matrix/Tensor to locate the most effective Principles, or
2. Read every Principle and choose the most appropriate one.

At first glance, the second method seems ridiculous, but experience of many TRIZniks shows that a thorough knowledge of these Principles noticeably increases the creative potential of a problem solver, engineer, or inventor. After practice, you will remember the majority of the Principles (or at least those most important for you) and their meaning and will be able to apply them without the Matrix.

TABLE 13.1 Contradiction Matrix

Left - Down +	1	2	3	4	5	6	7	8	9	10	11	12	13	14	15	16
1	PhC		15.8. 29.34		29.17. 38.34		29.2. 40.28		2.8. 15.38	8.10. 18.37	10.36. 37.40	10.14. 35.40	1.35. 19.39	28.27. 18.40	5.34. 31.35	
2		PhC		10.1. 29.35		35.30. 13.2		5.35. 14.2		8.10. 19.35	13.29. 10.18	13.10. 29.14	26.39. 1.40	28.2. 10.27		2.27. 19.6
3	15.8. 29.34		PhC		15.17. 4		7.17. 4.35		13.4.8	17.10. 4	1.8.35	1.8. 15.34	1.8. 29.34	8.35. 29.34	19	
4		35.28. 40.29		PhC		17.7. 10.40		35.8. 2.14		28.1	1.14. 35	13.14. 15.7	39.37. 35	15.14. 28.26		1.40. 35
5	2.17. 29.4		14.15. 18.4		PhC		7.14. 17.4		29.30. 4.34	19.30. 35.2	10.15. 36.28	5.34. 29.4	11.2. 13.39	3.15. 40.14	6.3	
6		30.2. 14.18		26.7. 9.39		PhC				1.18. 35.36	10.15. 36.37		2.38	40		2.10. 19.30
7	2.26. 29.40		1.7. 35.4		1.7.4. 17		PhC		29.4. 38.34	15.35. 36.37	6.35. 36.37	1.15. 29.4	28.10. 1.39	9.14. 15.7	6.35.4	
8		35.10. 19.14	19.14	35.8. 2.14				PhC		2.18. 37	24.35	7.2.35	34.28. 35.40	9.14. 17.15		35.34. 38
9	2.28. 13.38		13.14. 8		29.30. 34		7.29. 34		PhC	13.28. 15.19	6.18. 38.40	35.15. 18.34	28.33. 1.18	8.3. 26.14	3.19. 35.5	
10	8.1. 37.18	18.13. 1.28	17.19. 9.36	28.1	19.10. 15	1.18. 36.37	15.9. 12.37	2.36. 18.37	13.28. 15.12	PhC	18.21. 11	10.35. 40.34	35.10. 21	35.10. 14.27	19.2	
11	10.36. 37.40	13.29. 10.18	35.10. 36	35.1. 14.16	10.15. 36.28	10.15. 36.37	6.35. 10	35.34	6.35. 36	36.35. 21	PhC	35.4. 15.10	35.33. 2.40	9.18. 3.40	19.3. 27	
12	8.10. 29.40	15.10. 26.3	29.34. 5.4	13.14. 10.7	5.34. 4.10		14.4. 15.22	7.2. 35	35.15. 34.18	35.10. 37.40	34.15. 10.14	PhC	33.1. 18.4	30.14. 10.40	14.26. 9.25	
13	21.35. 2.39	26.39. 1.40	13.15. 1.28	37	2.11. 13	39	28.10. 19.39	34.28. 35.40	33.15. 28.18	10.35. 21.16	2.35. 40	22.1. 18.4	PhC	17.9. 15	13.27. 10.35	39.3. 35.23
14	1.8. 40.15	40.26. 27.1	1.15. 8.35	15.14. 28.26	3.34. 40.29	9.40. 28	10.15. 14.7	9.14. 17.15	8.13. 26.14	10.18. 3.14	10.3. 18.40	10.30. 35.40	13.17. 35	PhC	27.3. 26	
15	19.5. 34.31		2.19. 9		3.17. 19		10.2. 19.30		3.35.5	19.2. 16	19.3. 27	14.26. 28.25	13.3. 35	27.3. 10	PhC	
16		6.27. 19.16		1.40. 35				35.34. 38					39.3. 35.23			PhC
17	36.22. 6.38	22.35. 32	15.19. 9	15.19. 9	3.35. 39.18	35.38	34.39. 40.18	35.6. 4	2.28. 36.30	35.10. 3.21	35.39. 19.2	14.22. 19.32	1.35. 32	10.30. 22.40	19.13. 39	19.18. 36.40
18	19.1. 32	2.35. 32	19.32. 16		19.32. 26		2.13. 10		10.13. 19	26.19. 6		32.30	32.3. 27	35.19	2.19.6	
19	12.18. 28.31		12.28		15.19. 25		35.13. 18		8.15. 35	16.26. 21.2	23.14. 25	12.2. 29	19.13. 17.24	5.19. 9.35	28.35. 6.18	
20		19.9. 6.27								36.37			27.4. 29.18	35		
21	8.36. 38.31	19.26. 17.27	1.10. 35.37		19.38	17.32. 13.38	35.6. 38	30.6. 25	15.35. 2	26.2. 36.35	22.10. 35	29.14. 2.40	35.32. 15.31	26.10. 28	19.35. 10.38	16
22	15.6. 19.28	19.6. 18.9	7.2.6. 13	6.38. 7	15.26. 17.30	17.7. 30.18	7.18. 23	7	16.35. 38	36.38			14.2. 39.6			
23	35.6. 23.40	35.6. 22.32	14.29. 10.39	10.28. 24	35.2. 10.31	10.18. 39.31	1.29. 30.36	3.39. 18.31	10.13. 28.38	14.15. 18.40	3.36. 37.10	29.35. 3.5	2.14. 30.40	35.28. 31.40	28.27. 3.18	27.16. 18.38
24	10.24. 35	10.35. 5	1.26	26	30.26	30.16		2.22	26.32						10	10
25	10.20. 37.35	10.20. 26.5	15.2. 29	30.24. 14.5	26.4. 5.16	10.35. 17.4	2.5. 34.10	35.16. 32.18		10.37. 36.5	37.36. 4	4.10. 34.17	35.3. 22.5	29.3. 28.18	20.10. 28.18	28.20. 10.16
26	35.6. 18.31	27.26. 18.35	29.14. 35.18		15.14. 29	2.18. 40.4	15.20. 29		35.29. 34.28	35.14. 3	10.36. 14.3	35.14	15.2. 17.40	14.35. 34.10	3.35. 10.40	3.35. 31
27	3.8. 10.40	3.10. 8.28	15.9. 14.4	15.29. 28.11	17.10. 14.16	32.35. 40.4	3.10. 14.24	2.35. 24	21.35. 11.28	8.28. 10.3	10.24. 35.19	35.1. 16.11		11.28	2.35. 3.25	34.27. 6.40
28	32.35. 26.28	28.35. 25.26	28.26. 5.16	32.28. 3.16	26.28. 32.3	26.28. 32.3	32.13. 6		28.13. 32.24	32.2	6.28. 32	6.28. 32	32.35. 13	28.6. 32	28.6. 32	10.26. 24
29	28.32. 13.18	28.35. 27.9	10.28. 29.37	2.32. 10	28.33. 29.32	2.29. 18.36	32.28. 2	25.10. 35	10.28. 32	28.19. 34.36	3.35	32.30. 40	30.18	3.27	3.27. 40	
30	22.21. 27.39	2.22. 13.24	17.1. 39.4	1.18	22.1. 33.28	27.2. 39.35	22.23. 37.35	34.39. 19.27	21.22. 35.28	13.35. 39.18	22.2. 37	22.1. 3.35	35.24. 30.18	18.35. 37.1	22.15. 33.28	17.1. 40.33
31	19.22. 15.39	35.22. 1.39	17.15. 16.22		17.2. 18.39	22.1. 40	17.2. 40	30.18. 35.4	35.28. 3.23	35.28. 1.40	2.33. 27.18	35.1	35.40. 27.39	15.35. 22.2	15.22. 33.31	21.39. 16.22
32	28.29. 15.16	1.27. 36.13	1.29. 13.17	15.17. 27	13.1. 26.12	16.4	13.29. 1.40	35	35.13. 8.1	35.12	35.19. 1.37	1.28. 13.27	11.13. 1	1.3. 10.32	27.1. 4	35.16
33	25.2. 13.15	6.13. 1.25	1.17. 13.12		1.17. 13.16	18.16. 15.39	1.16. 35.15	4.18. 31.39	18.13. 34	28.13. 35	2.32. 12	15.34. 29.28	32.35. 30	32.40. 3.28	29.3. 8.25	1.16. 25
34	2.27. 35.11	2.27. 35.11	1.28. 10.25	3.18. 31	15.32. 13	16.25	25.2. 35.11	1	34.9	1.11. 10	13	1.13. 2.4	2.35	1.11. 2.9	11.29. 28.27	1
35	1.6. 15.8	19.15. 29.16	35.1. 29.2	1.35. 16	35.30. 29.7	15.16	15.35. 29		35.10. 14	15.17. 20	35.16	15.37. 1.8	35.30. 14	35.3. 32.6	13.1. 35	2.16
36	26.30. 34.36	2.26. 35.39	1.19. 26.24	26	14.1. 13.16	6.36	34.26. 6	1.16	34.10. 28	26.16	19.1. 35	29.13. 28.15	2.22. 17.19	2.13. 28	10.4. 28.15	
37	27.26. 28.13	6.13. 28.1	16.17. 26.24	26	2.13. 18.17	2.39. 30.16	29.1. 4.16	2.18. 26.31	3.4. 16.35	36.28. 40.19	35.36. 37.32	27.13. 1.39	11.22. 39.30	27.3. 15.28	19.29. 25.39	25.34. 6.35
38	28.26. 18.35	28.26. 35.10	14.13. 28.17	23	17.14. 13		35.13. 16		28.10	2.35	13.35	15.32. 1.13	18.1	25.13	6.9	
39	35.26. 24.37	28.27. 15.3	18.4. 28.38	30.7. 14.26	10.26. 34.31	10.35. 17.7	2.6. 34.10	35.37. 10.2		28.15. 10.36	10.37. 14	14.10. 34.40	35.3. 22.39	29.28. 10.18	35.10. 2.18	20.10. 16.38
40	31.1.8	27	28	7			15	8		3	10.34					
1956-1999	1	2	3	4	5	6	7	8	9	10	11	12	13	14	15	16

17	18	19	20	21	22	23	24	25	26	27	28	29	30	31	32	33
6.29 / 4.38	19.1 / 32	35.12 / 34.31		12.36 / 18.31	6.2 / 34.19	5.35 / 3.31	10.24 / 35	10.35 / 20.28	3.26 / 18.31	3.11 / 1.27	28.27 / 35.26	28.35 / 26.18	22.21 / 18.27	22.35 / 31.39	27.28 / 1.36	35.3 / 2.24
28.19 / 32.22	35.19 / 32		18.19 / 28.1	15.19 / 18.22	18.19 / 28.15	5.8 / 13.30	10.15 / 35	10.20 / 35.26	19.6 / 18.26	10.28 / 8.3	18.26 / 28	10.1 / 35.17	2.19 / 22.37	35.22 / 1.39	28.1.9	6.13 / 1.32
10.15 / 19	32	8.35 / 24		1.35	7.2 / 35.39	4.29 / 23.10	1.24	15.2 / 29	29.35	10.14 / 29.40	28.32 / 4	10.28 / 29.37	1.15 / 17.24	17.15	1.29 / 17	15.29 / 35.4
3.35 / 38.18	3.25			12.8	6.28	10.28 / 24.35	24.26	30.29 / 14		15.29 / 28	32.28 / 3	2.32 / 10	1.18		15.17 / 27	2.25
2.15 / 16	15.32 / 19.13	19.32		19.10 / 32.18	15.17 / 30.26	10.35 / 2.39	30.26	26.4	29.30 / 6.13	29.9	26.28 / 32.3	2.32	22.33 / 28.1	17.2 / 18.39	13.1 / 26.24	15.17 / 1316
35.39 / 38				17.32	17.7 / 30	10.14 / 18.39	30.16	10.35 / 4.18	2.18 / 40.4	32.35 / 40.4	26.28 / 32.3	2.29 / 18.36	27.2 / 39.35	22.1 / 40	40.16	16.4
34.39 / 10.18	10.13 / 2	35		35.6 / 13.18	7.15 / 13.16	36.39 / 34.10	2.22	2.6 / 34.10	29.30 / 7	14.1 / 40.11	25.26 / 28	25.28 / 2.16	22.21 / 27.35	17.2 / 10	29.1 / 40	15.13 / 30.12
35.6.4				30.6		10.39 / 35.34		35.16 / 32.18	35.3	2.35 / 16		35.10 / 25	34.39 / 19.27	30.18 / 35.4	35	
28.30 / 36.2	10.13 / 19	8.15 / 35.38		19.35 / 38.2	14.20 / 19.35	10.13 / 28.38	13.26		10.19 / 29.38	11.35 / 27.28	28.32 / 1.24	10.28 / 32.25	1.28 / 35.23	2.24 / 32.21	35.13 / 8.1	32.28 / 13.12
35.10 / 21		19.17 / 10	1.16 / 36.37	19.35 / 18.37	14.15	8.35 / 40.5		10.37 / 36	14.29 / 18.36	3.35 / 13.21	35.10 / 23.24	28.29 / 37.36	1.35	13.3 / 36.24	15.37 / 18.1	1.28 / 3.25
35.39 / 19.2		14.24 / 10.37		10.35 / 14	2.36 / 25	10.36 / 37		37.36 / 4	10.14 / 36	10.13 / 19.35	6.28 / 25	3.35	22.2 / 37	2.33 / 27.18	1.35 / 16	11
22.14 / 19.32	13.15 / 32	2.6 / 34.14		4.6.2	14	35.29 / 3.5		14.10 / 34.17	36.22	10.40 / 16	28.32 / 1	32.30 / 40	22.1 / 2.35	35.1	1.32 / 17.28	32.15 / 26
35.1 / 32	32.3 / 27.15	13.19	27.4 / 29.18	32.35 / 27.31	14.2 / 39.6	2.14 / 30.40		35.27	15.32 / 35		13	18	35.23 / 18.30	35.40 / 27.39	35.19	32.35 / 30
30.10 / 40	35.19	19.35 / 10	35	10.26 / 35.28	35	35.28 / 31.40		29.3 / 28.10	29.10 / 27	11.3	3.27 / 16	3.27	18.35 / 37.1	15.35 / 22.2	11.3 / 10.32	32.40 / 28.2
19.35 / 39	2.19 / 4.35	28.6 / 35.18	35	19.10 / 35.38		28.27 / 3.18	10	20.10 / 28.18	3.35 / 10.40	11.2 / 13	3	3.27 / 16.40	22.15 / 33.28	21.39 / 16.22	27.1.4	12.27
19.18 / 36.40				16		27.16 / 18.38	10	28.20 / 10.16	3.35 / 31	34.27 / 6.40	10.26 / 24		17.1 / 40.33	22	35.10	1
PhC	32.30 / 21.16	19.15 / 3.17		2.14 / 17.25	21.17 / 35.38	21.36 / 29.31		35.28 / 21.18	3.17 / 30.39	19.35 / 3.10	32.19 / 24	24	22.33 / 35.2	22.35 / 2.24	26.27	26.27
32.35 / 19	PhC	32.1 / 19	32.35 / 1.15	32	19.16 / 1.6	13.1	1.6	19.1 / 26.17	1.19		11.15 / 32	3.32	15.19	35.19 / 32.39	19.35 / 28.26	28.26 / 19
19.24 / 3.14	2.15 / 19	PhC		6.19 / 37.18	12.22 / 15.24	35.24 / 18.5		35.38 / 19.18	34.23 / 16.18	19.21 / 11.27	3.1.32		1.35 / 6.27	2.35.6 / 30	28.26 / 30	19.35
	19.2 / 35.32	19.2	PhC			28.27 / 18.31			3.35 / 31	10.36 / 23			10.2 / 22.37	19.22 / 18	1.4	
2.14 / 17.25	16.6 / 19	16.6 / 19.37		PhC	10.35 / 38	28.27 / 18.38	10.19	35.20 / 10.6	4.34 / 19	19.24 / 26.31	32.15 / 2	32.2	19.22 / 31.2	2.35 / 18	26.10 / 34	26.35 / 10
19.38 / 7	1.13 / 32.15	1.13		3.38	PhC	35.27 / 2.37	19.10	10.18 / 32.7	7.18 / 25	11.10 / 35	32		21.22 / 35.2	21.35 / 2.22		35.32 / 1
21.36 / 39.31	1.6.13	35.18 / 24.5	28.27 / 12.31	28.27 / 18.38	35.27 / 2.31	PhC		15.18 / 35.10	6.3 / 10.24	10.29 / 39.35	16.34 / 31.28	35.10 / 24.31	33.22 / 30.10	10.1 / 15.34	15.34 / 33	32.28 / 2.24
	19			10.19	19.10		PhC	24.26 / 28.32	24.28 / 35	10.28 / 23			22.10 / 1	10.21 / 22	32	27.22
35.29 / 21.18	1.19 / 21.17	35.38 / 19.18	1	35.20 / 10.6	10.5 / 18.32	35.18 / 10.39	24.26 / 28.32	PhC	35.38 / 18.16	10.30 / 4	24.34 / 28.32	24.26 / 28.18	35.18 / 34	35.22 / 18.39	35.28 / 34.4	4.28 / 10.34
3.17 / 39		34.29 / 16.18	3.35 / 31	35	7.18 / 25	6.3 / 10.24	24.28 / 35	35.38 / 18316	PhC	18.3 / 28.40	3.2.28	33.30	35.33 / 29.31	3.35 / 40.39	29.1 / 35.27	35.29 / 10.25
3.35 / 10	11.32 / 13	21.11 / 27.19	36.23	21.11 / 26.31	10.11 / 35	10.35 / 29.39	10.28	10.30 / 4	21.28 / 40.3	PhC	32.3 / 11.23	11.32 / 1	27.35 / 2.40	35.2 / 40.26		27.17 / 40
6.19 / 28.24	6.1.32	3.6.32		3.6.32	26.32 / 27	10.16 / 31.28		24.34 / 28.32	2.6.32	5.11 / 1.23	PhC		28.24 / 22.26	3.33 / 39.10	6.35 / 25.18	1.13 / 17.34
19.26	3.32	32.2		32.2	13.32 / 2	35.31 / 10.24		32.26 / 28.18	32.30	11.32 / 1		PhC	26.28 / 10.36	4.17 / 34.26		1.32 / 35.23
22.33 / 35.2	1.19 / 32.13	1.24 / 6.27	10.2 / 22.37	19.22 / 31.2	21.22 / 35.2	33.22 / 19.40	22.10 / 2	35.18 / 34	35.33 / 29.31	27.24 / 2.40	28.33 / 23.26	26.28 / 10.18	PhC		24.35 / 2	2.25 / 28.39
22.35 / 2.24	19.24 / 39.32	2.35.6	19.22 / 18	2.35 / 18	21.35 / 22.2	10.1 / 34	10.21 / 29	1.22	3.24 / 39.1	24.2 / 40.39	3.33 / 26	4.17 / 34.26		PhC		
27.26 / 18	28.24 / 27.1	28.26 / 27.1	1.4	27.1 / 12.24	19.35	15.34 / 33	32.24 / 18.16	35.28 / 34.4	35.23 / 1.24		1.35 / 12.18		24.2		PhC	2.5 / 13.16
26327.13	13.17 / 1.24	1.13 / 24		35.34 / 2.10	2.19 / 13	28.32 / 2.24	4.10 / 27.22	4.28 / 10.34	12.35	17.27 / 8.40	25.13 / 2.34	1.32 / 35.23	2.25 / 28.39		2.5.12	PhC
4.10	15.1 / 13	15.1 / 28.16		15.10 / 32.2	15.1 / 32.19	2.35 / 34.27		32.1 / 10.25	2.28 / 10.25	11.10 / 1.16	10.2 / 13	25.10	35. / 102.16		1.35 / 11.10	1.12 / 26.15
27.2 / 3.35	6.22 / 26.1	19.35 / 29.13		19.1 / 29	18.15 / 1	15.10 / 2.13		35.28	3.35 / 15	35.13 / 8.24	35.5 / 1.10		35.11 / 32.31		1.13 / 31	15.34 / 1.16
2.17 / 13	24.17 / 29.28	27.2 / 29.28		20.19 / 30.34	10.35 / 13.2	35.10 / 28.29		6.29	13.3 / 27.10	13.35 / 1	2.26 / 10.34	26.24 / 32	22.19 / 29.40	19.1	27.26 / 1.13	27.9 / 26.24
3.27 / 35.16	2.24 / 26	35.38	19.35 / 16	19.1 / 16.10	35.3 / 15.19	1.18 / 10.24	35.33 / 27.22	18.28 / 32.9	3.27 / 29.18	27.40 / 28.8	26.24 / 32.28		22.19 / 29.28	2.21	5.28 / 11.29	2.5
26.2 / 19	8.32 / 19	2.32 / 13		28.2 / 27	23.28	35.10 / 18.5	35.33	24.28 / 35.30	35.13	11.27 / 32	28.26 / 10.34	28.26 / 18.23	2.33	2	1.26 / 13	1.12 / 34.3
35.21 / 28.10	26.17 / 19.1	35.10 / 38.19	1	35.20 / 10	28.10 / 29.35	28.10 / 35.23	13.15 / 23		35.38	1.35 / 10.38	1.10 / 34.28	1.32 / 18.10	22.35 / 13.24	35.22 / 18.39	35.28 / 2.24	1.28 / 7.19
36	32						32	28.19	22	11.3 / 35	22		22	2.3 / 35.22		35.9 / 28
17	18	19	20	21	22	23	24	25	26	27	28	29	30	31	32	33

34	35	36	37	38	39	40	GSA SDS
2. 27. 28. 11	29. 5. 15. 8	26. 30. 36. 34	28. 29. 26. 32	26. 35. 18. 19	35. 3. 24. 37		1
2. 27. 28. 11	19. 15. 29	1. 10. 26. 39	25. 28. 17. 15	2. 26. 35	1. 28. 15. 35	31	2
1. 28. 10	14. 15. 1. 16	1. 19. 26. 24	35. 1. 26. 24	17. 24. 26. 16	14. 4. 28. 29	28	3
3	1. 35	1. 26	26		30. 14. 7. 26	7	4
15. 13. 10. 1	15. 30	14. 1. 13	2. 36. 26. 18	14. 30. 28. 23	10. 26. 34. 2		5
16	15. 16	1. 18. 36	2. 35. 30. 18	23	10. 156. 17. 7		6
10	15. 29	26. 1	29. 26. 4	35. 34. 16. 24	10. 6. 2. 34		7
1		1. 31	2. 17. 26		35. 37. 10. 2		8
34. 2. 28. 27	15. 10. 26	10. 28. 4. 34	3. 34. 27. 16	10. 18		15. 21	9
15. 1. 11	15. 17. 18. 20	26. 35. 10. 18	36. 37. 10. 19	2. 35	3. 28. 35. 37	3. 17	10
2	35	19. 1. 35	2. 36. 37	35. 24	10. 14. 35. 37		11
2. 13. 1	1. 15. 29	16. 29. 1. 28	15. 13. 39	15. 1. 32	17. 26. 34. 10	7	12
2. 35. 10. 16	35. 30. 34. 2	2. 35. 22. 26	35. 22. 39. 23	1. 8. 35	23. 35. 40. 3		13
27. 11. 3	15. 3. 32	2. 13. 28	27. 3. 15. 40	15	29. 35. 10. 14	40	14
29. 10. 27	1. 35. 13	10. 4. 29. 15	19. 29. 39. 35	6. 10	35. 17. 14. 19		15
1	2		25. 34. 6. 35	1	20. 10. 16. 38		16
4. 10. 16	2. 18. 27	2. 17. 16	3. 27. 35. 31	23. 2. 19. 16	15. 28. 35		17
15. 17. 13. 16	15. 1. 19	6. 32. 13	32. 15	2. 26. 10	2. 25. 16	32	18
1. 15. 17. 28	15. 17. 13. 16	2. 29. 27. 28	35. 38	32. 2	12. 28. 35		19
			19. 35. 16. 25		1. 6		20
35. 2. 10. 34	19. 17. 34	20. 19. 30. 34	19. 35. 16	28. 2. 17	28. 35. 34		21
2. 19		7. 23	35. 3. 15. 23	2	28. 10. 29. 35		22
2. 35. 34. 27	15. 10. 2	35. 10. 28. 24	35. 18. 10. 13	35. 10. 18	28. 35. 10. 23		23
			35. 33	35	13. 23. 15	32	24
32. 1. 10	35. 28	6. 29	18. 28. 32. 10	24. 28. 35. 30		19	25
2. 32. 10. 25	15. 3. 29	3. 13. 27. 10	3. 27. 29. 18	8. 35	13. 29. 3. 27	22	26
1. 11	13. 35. 8. 24	13. 35. 1	27. 40. 28	11. 13. 27	1. 35. 29. 38	11	27
1. 32. 13. 11	13. 35. 2	27. 35. 10. 34	26. 24. 32. 28	28. 2. 10. 34	10. 34. 28. 32	22	28
25. 10		26. 2. 18		26. 28. 18. 23	10. 18. 32. 39		29
35. 10. 2	35. 11. 22. 31	22. 19. 29. 40	22. 19. 29. 40	33. 3. 34	22. 35. 13. 24	22	30
		19. 1. 31	2. 21. 27. 1	2	22. 35. 18. 39	3. 22	31
35. 1. 11. 9	2. 13. 15	27. 26. 1	6. 28. 11. 1	8. 28. 1	35. 1. 10. 28		32
12. 26. 1. 32	15. 34. 1. 16	32. 25. 12. 17		1. 34. 12. 3	15. 1. 28	11. 28. 9	33
PhC	7. 1. 4. 16	35. 1. 13. 11		34. 35. 7. 13	1. 32. 10	13	34
1. 16. 7. 4	PhC	15. 29. 37. 28	1	27. 34. 35	35. 28. 6. 37		35
1. 13	29. 15. 28. 37	PhC	15. 10. 37. 28	15. 1. 24	12. 17. 28		36
12. 26	1. 15	15. 10. 37. 28	PhC	34. 21	35. 18		37
1. 35. 13	27. 4. 1. 35	15. 24. 10	34. 27. 25	PhC	5. 12. 35. 26		38
1. 32. 10. 25	1. 35. 28. 37	12. 17. 28. 24	35. 18. 27. 2	5. 12. 35. 26	PhC	10. 14. 25. 35. 38	39
13	15		28		14. 15. 28. 35	PhC	40
34	35	36	37	38	39	40	Right - Up +

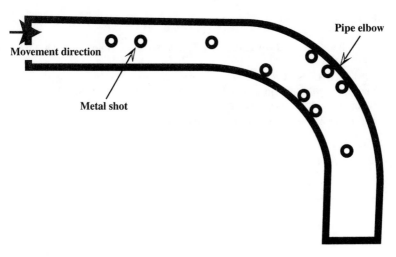

FIGURE 13.2 Initial situation.

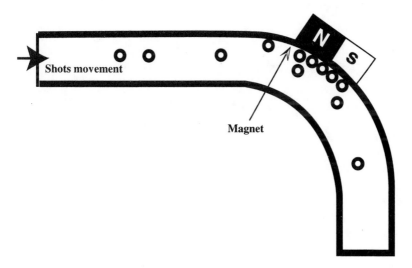

FIGURE 13.3 Possible solution.

The Contradiction Matrix and the Principles (as well as other TRIZ heuristics and instruments) are like musical instruments — they do not work by themselves; they only suggest the most promising directions for you to search for a solution. The problem solver has to interpret these suggestions and determine how they apply to the particular situation. Nevertheless, TRIZ heuristics are very valuable because a problem solver does not have to study all patents from all disciplines if he or she can reformulate the problem in the format of the Contradiction Matrix, thereby increasing the efficiency of solving.

The Matrix is a part of modern ARIZ and could be used in combination with other TRIZ heuristics presented in the next chapters.

REFERENCES

1. Altshuller, G. S., *40 Principles: TRIZ Keys to Technical Innovation*, TIC, Worchester, 1998.
2. Altshuller, G. S., *Creativity as an Exact Science: The Theory of the Solution of Inventive Problems,* Gordon and Breach Science Publishing, New York, 1984.
3. Sklobovsky, K. A. and Sharipov, R. H., Eds., *Theory, Practice and Applications of the Inventive Problems Decision*, Protva-Prin, Obninsk, 1995 (in Russian).

14 Physical Point Contradictions: Ontology and Resolution

14.1 INTRODUCTION

The term "physical contradiction" is common in TRIZ; it came from the original works of Genrich S. Altshuller [1]. A physical contradiction is neither more nor less scientific than the technical contradictions described in the previous chapter. Both contradictions can appear in any technique. The mutually exclusive requirements are demanded from the same subsystem (function, characteristic or property, parameter, etc.) of the technical system or technological process in the case of physical contradictions. That is why a physical contradiction represents the *point* problem in the so-called *key* subsystem (function, characteristic, etc.). Often it is easier to operate with the opposite characteristics instead of working with the mutually exclusive requirements or two unequal values of a parameter for this characteristic. For example, an electron emitter should have a needle-like shape to emit a large electrical current in field emission flat panel displays, but with such a shape the emitter cannot maintain the electrical load and burns down. We can formulate the physical contradiction as follows: The edge of an electron emitter should be thick in order not to burn down and should be sharp to emit a large electrical current. In this case, the key subsystem is the edge of an electron emitter and the opposite parameters are sharp vs. thick. Such mutually exclusive or opposite requirements are often labeled as "positive" and "negative" characteristics.

The transition from a technical, pair contradiction to a physical, point one is often possible due to the identification of the key subsystem or characteristic of the desired result that causes the negative impact. This key subsystem **A** (or its characteristic) that should have the opposite parameters ± **B** becomes the object of the point contradiction. The last can be formulated as

A has/should be (+B) AND A has/should be (–B)

or

(+A) acts as B AND (–A) acts as B

The technical contradiction is that some subsystem "Y" should be hot to work correctly but the heating causes degradation of the neighbor subsystem "X." The key parameter A here is temperature. Therefore, the physical point contradiction is that the temperature is high and it is low. Both contradictions are identical: A high temperature improves "Y" but degrades "X," while a low temperature does not improve "Y" and does not degrade "X." Sometimes it is easier to formulate a physical contradiction than a technical one. For example, a flow of reagents should be high to increase the rate of a chemical reaction and low to avoid waste of energy and destruction of submicron features of IC chips.

14.2   OPERATIVE ZONE AND PERIOD

TRIZ assumes that TS and TP exist in time and space without Einstein's relativistic effects (which is a good approximation for the majority of TS and TP). Therefore, the two peculiarities of contradiction — spatial and temporal — are considered in TRIZ. A contradiction exists in the operative zone and during the operative period that shows exactly where the contradiction arises, and exactly when.

Operative period T consists of the interval of contradiction T1 itself, the term prior to the contradiction T2 and the interim after the contradiction T3. A value of any Ti is not predetermined in TRIZ. In the general case, the sum T = T1 + T2 + T3 represents the available time resources.

Operative zone Z consists of the volume Z1 of the key subsystem in which contradiction occurs and the available region Z2 around this area Z1 (in the key subsystem, in other subsystems, in environment). In the general case, Z = Z1 + Z2 represents the available space resources.

In order to decide whether T and Z (often also called conflict time and conflict zone) are selected correctly, we need to point to a tool and to a product. Both operative zone and operative period ideas are applicable to a tool (**Zt** and **Tt**) and to a product (**Zp** and **Tp**), as well as to positive (+B) or negative (–B) requirements (**Z+, Z–** and **T+, T–**) and to the key subsystem. As a rule, PF or UF are sources of the positive requirements, while HF are sources of the negative requirements.

TRIZ assumes that the operative zone and period show when and where a working tool of a technique interacts with a product. Otherwise, it is the selection error of the product/tool pair or one or both of the operative's peculiarities. The tool and product can often interact directly or via some field within the operative zone Z1 during the operative period T1. Identification of the operative's peculiarities T and Z can help distinguish the kind of contradiction in a problem and therefore lead

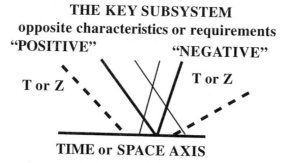

THE KEY SUBSYSTEM
opposite characteristics or requirements
"POSITIVE"   "NEGATIVE"

T or Z   T or Z

TIME or SPACE AXIS

FIGURE 14.1 Illustration of possible "interactions" of operative times and operative zones: separate (dotted line), touching (thick solid line), and intersecting (thin solid line).

to choosing an appropriate TRIZ heuristic for a search for the problem's solution. The key subsystem, as a rule, is a tool because we usually cannot change requirements for a product. Two opposite physical, geometric, or functional characteristics are usually requested from a tool in the point problem (see subsection 14.2.1).

Therefore, TRIZ considers only a few possibilities for operative periods and zones of the key subsystem's opposite parameters to be *separate, touching,* or *intersecting.* Figure 14.1 illustrates schematically three variants of "interactions" between operative zones and operative periods, shown for simplicity as straight lines. Any point at the horizontal axis represents a unique moment of time or a unique space element (e.g., point), so the exact direction of the horizontal axis is not important although it should be the same for the positive and negative parameter.

The one-dimensional illustration is sufficient here (and actually is used almost always in TRIZ analysis) because it is assumed that no operative zone intersects with any time (i.e., classical Euclid-Newton geometry works for a technique). Hence, there is no vertical axis in Figure 14.1. The two-dimensional picture is used in Figure 14.1 for visual ease. Representation of the separate, touching, and intersecting cases is by the simple symbols \/, **V**, and **X**, existing on any computer keyboard.

A total of six cases exists:

Three cases for operative periods

$X\tau$ — the properties +B and –B exist at the same time interval (or even strongly for some problems: always +B and always –B);

$V\tau$ — +B till moment $\tau \leq \tau 0$, then –B immediately from $\tau \geq \tau 0$;

$\backslash/\tau$ — +B till the moment $\tau 1$, then –B from the moment $\tau 2$ that is later than $\tau 1$.

Three cases for operative zones

Xz — the properties +B and –B exist in the same space segment (or even strongly for some problems: everywhere +B and everywhere –B);

Vz — +B till the coordinate $\zeta \leq \zeta 0$ (before $\zeta 0$) and –B at the point $\zeta \geq \zeta 0$ (after $\zeta 0$) (note that in this case the dimension of the boundary between Z+ and Z- is not higher than the lowest dimension of these zones);

$\backslash/z$ — +B till the coordinate $\zeta 1$, then –B from the coordinate $\zeta 2$ that is higher or smaller than $\zeta 1$, $\zeta 1 \neq \zeta 2$.

Unfortunately, some technical problems are rather difficult to illustrate in the way seen in Figure 14.1; hence, practice is recommended for beginners in TRIZ. According to Nikolay N. Khomenko and Simon S. Litvin, these six possibilities lead to different methods of resolution of the physical contradictions, which will be considered later in this chapter.

14.2.1 SHORT LIST OF OPPOSITE CHARACTERISTICS

As a rule, two opposite physical, geometric, or functional characteristics are requested from the tool in the formation of a physical contradiction. Any characteristic can have a wide spectrum of parameter values. The TRIZ roster of technical fields and states of matter have a wider classification compared to natural sciences such as physics or chemistry, where only "fundamental" fields and states are considered; however, the TRIZ roster of shapes is the same as what we know from geometry. For example, the basic (gravity, electrical, etc.) and composite (chemical, acoustic, etc.) fields are considered in TRIZ equally, as are substances which can exist in fundamental (solid, gas) and mixed (suspension, viscous gel, smog, etc.) states.

The following brief list shows some characteristics reflecting various possible physical contradictions.

Geometric	Material and Energetic	Functional
Long vs. Short	(Big vs. Small)	Throw vs. Catch
Symmetrical vs. Asymmetrical	Density	Pull vs. Push
Parallel vs. Intersecting	Conductivity	Hot vs. Cold
Thin vs. Thick	Temperature	Fast vs. Slow
Round vs. Non-Circle	Time	Move vs. Stay
Sharp vs. Dull	Viscosity	Strong vs. Weak
Narrow vs. Wide	Power	Soft vs. Hard
Horizontal vs. Vertical	Friction	Cheap vs. Expensive
etc.	etc.	Use Antonyms Dictionary

You can expand the list and add other characteristics that are important for your technical system or technological process.

14.3 TYPES OF PHYSICAL CONTRADICTIONS

Any physical contradiction corresponds to a situation when opposite requirements are applied to the same subsystem A. The key subsystem A can be any substance or field as these conceptions are defined in TRIZ.

Based on the results described in the previous sections, we can formulate the physical (point) contradiction in the following form:

> The key subsystem (*name*) should be or has (*"positive" parameter*),
> in order to (*the first requirement for the tool*),

AND

> the key subsystem (*name*) should not be or not have (*"negative" parameter*),
> in order to (*the second requirement for the tool*).

The italicized words should be replaced with terms relevant to the problem.

This generic formulation includes many types of physical contradictions. To the best of the author's knowledge, three classifications for these types coexist.

A. The author constructed the following variants in 1982:

1. The key subsystem (or its parameter) A must exist and A must not exist.
2a. A has to have the characteristic B and the characteristic –B (the opposite characteristics). For example, B is superconductivity and –B is the dielectric resistivity; or heavy/light. (The list of opposite characteristics is given in the previous section.)
2ai. The field-substance interaction must be strong and must be weak.
2b. A must be in a phase state C and in another state C'; e.g., C is plasma and C' is solid; or C is gas and C' is liquid.
3a. A must be at a time period D and must not be at time period E (if E = D, see type 1).
3b. A must be constant (i.e., time independent) and A must change in time.

B. V. N. Glazunov argued in 1990 that the set of key subsystems with two nonequal parameter values for the opposite characteristic consists of the following elements [2]:

1. A parameter of a key subsystem has a spatially or temporally distributed value.
2. Two spatially or temporally related subsystems of a system have two values of the same parameters in the same and/or different space elements or at the same and/or different time (for example, different phases of matter).
3. A key subsystem has the same parameter as one of its elements or as another subsystem of the system or whole system.

C. J. Terninko with coworkers showed in 1998 that the physical contradictions can be separated into three groups in terms of harmful or desired effects [3]. They can be reformulated in terms of UF and HF as the following:

1. Performing the key function is necessary to achieve useful functions (UF), and not performing this function is necessary to avoid harmful functions (HF). For example, heat is applied to an integrated circuit chip pin that will be attached to a circuit board, but heat must not be applied because it damages the integrated circuit chip.
2. The characteristic of the key subsystem must be of one value (big, infinite) to achieve one UF and must be the opposite (small, zero) one to avoid HF or to achieve another UF. For example, the metal lines in an integrated

circuit chip should be wide for good electrical conductivity and should be narrow for miniaturization of the integrated circuit chip.

3. The key subsystem must be present to achieve one UF and must be absent to avoid HF or to achieve another UF. For example, a relatively thick wafer is necessary during integrated circuit chip production (mechanical properties are better) but undesirable during integrated circuit chip operation (thermal and and some electrical propertie are worse).

It is easy to see that these three classifications (A, B, and C) are mutually supplementary.

14.4   HEURISTICS FOR RESOLUTION OF PHYSICAL CONTRADICTIONS

In the 1970s Altshuller proposed 11 methods, shown in the following table, to resolve physical contradictions.

Methods of Solving Physical Contradictions

Generic Operative	Examples
1. Separation of contradicting properties in space	In order to suppress dust during mining work, drops of water should be small, but small drops create fog. It was proposed to surround small drops with a cone of large drops.
	Change of resistivity due to a nonuniform diffusion and creation of $n - p$ or $n^+ - n$ or $p^- - p$ regions in semiconductor devices.
2. Separation of contradicting properties in time	The width of a tape electrode is changed, depending on the width of the weld.
	The conductivity of a channel in MOSFET transistor is modulated by the gate AC voltage.
3. The joining of homogenous and/or heterogeneous systems or elements into a super-system	Slabs on a roller conveyer are transported butt-jointed, so the ends won't cool.
	Growing a thin layer of epitaxial Si on the surface of a silicon wafer allows it to receive very high electrical quality in a mechanically strong material for a relatively short time.
4. A changeover from system to anti-system or to a combination of system and anti-system	To stop bleeding, a tissue, soaked in blood of another group, is applied.
	Electrons and holes as opposite charge carriers in bipolar semiconductor devices.
5. The whole system is endowed with the property "B," and its subsystems with the property "anti-B"	Working parts of vices for fixing work pieces of a complicated form: each part (steel bushes) is hard, but on the whole the clamp is pliable, capable of changing its shape.
	Glass for an image intensifier of a night vision tube is an insulator while porous-like micro-channels inside this glass are good conductors.
6. A transition to a system, operating on the micro-level	Instead of a mechanical crane, a "thermo-crane" made of two materials with different coefficients of thermal expansion. When they are heated, a gap appears.

Generic Operative	Examples
	Microwave oven (heating of water molecules inside a food) instead of heating by hot air of food's surface as in traditional gas or electric ovens.
7. An alteration of the phase state of a part of the system or the environment	Method of supplying compressed gas in mines — transportation of liquid that evaporates into this gas once in the mines.
	Molecular beam epitaxy for growing of multi-component semiconductor films in which the gaseous environment of the film changes in accordance with the requirements to the film chemical composition.
8. A "dual" phase state of part of the system (transition of this part from one state to another depending on the conditions of work)	Heat exchanger has "petals" made of titanium nickelide, which are tightly pressed against it; when the temperature rises, the "petals" bend back, increasing the cooling area.
	Electro-chromic glass with transparency dependent on illumination level.
14. Use of phenomena that accompany the phase transition	A device for transporting frozen loads has supporting elements in the form of ice bars (reduction of friction due to melting).
	Memory elements from chalcogenide glasses or vanadium oxide that utilize changes of electrical properties during the phase transition between amorphous and crystalline states of these materials.
10. Substitution of double-phase matter for single-phase	Method of polishing products: the working medium consists of liquid (lead melt) and ferromagnetic abrasive particles.
	A bubble chamber for studies in the physics of particles.
11. Physical-chemical transition: formation/disappearance of matter due to their decomposition/compounding, ionization/recombination, etc.	In order to plasticize wood with ammonia, the wood is impregnated with salts of ammonia which decompose from friction.
	Fuses in electrical equipment.
	To increase frequency parameters, recombination centers are introduced in silicon semiconductor devices by Au or Pt diffusion.

Notes:

1. USSR Patents 256708, 258490, 722624, 523695, 510350, 179479, 256222, 958837, 601192, 722740, 342761, and some results of the author's projects were used for illustration of solutions.
2. V. A. Korolyev argued that methods 6 and 11 are not independent and that method 3 can be divided into two entries for the uniform and nonuniform (in the limit anti-system) systems.

Some TRIZniks (Gregory I. Frenklach, Yury V. Gorin) have worked to expand this toolbox of heuristics. The most extensive studies have been performed by Vitaly N. Glazunov in the late 1980s. Glazunov [2] counted 30 "theoretically possible" approaches for eliminating physical contradictions for the set of different key sub-systems with two mutually exclusive values of some parameter. However, among all these approaches, only 14, according to the author, have an independent practical importance. Those 14 approaches are listed below.

1. Replace the key subsystem by two different subsystems, each of which has the value of the parameter presented in the formula of physical contradiction.
2. Replace the key subsystem with another more complex subsystem, different parts of which have different values of the parameter in the physical contradiction formula.
3. Replace the key subsystem by two identical subsystems, each of which has the value of the parameter presented in the physical contradiction formula.
4. Replace the key subsystem with another subsystem characterized by two parameters similar to the parameter that should have two nonequal values, each having one of two values of the parameter in the physical contradiction formula.
5. Change the conditions under which the key subsystem exists, so that its different parts have different values of the parameter in the physical contradiction formula.
6. Change the conditions under which the key subsystem exists, so that it is characterized by different values of the parameter in the physical contradiction formula at different stages of the technique's life.
7. Replace the key subsystem with another more dynamic (or in other words, less stable) subsystem, which is characterized by different values of the parameter in the physical contradiction formula at different stages of the technique's life.
8. Replace the key subsystem with another subsystem which experiences a conversion (for example, phase transformation) into a third subsystem, both characterized by a single unique value of the parameter in the physical contradiction formula.
9. Include the key subsystem into another large subsystem characterized by one value of the parameter in the physical contradiction formula, while the key subsystem itself is characterized by another value.
10. Replace the key subsystem with another subsystem which is characterized by the parameter similar to the initial value, but with such a value that can be considered "different" as it is necessary for the physical contradiction formula with respect to dissimilar external systems.
11. Change the conditions under which the key subsystem exists, so that it is transformed (for example, due to a phase transition) into another state that is characterized by the value of the parameter in the physical contradiction formula other than the value before transformation.
12. Change the conditions under which the key subsystem exists, so that one of its parts (e.g., an element or sub-subsystem) experiences transformations (for example, due to a phase transition) into another object characterized by one value of the parameter in the physical contradiction formula, while the remaining part of the key subsystem is characterized by another value.
13. Change the conditions under which the key subsystem exists, so that it is characterized by two different parameters similar to the key parameter, each having one of the values of the parameter in the physical contradiction formula.

14. Consider the key subsystem as a composite system characterized by one value of the parameter in the physical contradiction formula while one of its elements is characterized by another value.

V. N. Glazunov noted that approaches 3, 4, and 10 provide the simplest solution and the best results to a problem with a physical contradiction.

It was reasoned by V. A. Korolyev, G. I. Ivanov, S. D. Savransky, and some other TRIZniks that these methods for resolving physical contradictions are not independent of each other. On the other hand, some of the methods described above are equivalent to the Standards, which are discussed in the next chapter.

Modern TRIZ seeks to eliminate physical contradictions in the key subsystem by separating the mutually exclusive requirements

in space,
in time,
upon condition,
between parts and whole.

In the solving of numerous technical problems in various engineering fields, physical contradictions were resolved using these separation methods and/or internal resources. As we already know, internal resources, such as materials and energy sources, can be found inside the system that is under consideration. Sergey A. Faer argued that geometric or chemical effects should be useful for resolving physical contradictions if the key subsystem is a substance, while physical effects are more suitable when the key subsystem is a field. Sometimes a contradiction can be resolved by transition from a substance to a field, or vice versa.

Although the first two separations often appear because of selection errors, it is useful to consider these cases in order to reduce the solver's psychological inertia. The next section demonstrates how the separation's methods help resolve physical contradictions.

14.5 SEPARATION HEURISTICS

This section details, through several problems resolved by TRIZ experts, the four separation heuristics that are used for resolving physical contradictions. It also links Altshuller's and Glazunov's ideas with these separation heuristics. The level of detail in the examples' explanations vary; the author suggests that the reader analyze the examples without detailed explication because such an approach usually helps in learning how to handle physical contradictions and other TRIZ heuristics.

14.5.1 SEPARATION IN SPACE

If mutually exclusive requirements are demanded from the key subsystem, separation in space is possible when a requirement exists (or is made larger) in one place and is absent (or made smaller) in another. To apply this separation heuristic, answer the following question: Do we need this parameter to be "positive" and "negative" everywhere, or is there a place in space where it is not necessary? If such a place exists, it might be possible to separate the opposite requirements to the key subsystem in space.

Example
A submarine pulls sonar detectors to get information about the outside world in a dark sea. It drags the detectors at the end of several thousand feet of cable in order to separate the detector from the noise of the submarine. Thus, a submarine and its sonar are separated in space.

If parameters of the physical contradiction formula are inherent in the key subsystem in the same space, then application of Glazunov's approaches 1, 2, 5, and 12 fits Altshuller's method 1 for physical contradiction elimination "in space."

Example
Problem — Metallic parts are placed in a container with a metal salt suspension (nickel, cobalt, chromium in a liquid) for chemical coating of parts' surfaces. During the reduction reaction, metal from the suspension precipitates onto the parts' surface. Traditionally, the process is performed in hot suspension. The higher the temperature, the faster the process, but the suspension decomposes at high temperatures. As much as 75% of the chemicals settle on the bottom and walls of the container. Adding stabilizers is not effective and conducting the process at a low temperature sharply decreases productivity of this technological process.

Analysis — In this situation the metal salt suspension is the tool and the surface is the product. Requirements for the tool are to be hot and to be cold.

Contradiction — The process must be hot for fast, effective coating and cold to efficiently utilize the metallic salt solution.

Analysis (continued) — Both $T1t$ and $T1p$ correspond to the time when the metal salt suspension is hot, so $T1t$ and $T1p$ are "touching" or "intersecting" and the same is true for the requirements of the tool. Operative zones are different because $Z1t$ is volume away from the part surface and $Z1p$ is the parts surface area, so $Z1t$ and $Z1p$ are separated. The requirements for the tool also do not specify that it must have opposite temperatures in the whole volume, hence $Z+$ and $Z-$ are separated. The answer becomes apparent with a succinct rephrasing of this $\backslash/z$ problem — only the suspension's areas around the part must be hot.

Solution — The product is heated to a high temperature before it is immersed in a cool suspension. In this case, the suspension is hot where it is near the product, but cold elsewhere. To accomplish this, the parts themselves, rather than the suspension, may be heated.

Practical suggestion — In the nickel-plating of parts, increased temperature is necessary only in proximity to the parts. One way to keep the product hot is an electric current application for inductive heating of metallic parts during the coating process.

In summary, for a $\backslash/z$ point problem, a solver should

a. try to partition (actually or theoretically) the key subsystem into two or more subsystems, and then
b. assign each contradictory function or opposite requirement to a different subsystem.

Note: The available resources of a tool and/or a product or raw object in Z2 can often be used for separation in space.

Therefore, it is possible to separate the opposite requirements to the key subsystem in space when Z– and Z+ are *separated* from each other.

14.5.2. SEPARATION IN TIME

If mutually exclusive requirements are demanded from the key subsystem, the separation in time is possible when a requirement exists (or made larger) at one period and absent (or made smaller) at another time interval. To apply this separation heuristic, answer the following question: Do we need the parameter to be "positive" and "negative" at all times, or is there some time interval(s) during which it is not necessary? If such an interval exists, it might be possible to separate the opposite requirements to the key subsystem in time.

Examples
    Concrete piles must be pointed for easy driving but not pointed in order to support a load. The piles are made with pointed tips, which after driving are destroyed via an embedded explosive. Thus, the piles' sharpness is time-dependent.
    Aircraft wings are longer for takeoff, and then pivot back for high-speed flight. Thus, the geometry of the wings is time-dependent.

If parameters of the physical contradiction formula characterize the initial system at different stages and phases of its life, then application of Glazunov's approaches 6, 7, 8, and 11 fit Altshuller's method 2 for physical contradiction elimination "in time."

Example
Problem — When a wire for electrotechnical applications is manufactured, it passes through a liquid enamel bath and then through a die which removes excess enamel and controls the diameter of the wire. The die must be hot to ensure reliable calibration. If the wire feed is interrupted for ten or more minutes, the enamel bakes in the hot die and the die firmly grips the wire. The process must then be halted to cut the wire and clean the die.
    Analysis — In this situation the die is the tool and the enamel on wire is the product. The mutually exclusive requirements to the tool are to be hot and to be cold.
    Contradiction — The die should be hot and cold which is forbidden by laws of physics.
    Analysis (continued) — Z+ and Z– intersection does exist because the whole die must have opposite temperatures. The intersection also exists for Zt and Zp because the wire and die should be in the same place. Is it possible to interchange the tool and the product? Request for the product: The wire should be enameled and the diameter of the enamel should be constant. The existing process allows enamel to touch the wire with random volume of insulator on the wire's surface. We cannot change the product or modify the process. Let us consider the operative periods. T1p is the whole interval when excess enamel exists on the wire surface (regardless of wire movement); T1t is the time interval for the die to be cold when wire with excess enamel is not moving. Tt and Tp do not cross. The wire should be hot all the time, and the die should be hot only at the period when it sizes the diameter of wire (removes excess enamel), and therefore T+ and T– are separated.

Detailed Contradiction — The die should be hot when the wire is being drawn and cold when the wire is not moving.

Simplified Problems — How do we keep the die hot when the freshly enameled wire moves through the die and cold when the wire with enamel does not move? Is there a way to have the die heated and not heated automatically? While the wire is being drawn on the die, there is a significant force pulling the die in the direction of the wire pull. When the wire stops there will be no pull.

Solution — The die can be fixed to a spring. When the wire moves, it pulls the die, which compresses the spring into a heating zone. The die is heated either by induction or by contact with the hot chamber walls. When the wire stops moving, the spring pushes the die back into the cold zone.

In summary, for a $\backslash/\tau$ point problem, a solver should

a. try to schedule performing (actually or theoretically) the technique UF in such a way that the opposite requirements (or conflicting functions) take effect at different times, and/or

b. change the key subsystem parameter(s) or environment if the contradiction is not initially stated as a time constraint.

Note: The available resources T2 and T3 can be employed for separation in time. T2 is used more often than T3.

Therefore, it is possible to separate the opposite requirements to the key subsystem in time when T- and T+ are *separated* from each other.

These two separation possibilities link the physical and technical contradictions. Simon S. Litvin provided examples of many patented solutions that are based on some Inventive Principles (see the previous chapter) for use when the physical contradictions appear in the "separate" cases. These links are summarized in Appendix 6.

14.5.3 SEPARATION UPON CONDITION

If mutually exclusive requirements are demanded from the key subsystem, separation upon condition is possible if a requirement exists (or is high) under one condition and is absent (or is low) under another.

Examples
Water is a "soft" and a "hard" substance depending on the velocity of a solid body entered in or on the speed of a water jet. Thus, speed is the condition to be considered when properties of water are discussed.

A kitchen sieve is porous with regard to water and solid with regard to food, e.g., pasta. Thus, the dimension and flexibility of the substance are the condition to be considered when a sieve is used to work with this substance.

If replacement of the key subsystem is not allowed by problem conditions, then application of Altshuller's methods 7–10 and Glazunov's approaches 5, 6, 9, 11, 12, and 13 fit the generic TRIZ heuristic for eliminating physical contradictions by "changing conditions."

Example
Problem — Pure water flows through a pipeline. Water freezes in the pipeline during winter and the pipes burst.

Analysis — In this situation the tool and the product seem interchangeable; both pipe and water look like a tool and can be considered a product. Let us formulate the opposite requirements to the pipe and water first. The mutually exclusive requirements of the pipe are (1) pipe must be soft and pipe must be rigid and (2) the pipe's walls must be continuously solid to conduct water and must have holes to allow removal of excess water created by strong thermal expansion of water near the freezing temperature. The mutually exclusive requirements of water are to become icy at temperatures well below 0°C during winter and not to expand during the phase transition. The requirements for water are inconsistent with the nature of this liquid and we cannot change its physical and/or chemical properties (we have to have pure water). On the other hand, the second set of requirements of the pipe is almost against the natural behavior of common pipe materials (usually plastics or metals). Plastics and metals can have holes, but their openings would not increase when temperature decreases because of the positive coefficient of thermal expansion of these materials. Therefore, it is more convenient to assume that the pipe is the tool and water is the product and to use the first set of opposite requirements to formulate the physical contradiction.

Contradiction — Pipe should be rigid in order not to bend, and should be soft in order to concede to ice.

Analysis (continued) — The operative periods T+ and T– as well as T1t and T1p coincide ("touch" or "intersect") when water in the pipeline freezes in winter. The operative zones for the tool Z+ (rigid pipe) and Z– (soft pipe) cannot be separated (pipe should be unbroken) and Z+ and Z– do not necessarily intersect along all lengths of the pipe. On the other hand, the operative zones for the tool and product are different because Z1t is a wall of pipeline and Z1p is the space occupied by water in the pipeline; they touch each other only at the inner surface of the pipeline wall.

Solution — An elastic substance is laid in a pipe.

In summary, for a $V\tau$ (or Vz) point problem, a solver should

a. try to find the special characteristics or features of the key subsystem,
b. determine and apply a stimulus to initiate or terminate these characteristics or features of the key subsystem.

In some cases, such contradictions can be resolved by applying physical transitions at the boundary of Z– and Z+ operative zones or in the moment of "collision" of T– and T+ operative periods. Therefore, sometimes it is possible to separate opposing requirements to the key subsystem upon conditions when T– and T+ or Z+ and Z– *touch* each other.

14.5.4 SEPARATION BETWEEN PARTS AND WHOLE

If mutually exclusive requirements are demanded from the key subsystem, the separation between parts and whole is possible when a requirement exists (has one value) at the key subsystem level but does not exist (has the opposite value) at the sub-subsystem, and/or system, and/or super-system level.

Examples

A bicycle chain is rigid at the micro-level for strength and flexible at the macro-level for a bicycle's manufacturer. Thus, the macro → micro transition is to be considered when properties of the bicycle chain are discussed.

Epoxy resin and hardener are liquid but both solidify when mixed and form an epoxy compound. Thus, the subsystems → system (or system → super-system) transition is to be considered when properties of the system are discussed.

If requirements to the key subsystem are formulated from the viewpoint of different external objects (or starting from different frames of reference), then application of Altshuller's methods 3–6 and 11 and Glazunov's approaches 10 and 14 fit the generic TRIZ heuristic for physical contradiction elimination "in relations."

Example

Problem — Work pieces having complex shapes can be difficult to grip with an ordinary vise.

Contradiction — The main function of the vise is to provide an evenly distributed clamping force (a firm, flat grip face). The subsystem requires some means of conforming to the irregular shape of the object (a flexible grip face).

Solution — The hard bushings stand on end between the flat surface of the vise jaws and the irregular surface. Each bushing is free to move horizontally in order to conform to the shape of the piece as pressure increases, while distributing even gripping force on the object.

In summary, for an $X\tau$ (or Xz) point problem, a solver should

a. try to partition the key subsystem and assign one of the contradictory functions or opposite requirements to this subsystem's element, or
b. let the technique as a whole (or a few other subsystems of this TS or TP) retain the remaining functions and requirements.

Note: The available resources are the internal structure of the TS or TP (because the division of a technique on subsystem is often arbitrary) and environment.

In some cases such contradictions can be resolved by artificial separation of the operative zones or periods. Therefore, sometimes it is possible to separate the opposite requirements of the key subsystem inside the technique when T– and T+ or Z+ and Z– *intersect* each other.

I specifically use the word *sometimes* in the last two separation heuristics. It is often hard to separate "touch" and "intersection" cases for operative periods T+ and T– because the term moment $\tau 0$ is uncertain for many TS and TP. (Perhaps we should compare the longevity of $\tau 0$ with some characteristic time of the technique.) Unfortunately, it is impossible to predict which of the separation heuristics would work for the problem under consideration. It is usually suggested to try both heuristics in this situation. Thus, the trial and error method is still present in TRIZ here, although the set for possible attempts is rather small and can be handled easily by a solver.

Occasionally all separation heuristics work fairly well for the same technique. Let us demonstrate it by a well-known example.

Problem — Some people have two types of bad vision.
Contradiction — Eyeglasses are used to see near and far.
Solutions —
  Separation in Space: Two different lenses (bifocals).
  Separation in Time: Two pairs of glasses, changing back and forth as the need arises.
  Separation Between Parts and Whole: Plastic lenses with possibility to change curvature and focus.
  Separation upon Condition: The self-focusing camera-type lens.

If inventors had used the separation methods a century ago, they would have designed today's eyeglasses a hundred years earlier. Clearly the use of the TRIZ heuristics for resolution of point or physical contradictions helps resolve various technical problems.

14.6 CONCLUSION

Often a single subsystem must perform contradictory functions or operate under incompatible requirements. In TRIZ, such a situation is called a physical (point) contradiction. Any key subsystem that should have at least two nonequal values of some parameter meets the requirements of the corresponding physical contradiction. A description of such a key subsystem with mutually exclusive requirements should involve these characteristics and conditions for their realization. The TRIZ heuristics for resolving point or physical contradictions are separation in space, separation in time, separation between parts and whole (in relations), and separation upon conditions. Although there are fewer heuristics to compare with pair or technical contradictions, they usually provide the possibility of finding stronger solutions. As a rule, each separation heuristic should be investigated because one cannot predict absolutely which will lead to the most significant breakthrough.

In this and previous chapters we considered various TRIZ heuristics for the simplest structures of problems. Such problems with physical (point) and technical (pair) contradictions have been studied extensively by TRIZniks during the last 50 years. A few heuristics have been developed for resolving problems with more complicated structures, but the volume and scope of this book do not allow presentation of those results here.

REFERENCES

1. Altshuller, G. S., *Creativity as an Exact Science: The Theory of the Solution of Inventive Problems,* Gordon and Breach Science Publishing, New York, 1984.
2. Glazunov, V. N., *Parametric Method for Resolution of Contradictions in Technique,* Rechnoj Transport, Moscow, 1990.
3. Terninko, J., Zusman, A., and Zlotin, B., *Systematic Innovation: An Introduction to TRIZ (Theory of Inventive Problem Solving),* Saint Lucie Press, Boca Raton, 1998.

15 Standard Solutions of Invention Problems

15.1   INTRODUCTION

Many years of Su-Field analysis of technical problem solutions in various engineering fields, as well as study of high-Level patents from the point of view of Su-Fields, has revealed the standard situations and standard methods for problem solution. The term "Standard," introduced by G. S. Altshuller, means that a common "trick" to solve different problems results in very similar problem models. Altshuller proposed a system of 76 Standards in 5 different classes and illustrated each Standard by a few inventions [1]. The titles of the Standards are given in Appendix 5.

15.2   STANDARDS

Standards are the precepts of synthesis and transformation with the aim to overcome or circumvent technical and physical contradictions. When a problem is solved with the help of Standards, there is *no* need to formulate a contradiction. Standards are used in those situations when a problem involves an *undesired interaction* between two or more subsystems:

- *missing* — some parameter of a subsystem or product has to be changed during operation, but we do not know how to change it;
- *harmful* — a subsystem produces a harmful effect;
- *excessive* — an action of one subsystem on another is too strong;
- *insufficient* — an action of one subsystem on another is too weak.

The usage of Standards requires thorough knowledge of how to identify the components and to construct the Su-Field model, how to perform the physical, chemical, and other effects for technical problems, and how to develop the concepts to support the solutions founded from the Su-Field Analysis. A continuous switching back and forth between the analytic and knowledge-based approaches occurs during work with Standards.

15.2.1 CRITIQUE OF ALTSHULLER'S STANDARDS SYSTEM

The system of Standards proposed by Altshuller is quite inhomogeneous (see Appendix 5). For example, Standard 5.1.1 actually comprises nine ideas while the last Standard is no more than an explanation of Standards 5.5.1 and 5.5.2.

Many Standards (e.g., all in class 3 and groups 2.3 and 4.5, and particularly 2.2.2–2.2.4, 4.3.2, and 4.3.3) are equal to the trends of technique evolution. In general, Standards classification matches a trend of technique evolution: simple Su-Fields → complex Su-Fields → forced Su-Fields → complex-forced Su-Fields → transition to super-system or subsystems.

Standards 2.4.7, 4.3.1, and 4.4.5 are obvious because any TS or TP is impossible without "action" of physical effects (they are the background of chemical, biological, materials science, and other phenomena). Practically any good invention can illustrate such Standards. For example, the third patent found during a quick search of the Internet was Patent USA 5001054: Organic contamination detector has an optical fiber in contact with the fluid (where organic contamination could exist), laser, and a receiver. Organic substrate oxidation causes a change of air composition. The optical fiber detects gases in such a manner that signal depends on the level of organic impurities in the fluid. It is obvious that this patent fits Standard 4.3.1: "If given is a Su-Field system, the efficiency of measurement and detection in it can be increased by using physical effects."

Some Standards (e.g., 2.2.3) present the special cases of more general Standards (e.g., 2.2.2). Too much attention is given to Su_M-Fields, which are presented not only in subclasses 2.4 and 4.1, but also along Standards in other classes. For example, there is no constructive answer to why Standard 1.2.5 deals only with a magnetic field. I would like to stress that introduction of a magnetic field can give interesting solutions, but one should remember that incorporation of the subsystems for magnetic fields creation in many kinds of techniques would decrease the Ideality of these techniques. Some Standards (e.g., 4.1.2, 5.1.4) repeat the Inventive Principles, while others (e.g., all in the group 5.3, 3.1.5, and 5.4.1) repeat Altshuller's methods to resolve physical contradictions (see Appendix 6). Most Standards can be modeled by Su-Fields, but a large group of Standards (e.g., all in classes 3 and 5 and subclass 4.1, as well as 2.3.2, 2.3.3, 4.3.1, and others) are beyond the scope of symbolic description of Su-Fields.

Formulation of various Standards are too different and inconsistent. For example, compare Standard 5.3.4 and Standard 2.1.1.

2.1.1. Chain Su-Field — If Su-Field is weakly controllable, and you need to make it more efficient, the problem is solved by transforming a Su-Field part into an independently controlled Su-Field and construction of a chain Su-Field. Chain Su-Fields can also be obtained by expanding relations in Su-Field. In this case, a new link F2-S1 is integrated into the relation S1-S2.

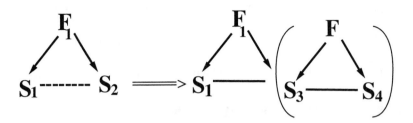

Example*
Patent UK 824047. Proposed is the device for rotating the transmission from one shaft to another one (coupling). The device includes inner and outer rotors enclosed by an electric magnet. The gap between rotors is filled with a magnetic liquid solidifying in the magnetic field. If the electric magnet is off, the rotors are free to rotate relative to each other. When the electric magnet is on, the liquid becomes solid and rigidly binds the rotors together, i.e., allows the transmission of torque.

5.3.4. Two-phase state — "Double" properties of a system can be provided by replacing the single-phase state with the two-phase one.

Example*
Patent USA 3589468. For noise suppression, as well as for trapping vapors, odors, and cuts in cutting, cutting zone is coated with foam. Foam is pervious to instrument, but not pervious to noise, vapors, etc.

Therefore, although many of the Standards are quite useful in problem solving and design, the set proposed by Altshuller does not represent a system, a set of orderly interacting elements possessing properties which cannot be reduced to properties of separate elements, and intended for execution of certain functions. In the "system" of Standards, the Standards

- do not have a strong order between the classes and sometimes among groups,
- only a few elements "interact" as "mother-child,"
- execute various incompatible functions (describe the evolution of technique, play a role in TRIZ heuristics, model behavior of a technique, etc.), and
- do not have synergetic properties, which cannot be reduced to properties of separate elements.

There have been some attempts to improve the situation with Standards, the most important being

---

* I replace USSR patents given by Altshuller with patents of Western countries in the examples because the latter are more available for the readers of this book.

- algorithm and flowchart that to help orient Altshuller's system and to use the right Standard,
- table of unification of the Standards that are most important for problem solving,
- lists of Standards for typical problems (see Appendix 7).

15.3 STANDARDS AS TRIZ HEURISTICS

Attempts to improve Altshuller's Standards were made independently by a few TRIZ experts, notably A.M. Pinyaev, B.I. Goldovsky, and G. Frenklach with the author; some of their results are presented in Appendix 7.

15.3.1 STANDARDS SYSTEM USAGE ALGORITHM

It is not easy to orient in Altshuller's Standards, so Russian TRIZniks designed an algorithm for choosing the right Standard from the Altshuller system. A flowchart of this algorithm is shown in Figure 15.1.

1. Construct a model of the problem.
2. Transform the model of the problem to the Su-Field form.
   Note-0: Complete model should have a product (S1), a tool (S2), and an interaction of a product and tool (F).
3. Check if it is a measurement problem.
   If *yes*, go to step 3.1.
   If *no*, go to step 4.1.
3.1. Check if a replacement of the initial problem in measurement or detection tasks is accessible.
   If *yes*, apply the Standards of group 4.1.
   If *no*, go to step 4.
   Note-1: If the direct transition is too complicated, first transfer the problem to a detection task, and then translate it to a measurement task.
4. Check the completeness of the Su-Field.
   If the Su-Field is incomplete (or *no*), complete step 4.1, then go to step 5.
   If the Su-Field is complete, go directly to step 5.
4.1. Check presence of harmful links. If such a link is absent, go to step 4.2.
   4.1.1. Check if the introduction of substances and fields is allowable.
      If *yes*, apply Standards 1.1.1–1.1.6 or Standards of group 4.2.
      If *no*, apply the Standards of group 5.1, 5.2, 5.5.
4.2. Check if introduction of substances and fields is allowable.
   If *yes*, apply Standards 1.1.7, 1.1.8, 1.2.3.
   If *no*, apply the Standards of groups 5.1, 5.2, 5.5.
5. Check presence of harmful links.
   If *yes*, go to step 5.1.
   If *no*, go to step 6.

5.1. Check if the introduction of substances and fields is allowable.

If *yes*, apply Standards 1.2.1, 1.2.2, 1.2.4, 1.2.5.

If *no*, apply the Standards of groups 5.1, 5.2, 5.5.

6. Check presence of ferromagnetic substances in the Su-Field.

If *yes*, go to step 7.

If *no*, go to step 8.

Note-2: Check presence of any ferromagnetic substance in subsystems which could be included in the Su-Field under consideration.

7. Check if introduction of a magnetic field is allowable.

If *yes*, go to step 17.

If *no*, go to step 8.

8. Check if formation of the complex Su-Fields is allowable.

If *yes*, apply the Standards of group 2.1.

If *no*, go to step 9.

Note-3: If the complication of the system is *not* restricted in conditions of the problem, it is often possible to solve the problem by formation of complex Su-Fields.

9. Check if replacement of the Su-Field is allowable.

If *yes*, apply Standard 2.2.1.

If *no*, go to step 10.

Note-4: Replace any field except magnetic and electrical.

Note-5: Replacement of a field is inadmissible if the replacing field is a source of hindrances.

10. Check if the system is dynamic.

If *yes*, go to step 11.

If *no*, apply Standards 2.2.2–2.2.4.

Note-6: Remember the principle of increased dynamism of the technique.

11. Check if the structure of components of the Su-Field is coordinated.

If *yes*, go to step 12.

If *no*, apply Standards 2.2.5, 2.2.6, or 4.3.1 and of groups 5.3 and 5.4.

Note-7: Remember duality of this law! It may be necessary to misbalance consciously the components.

12. Check if dynamics of components of the Su-Field are coordinated.

If *yes*, go to step 13.

If *no*, apply Standards 2.3.1–2.3.3 or 4.3.2 and 4.3.3.

13. Check if introduction of ferromagnetic substances and magnetic fields is allowable in Su-Field instead of current components.

If *yes*, apply Standards 2.4.1 or 4.4.1.

If *no*, go to step 14.

14. Check if introduction of the ferromagnetic additives is allowable in available substances.

If *yes*, apply Standards 2.4.5 or 4.4.3.

If *no*, go to step 15.

Algorithm for Use of Standards

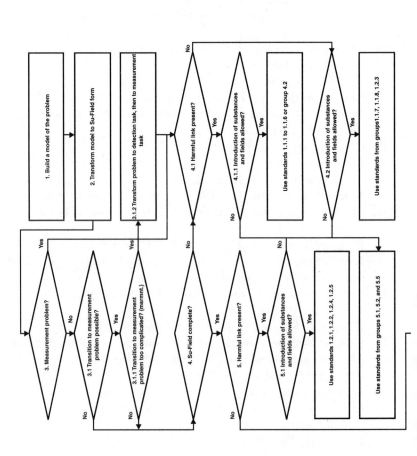

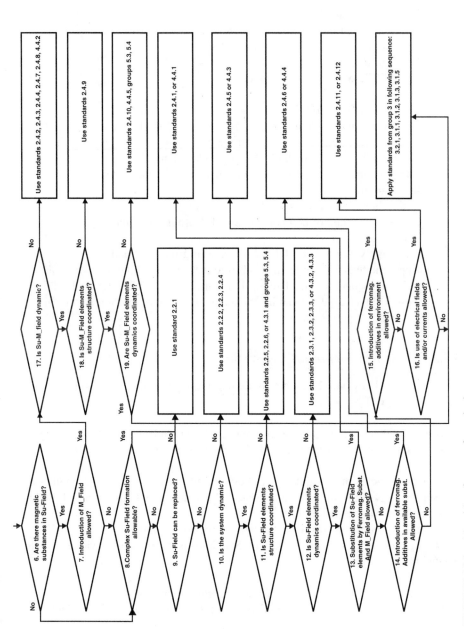

FIGURE 15.1 Algorithm for use of standards.

15. Check if introduction of the ferromagnetic additives is allowable in the environment.
    If *yes*, apply Standard 2.4.6 or 4.4.4.
    If *no*, go to step 16.
16. Check if use of electrical fields and/or currents is allowable.
    If *yes*, apply Standards 2.4.11 and 2.4.12.
    If *no*, go to step 20.
17. Check if Su-M_Field is dynamic.
    If *yes*, go to step 18.
    If *no*, apply Standards 2.4.2, 2.4.3, 2.4.4, 2.4.7, 2.4.8, and 4.4.2.
    Note-8: At step 7 we introduce only a magnetic field, and at step 17 we come to Su-M_Field, making ferromagnetic substance dynamic (Standards 2.4.2–2.4.4) or making all components dynamic.
18. Check if structure of components Su-M_Field is coordinated.
    If *yes*, go to step 19.
    If *no*, apply Standard 2.4.9.
19. Check if dynamic of components Su-M_Field is coordinated.
    If *yes*, go to step 20.
    If *no*, apply Standards 2.4.10, 4.4.5, and of groups 5.3 and 5.4.
20. Apply the Standards of the third class to the solution of the problem in the following sequence: Standard 3.2.1, and then 3.1.1, 3.1.2, 3.1.3, and 3.1.5.
    Note-9: Standard 3.1.4 can be applied at any stage of development of bi-systems and poly-systems.

15.3.2 TABLE OF STANDARDS UNIFICATION

Valeri Souchkov and the author argue that most Standards can be presented in a simple IF…THEN form:

IF a problem of *goal* is given as Su-Field *conditions* and *constraints* according to the problem circumstances, **THEN** such problems are solved by *action*.

All constituents of a Standard are presented in Table 15.1 with the number of the Standard given by Altshuller and some of his notes. Standards for Su-M_Fields have the letter M after the number.

Table 15.1 demonstrates that Standards unification is possible.

Using Table 15.1, one of Altshuller's Standards is used to make evident how to expand the pattern:

IF a problem of *goal* is given as Su-Field *conditions* and *constraints* according to the problem circumstances, **THEN** such problems are solved by *action*.

Standard 4.2.3 — If Su-Fields for measurement should be created from a poorly measurable or detectable complete Su-Field, and no substances can be added to the

TABLE 15.1
Standard Solutions of Technical Problems

Aim/Conditions	Constraints	Action	Altshuller's Numbers and Notes
Aim: Optimization of Su-Fields			
Minimal (dosed, optimal) mode	Hard, or even impossible, to achieve	Use the maximal mode followed by removal of surplus part	1.1.6
UF maximal mode	Maximal mode is intolerable on one substance (e.g., S1)	Retain maximal mode maintenance but direct it to another substance (e.g., S2) related to the first one (e.g., S1).	1.1.7
Selective mode	No restrictions on F value	Add a protective substance where minimal mode is needed, and add a substance giving a local field where maximal mode is needed.	1.1.8 F is maximal in some sectors and minimal in other sectors.
Aim: Destruction of Su-Fields			
Both UF and HF take place between substances in Su-Field	The substances must not necessarily be in direct contact	Add a new, free, or sufficiently inexpensive substance S3 between the substances S1 and S2.	1.2.1 Take S3 from the outside in the finished form or made of substances available under the action of fields; e.g, S3 is bubbles, "emptiness," foam, etc.
The same conditions as above	1.2.1 + the usage of foreign S3 is barred.	Add a new, free, or sufficiently inexpensive substance S3 between S1 and S2, and this third substance is a modification of the first two.	1.2.2 S3 is already available in a technique; S3 is just modified for performing new functions.
The same conditions as above	S1 and S2 must be in direct contact	Pass to double Su-Field, where available field F1 retains its UF, and added field F2 neutralizes (compensates) HF (or transforms it into useful one).	1.2.4
HF of a field on substance exists	No restrictions	Introduce a substance that will eliminate HF itself.	1.2.3, 1.2.5M

TABLE 15.1 (continued)
Standard Solutions of Technical Problems

Aim/Conditions	Constraints	Action	Altshuller's Numbers and Notes
		Aim: Construction of Su-Fields	
The given substance is hardly changeable in the needed direction	No restrictions on adding new substances and fields	Completion (synthesis) of Su-Field due to introduction of new (missing) components.	1.1.1 When performing operations with thin, fragile, and easily deformable substance, a subsystem is joined during these operations with a substance making it hard (strong). Then this subsystem can be removed by dissolving, evaporation, etc.
The same conditions as above	No restrictions on adding new substances into existing subsystem	Transition (constant or temporal) to internal complex Su-Field, introducing additions into available substances S1 or S2. Such additions must increase Su-Field controllability or add needed properties to it.	1.1.2 Sometimes one and the same solution, depending on the statement of a problem, can be obtained by constructing (complex) Su-Field. S3 is an addition to the tool S2.
The same conditions as above	Restrictions on adding new substances to available ones S1 or S2	Transition (constant or temporal) to external complex Su-Field, joining outer substance S3 with S1 or S2. The S3 must increase Su-Field controllability or give it needed properties.	1.1.3, 2.4.5M
The same conditions as above	Restrictions on adding or joining new substances	Completion (synthesis) of Su-Field using the available environment as a substance to be added.	1.1.4, 2.4.6M In particular, if a weight of a moving subsystem needs to change, and it is impossible, the subsystem must be shaped as a wing. Changing the angle of wing inclination about the movement direction, one obtains the additional upward or downward force.
The same conditions as above	1.14 + no substances in the environment	Substances can be obtained by replacement of the environment, its decomposition, or addition of new substances into it.	1.1.5

TABLE 15.1 (continued)
Standard Solutions of Technical Problems

Aim/Conditions	Constraints	Action	Altshuller's Numbers and Notes
Aim: Increase the Su-Field Efficiency Due to Resources			
Su-Field is weakly controllable and its efficiency should increase	No restrictions	Transformation of a Su-Field component into independently controlled Su-Field and construction of chain Su-Fields. (Analogies: 2.4.1 for Su_M_Fields and 2.4.11 for Su_E_Fields).	2.1.1, 2.4.1M A chain Su-Field can be obtained by expanding relations in Su-Field. In this case, a new link F2-S1 is integrated into the relation S1-S2.
The same conditions as above	No restrictions	Increase the degree of dispersion of a substance operating as a tool. Increase the degree of flexibility of the Su-Field.	2.2.2, 2.4.2M, 2.2.4, 2.4.3M, 2.4.8M Standards reflect the technique evolution trends.
The same conditions as above	No restrictions	Transition from homogeneous fields (substances) or fields (substances) with unordered structure to inhomogeneous fields (substances) or fields (substances) with a certain spatial structure (constant or variable).	2.2.5. For field organization 2.2.6. For substances organization 2.4.9M For ferromagnets and magnetic fields
The same conditions as above	Su-Field components cannot be replaced (2.1.2) by adding new F and S (2.2.1)	Construct a double Su-Field due to introduction of the second well controllable field. (2.1.2) Replace uncontrollable (or weakly controllable) working field with controllable (well controllable) one (2.2.1).	2.1.2, 4.4.2M 2.2.1, 2.4.1M For example, a mechanical field can be replaced with an electric one, etc. Analogs are 4.4.2M, 2.4.1M
Aim: Growth of Su-Fields Efficiency by Phase Transitions			
Contradictory requirements to introduce S and	Restriction to add substances	Change the phase state of the available substance instead of adding a new substance.	5.3.1
F can be met only by using phase transitions	Opposite properties for existing substances	Use the substances capable of transition from one phase state to another one, depending on the operation conditions	5.3.2 The phase transition of the second type is preferable.
The same conditions	See the conditions	Use phenomena accompanying the phase transition.	5.3.3
The same conditions	The same restrictions	Replace the single-phase state of a substance with a two-phase.	5.3.4 See Standard 5.4.1.
The same conditions	The conditions are the restrictions	Introduce an interaction (physical, chemical) between phases of the substance (obtained by 5.3.4).	5.3.5

TABLE 15.1 (continued)
Standard Solutions of Technical Problems

Aim/Conditions	Constraints	Action	Altshuller's Numbers and Notes
Aim: Formation of Su-Fields for Measurement			
Poorly measurable or detectable incomplete Su-Field	No restrictions	Construct a simple or double Su-Field using a field passing through the system and carrying out the information about its state	4.2.1 The synthesis of measuring Su-Fields is distinguished by the fact that they must ensure obtaining a field at output. (Compare Standard 1.1.1.)
Poorly measurable or detectable complete Su-Field	No restrictions	Change the system in such a way that there will be no necessity for detection and measurement.	4.1.1 PF of some subsystems is measurements and detection. It is desirable to exclude (or minimize) such PF, without prejudice to technique accuracy and performance.
The same conditions as above	No restrictions	Transition to internal or external complex Su-Field, adding easy-to-detect substances to the system.	4.2.2, 4.4.3M Can be applied to a component of any complete Su-Field.
The same conditions as above	Standard 4.1.1 cannot be applied	Replace direct operations with a subsystem by operations with its copy or picture.	4.1.2 Such copy (picture) can have the opposite colors to the subsystem's colors.
The same conditions as above	Standards 4.1.1 and 4.1.2 cannot be applied	Perform the sequential detection of changes.	4.1.3 The change from the indistinct concept "measurement" to the clear model "two sequential detections" simplifies many problems.
The same conditions as above	No substances can be added	Add the substances generating easy-to-detect and easy-to-measure field to environment.	4.2.3, 4.4.4M The state of the technique can be judged from the state of environment.
The same conditions as above	Restriction for adding the substances according to Standard 4.2.3	Obtain the substances generating easy-to-detect and easy-to-measure field in the environment itself	4.2.4 Such substances can be obtained by decomposition of environment or change of the aggregate state of matter.

TABLE 15.1 (continued)
Standard Solutions of Technical Problems

Aim/Conditions	Constraints	Action	Altshuller's Numbers and Notes
Aim: Substances Management in Su-Fields			
Complete Su-Field	Restriction to add new substances	1. "Emptiness" and/or a field is used in spite of substance. 2. External addition is used in spite of internal one. 3. Substance is added in the form of chemical compound giving off the needed substance. 4. Particularly active addition in very small doses is used. 5. Usual substance in very small doses is added but only at certain points of a subsystem. 6. Addition is used for a while. 7. Technique model, to which substances can be added, is used in spite of the technique. 8. Addition is obtained from the technique itself, its subsystems, or environment by decomposing it using, for example, changing the aggregate state of matter.	5.1.1
Complete Su-Field	Substance's direct production is impossible	Destroy substance of the closest higher ("full" or "excessive") structure level (e.g., molecules) to obtain its parts (e.g., ions).	5.5.1
Complete Su-Field	Substance's direct production and destruction are impossible	Integrate a substance of the closest lower ("non-full") structure level (for example, ions).	5.5.2
Complete Su-Field	A technique is unchangeable and tool replacement or addition of substances is not allowed	Separate substance(s) into parts interacting with each other and use them as a tool.	5.1.2 Separation into parts charged positively and negatively. If all substance's parts have the same electrical charge, another substance should have the opposite charge.
Complete Su-Field	Added substance must disappear after being used	Make additive substance indistinguishable from the technique substance or in environment.	5.1.3

TABLE 15.1 (continued)
Standard Solutions of Technical Problems

Aim/Conditions	Constraints	Action	Altshuller's Numbers and Notes
Add a lot of substance	Much of substance cannot be added	Use "emptiness" substance as inflatable constructions (macro-level) or foam (micro-level).	5.1.4 Standard 5.1.4 is often used along with other Standards.

Aim: Add Fields in Su-Fields

Aim/Conditions	Constraints	Action	Altshuller's Numbers and Notes
Complete Su-Field	No restrictions	Use already available ("hidden") fields carrying by substances existing in the technique.	5.2.1. Using existing fields
Complete Su-Field	Standard 5.2.1 is inapplicable	Use fields from an environment.	5.2.2. Fields from environment
Complete Su-Field	Standards 5.2.1 and 5.2.2 are inapplicable	Use fields that can be generated by the technique's substances or environment.	5.2.3. Substances as sources of fields Utilize magnetism of ferromagnetic substances used in the technique only mechanically for better interaction between subsystems, for revealing information, etc.

Aim: Forcing of Measuring Su-Fields

Aim/Conditions	Constraints	Action	Altshuller's Numbers and Notes
Complete Su-Field	Changes cannot be directly detected or measured. A field cannot be passed via the system	Excite resonance vibrations (in the whole system or its part), and changes in frequency of these vibrations serve as indications of changes taking place in the system itself.	4.3.2
Complete Su-Field	Same as above + Standard 4.3.2 cannot be applied	Obtain information about the technique from the changes in intrinsic frequency of a subsystem (environment) related/added to the monitored technique.	4.3.3

Aim: Growth of Efficiency for Physical Effects Applications

Aim/Conditions	Constraints	Action	Altshuller's Numbers and Notes
Su-Field's component must be in various states	Periodically, from time-to-time, or occasionally	Use reversible physical transformations (e.g., phase transitions).	5.4.1 Transition by the subsystem itself is due to ionization-recombination, dissociation–association, etc. Also Standard 5.3.4.

TABLE 15.1 (continued)
Standard Solutions of Technical Problems

Aim/Conditions	Constraints	Action	Altshuller's Numbers and Notes
Su-Field has a "weak" input	Cannot increase input, but a "strong" output is needed	Use the substance-transformer into the state close to the critical one. Energy is accumulated in the substance, and an input signal plays a part of "trigger."	5.4.2 Goal here is to obtain a "strong" output, usually in the form of a field.

technique according to the problem circumstances, then such problems are solved by adding the substances generating easy-to-detect and easy-to-measure fields to the environment.

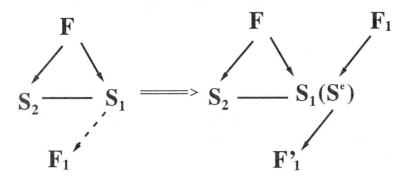

Notes:

1. The state of the technique can be judged from the state of the environment, noted by the letter "e" in the figure above.
2. The formulation above is identical to the original one:

"If a system is hard-to-detect or hard-to-measure at some instant and no substances can be added to an object, then these substances generating easy-to-detect and easy-to-measure field should be added to ambient medium, and the state of the object can be judged from the state of ambient medium" [1].

3. This Standard can be illustrated by the following invention:

To monitor engine wear, it is necessary to determine the quantity of "wiped off" metal. Metal particles fall into the ambient medium — an oil. It was proposed to add lumi-nophores to oil: metal particles stop the glow (USSR Patent 260249).

15.4   A CASE STUDY

We now return to the problem of metal shots or pellets in a plastic pipe, considered originally in Chapter 13 to show how the Standards work. As we already know, the PF is to change the direction of shots movement. We have a simple Su-Field with undesired (harmful) interaction: high-speed metal shots damage the elbow.

S1 — metal shots,
S2 — elbow,
F — mechanical.

Using the table from subsection 15.3.2 we can easily find

Standard 1.2.1 — Place metal shots at inner surface of elbow to form a protective layer (S3).
Standard 1.2.2 — Reinforce the elbow. Redesign shape of elbow.
Standard 1.2.4 — Use magnetic field to form a blanket of shots at the elbow.
Standard 1.2.4 — Electrically charge the metal shots and the elbow with the same polarity.

These Standards lead to known and new solutions.

15.5   CONCLUSION

In conclusion, note that while the Inventive Principles (see Chapter 13) operate with generalized technical parameters, the Standards are more formal and context-dependent since they operate with a specific model of a technique and/or product or raw object. This often makes Standards more accurate than the Inventive Principles. The comparison of different TRIZ heuristics is given in Appendix 6.

Su-Field Analysis and Standards work quite well with some TRIZ instruments, including the Agents Method and ARIZ, as will be demonstrated in the next chapters.

REFERENCE

1. Altshuller, G. S., *To Find an Idea,* Nauka, Novosibirsk, 1986 (in Russian).

16 Energy Synthesis of Systems

16.1 INTRODUCTION

In this chapter we consider an important step of conceptual design — the creation of a TS structure based on consideration of energy/field fluxes in it. This step can be performed when the basic concept of TS and its initial functional structure are known. A few TRIZniks have participated in developing this instrument, termed the Structure-and-Energy Synthesis, notably Michail I. Vaynerman, Lev G. Goryainov and Nikolaj N. Matvienko. This instrument nicely illustrates Su-Field Analysis, described in the previous chapter, and uses our knowledge about technical systems and technological processes (Chapter 3). In this chapter I follow Vaynerman's version of this TRIZ instrument for TS synthesis [1].

16.2 SYNTHESIS OF TECHNICAL SYSTEMS — OVERVIEW

The direct synthesis of a technical system given all the requirements imposed on it, including the selected fields at input and output, consists of the following steps.

1. Analyze functioning principles and constraints to the system or subsystems.
2. Determine the raw object and product of the system, i.e., that object that is changed or about which information is obtained. Then determine the type of system: whether it is changing or measuring, manipulated or controlling.
3. Outline possible structures at which the given inputs (including the raw object) can be processed into the given outputs (including the product).
4. Choose the appropriate source of energy that can transfer the raw object into the final product. (Different energies are listed in Appendix 4).

5. Construct energy chains starting from the product.
6. Compare the type of field, which should come to the product or with which the product affects energy flux, with the type of field at the input and output. Do not include too many transformations. It is always desirable to obtain minimal structures, and they always are the best starting point.
7. Having outlined the structures, determine the physical principles at which these structures can be put into practice.
8. Construct these structures from the four elementary energy elements and generic energy parts (to be discussed in the next two sections of this chapter).

Transition from function to energy is easy because the formulation of function itself involves prompts about the type of energy or its flux. To make the transition from energy fluxes to structure is not so straightforward; thus, function and energy should be described specifically. The more specific the descriptions are of all UF, the narrower is the variety of energy means for its realization, and the more definite are TS and its structure. In order to do the next steps of synthesis we will need to outline first elementary energy elements and energy chains.

16.3   ELEMENTARY ENERGY ELEMENTS

While passing through subsystems, energy transforms — its value or characteristics are changed. Unfortunately, there is always some dissipation of energy (i.e., HF), which is one of the reasons why the absolutely Ideal technique to perform only UF is impossible. We can classify four types of energy transformations:

Emission of energy (radioactive emission of uranium, discharge of a capacitor, heat release from a heated body, expansion of a compressed gas). Various subsystems to generate energy flow are known.

Energy flux transformation by a certain program (electric current traversing a conductor, heat and mass exchange, pressure transfer in a liquid or gas jet, mechanical exchange of forces and displacements, polarization of light). In this case, the changing parameter is not the type of energy but its organization in space and time: a value of energy and relevant parameters, direction of the energy flux, etc.

Conversion of energy (transformation of electrical energy into heat as current traverses a conductor, transformation of alternating current energy into magnetic energy, transformation of mechanical energy into electrical in a piezocrystal). In this case, the changing parameter is the type of energy.

Absorption of energy (thawing of ice, deformation, gas compression, absorption of radioactive radiation). Different subsystems accumulate and store energy over time.

This classification of energy transformation is relative. All types can be reduced to energy emission and absorption. Then energy transformation corresponds to unity of absorption and emission in one subsystem. Another way is to consider only transformations. Then emission corresponds to transformation of the subsystem's internal energy into emitted energy, while absorption corresponds to transforming the absorbed energy into the internal energy of the subsystem. This approach is even more accurate from the physical point of view. However, because energy sources and absorbers (receivers) are considered in engineering as independent subsystems, it is more convenient to consider four types of energy transformations. One should keep in mind that energy emission and absorption represent conversion of the inner energy of a substance.

We can define four elementary energy elements (EEEs) for any technique and generic technical problem that is associated with such EEEs:

EEE1 — "source" of a field/energy (such as accumulator in the discharge mode, nuclear reactor). Generic problem is to make a substance which provides the field generation: $P_1 \equiv S \Rightarrow S \to F$.

EEE2 — field/energy transformation by a program, including a change of field/energy intensity up to compensation for it or complete elimination without any change in field/energy type (such as communication line, heat supply line, transformer). Generic problem: $P_2 \equiv F \Rightarrow F \to F'$.

EEE3 — converter of a field/energy type (for example, electric motor, refrigerator, radio receiver), including the change in the number of types of fields/energies at the output. Generic problems: $P_3 \equiv F \Rightarrow F_1 \to F_2$, and $P_3 \equiv F \Rightarrow F_1 \to F_2$ and F_3.

EEE4 — field/energy "receiver" ("absolute" heat insulator, electrical capacitor, sports mats for high jumps, accumulator in the discharge mode, water container above ground level). Generic problem: $P_4 \equiv F \Rightarrow F \to S$.

Let us now see how to construct a structure using EEEs by example of the following typical problem.

Given are two substances S1 and S2. We need substance S1 to act upon substance S2, that is, S1 $--\to$ S2. However, we cannot change or transform these substances.

It is known that S1 emits field F_1 (S1 $\to F_1$) and S2 can perceive field F_2 ($F_2 \to$ S2). Since S1 and S2 must remain unchanged, we have only one way: transform F_1 into F_2 ($F_1 \Rightarrow F_2$). Assume it is known that the transformation can be performed by substance S3 in the excited state: $F_1 \to S'_3 \to F_2$, and the transition of substance S3 into state S'_3 is possible only under the action of field F_3: $S'_3 \equiv F_3 \to$ S3. Having joined four EEEs, we obtain the following structure:

$$S_1 \to F_1 \to S_3 \to F_2 \to S_2 \equiv S_1 \to S_3 \to S_2$$
$$\uparrow \qquad\qquad\qquad\quad \uparrow$$
$$F_3 \qquad\qquad\qquad\quad F_3$$

This Su-Field model gives the concept of a solution. Su-Field representation can be more itemized; for example, the following representations of the TS structure are fully equivalent:

$$S_1 \rightarrow S_2 \Rightarrow S_3 \equiv S_1 \rightarrow F_1 \rightarrow S_2 \rightarrow F_2 \rightarrow S_4$$
$$\uparrow\downarrow \qquad\qquad\qquad \downarrow \quad \uparrow$$
$$S_3 \qquad\qquad\qquad S_3 \leftarrow F_3$$

Although the righthand representation of the structure is more detailed, the left, shortened, representation is also possible, especially if the form of interaction between substances is insignificant. The representations of the type S1S2 and S1 – S2 are also equivalent. The first representation is more convenient if one needs to emphasize that substances form a mixture.

It is clear from these Su-Fields models that in order to build a structure in the generalized physical-energy form, we should record the desired interactions and use them to determine those typical (or generic) structure problems of EEE construction, which should be solved in order to implement the desired interactions.

The following example illustrates this idea.

Problem — Piers of bridges are coated with ice in winter, which shortens their service life. Eliminate this drawback using the simplest and cheapest method.

Denote ice near a pier as S_1; water as S_2; pier as S_3. Their interaction can be shown as:

$$\tilde{S}_1 \rightarrow S_3$$

while the desired state required phase is $S_2 \leftrightarrow S_3$.

To eliminate undesired $\tilde{S}_1$ ($\tilde{S}_1 \Longrightarrow$), we should affect it with some field, i.e., solve P_4.

There are two possible fields which efficiently affect ice. Correspondingly, there are two possible ways to solve this generic problem:

mechanical (ME) interaction: $F_{ME} \rightarrow S_1$
thermal (T) interaction: $F_T \rightarrow S_1$

Ambient water can be taken as a mechanical energy source. If there is some stream, then F_{ME} exists. Moreover, lower water layers in winter are often warmer than upper ones, so they accumulate a significant store of F_T. The energy of a stream has a drawback: the energy may be low or even absent (for example, if the bridge spans a gulf or bay). Thermal energy of the upper layers always exists. Therefore, the solution oriented to F_T is more versatile. As a result, we have **EEE** $S'_2 \rightarrow F_T$ which should be joined with **EEE** $F_T \rightarrow S1$ to form a chain. Two ways are possible here:

1) $S'_2 \rightarrow F_T \rightarrow S1$, i.e., direct contact of warm water with ice
2) $S'_2 \rightarrow F_T \dashrightarrow F'_T \rightarrow S1$, i.e., heat transfer to ice from the lower water layers through the upper layers.

To implement the first way, water must be moved:

$$F_{ME} \rightarrow S'_2(X_1) \rightarrow F^{movement}{}_{ME} \Rightarrow S'_2(X_2)$$

This method requires additional energy to overcome the forces of gravity.

To implement the second way, it is sufficient to solve the problem

$$P_2 \equiv F_T(X_1) \Rightarrow F'_T(X_2)$$

This method is easy if a heat-conductive substance is introduced. According to the laws of thermodynamics, heat itself comes from the warm water layers to the cold ones. As a result, we have the following structure:

$$S'_2 \rightarrow F_T \rightarrow S_4 \rightarrow F'_T \rightarrow S_1$$

Solution — heat-conductive plates around a pier (USA Patent No. 170299).

The solution is actually simple and cheap, and it is acceptable if the plates around a pier will not be barriers to anything. If the solution with plates (S4) is prohibited, then the method of upward transport of the warm layers is preferred. In this case, we need to solve the problem

$$P_1 \equiv S \dashrightarrow F_{ME}$$

(i.e., what energy can be taken for transportation of water and from where it can be taken?).

Introducing the corresponding EEEs could readily solve the problem. However, it is possible only if the interaction of the given fields and substances is allowed by the laws of nature, and if an application of the needed field (substance) is not prohibited by technical restrictions.

The following generic structural solutions exist for transformation of the given substance (field):

Sol_1. Utilization of a "good" substance:
   a. $\Rightarrow S$ repeated introduction of a substance;
   b. $\tilde{S}_1 \Rightarrow S_2$ replacement of a "bad" substance with a "good" one.
Sol_2. Modification of a substance by breaking (to finely dispersed state):
   a. $S \Rightarrow S - \Delta S$ separation of a part of a substance;
   b. $S \Rightarrow S_P$ transition to particles of a substance (milling of a substance).
Sol_3. Modification of a substance by introducing an additional substance (addition):
   a. $S \Rightarrow SS1$ an addition and substance form a mixture;
   b. $S \Rightarrow S — S1$ an addition is attached as a separate element.
Sol_4. Modification of a substance by affecting it:

$$S \Rightarrow S \equiv S'$$
$$\uparrow$$

The usage of these generic structural solutions for the problems $P_1 \ldots P_4$ is summarized in Table 16.1.

Sol_1 corresponds to the case of "pure" synthesis, when there is the possibility of free use of necessary physical solutions corresponding to EEE. The other solutions

TABLE 16.1
Generic Solutions of Typical Structure Problems

Generic Solutions	Typical Problems			
	P_1	P_2	P_3	P_4
	$S \to F$	$F \Rightarrow F'$	$F_1 \Rightarrow F_2$	$F \to S$
Sol_1	$S \to F$	$F \to S \to F$	$F_1 \to S \to F_2$	$F \to S$
Sol_2	$S \to \Delta S \to F$	$F \to \Delta S \to S \to F'$	$F_1 \to \Delta S \to S \to F_2$	$F \to \Delta S \to S$
	$S_p \to F$	$F \to S_p \to F'$	$F_1 \to S_p \to F_2$	$F \to S_p$
Sol_3	$S_1 S_2 \to F$	$F \to SS_1 \to F'$	$F_1 \to SS_1 \to F_2$	$F \to SS_1$
	$S - S_1 \to F$	$F \to S - S_1 \to F'$	$F_1 \to S - S_1 \to F_2$	$F \to S1 - S$
Sol_4	$S \to F$ $\uparrow$	$F \to S \to F'$ $\uparrow$	$F_1 \to S \to F_2$ $\uparrow$	$F \to S$ $\uparrow$

are usually applied if there are some restrictions which prohibit replacement of a substance-processor or if there are special requirements imposed on it. These solutions add complexity to the structure.

Sol_2 is applied when the energy transformation must be *spatially organized* in some way. If the required spatial organization is not complex, then some part of a substance is separated out. For example, separation of an electrode into parts specially arranged in space allows an electric field of a complex configuration to be obtained:

$$P_1 \equiv [S \text{ --- } S(x)] \to F_E(x).$$

To provide for a complex spatial organization, it is worth increasing the degree to which the substance is broken into particles.

Examples

1. If a typical vise's gripping jaws are filled with small rings or balls, then pieces of a complex shape can be uniformly gripped in the vise (USSR Patent No. 510350):

$$P_2 \equiv F_{ME} \to S_p \to F'_{ME}(x).$$

An outer layer of the vise's gripping jaws is treated as though it is broken into particles.

2. The method of drawing a steel wire without a draw plate (USSR Patent No. 499 912), in which a ferromagnetic powder under the action of a magnetic field plays the role of a draw plate, illustrates application of S2 to solution of P_3:

$$P_3 \equiv F_M(x) \to S_p \to F^{Power}_{ME}(x)$$

In some cases breaking gives finer effects. A finely dispersed substance is, as a rule, more chemically active. For example, separation of a vibrating mass into parts allows the resonance to be obviated by changing the coefficient of transformation of the mechanical energy. A solid body, which conducts strength well, becomes an absorber of mechanical energy after being broken into small particles:

$$P_4 \equiv F_{ME} \rightarrow \Delta S(x) - S.$$

The better method for imparting a new, needed property to a substance is Sol_3. It consists in forming a composition material (SS1) or simply attaching (joining) two substances (S — S1).

Example
Presence of conductive admixtures in plastic allows the plastic to be used to conduct electric current and to transform it into heat:

$$P_2 \equiv F_E \rightarrow SS_1 \rightarrow F'_E$$

$$P_3 \equiv F_E \rightarrow SS_1 \rightarrow F_T$$

If lead admixtures are introduced into a plastic, the plastic can protect from radiation:

$$P_4 \equiv F_R \rightarrow SS_1$$

Application of Sol_2 for changes of substance properties requires knowledge of the natural sciences, as discussed in Chapter 9. Compared to Sol_2, the mechanism of Sol_3 is simpler. We should find a single EEE which solves the corresponding typical problem and is provided for by physical laws and simply attach this EEE to the substance, replacement of which is prohibited by restrictions. Thus, Sol_3 is very widely applied and gives strong solutions.

Sol_4 is also directed to changing properties of a substance, providing energy transformations. However, in contrast to Sol_2 and Sol_3, this change is performed by actively affecting the substance with some field, such as activation of emission by exposing a substance to gamma radiation after neutron bombardment (stimulated radioactivity):

$$P_1 \equiv S \rightarrow F'^{\gamma}_{EM}(t_2)$$
$$\uparrow$$
$$F_{N,P}(t_2)$$

and the glow of luminophores under mechanical action:

$$P_1 \equiv S \rightarrow F_{EM}(t_2)$$
$$\uparrow$$
$$F_{ME}(t_1).$$

Action of the fields stimulates chemical activity of substances. For example, sand processed in a disintegrator acquires the binding properties of cement. At percussion compression of water solutions, fast oxidizing-reduction reactions run in such solutions (usually such reactions run only in the presence of acid).

Strong magnetic fields change the conductor's resistance:

$$P_2 \equiv F'_E \rightarrow S \rightarrow F''_E(t1)$$
$$\uparrow$$
$$F_M(t1),$$

while a weak thermal action on a ferromagnetic substance changes its capability to solve P_3:

$$P_3 \equiv F_M \rightarrow S \rightarrow F_{ME}(t_1)$$
$$\uparrow$$
$$F_T(t1)$$

due to the variation of magnetic properties of the ferromagnetic substance. Moreover, in some cases, weak magnetic fields change properties of nonferromagnetic substances. For example, spin-dependent reactions lead to anomalous changes of viscosity, electrical properties, and optical transparency of some diamagnetic chalcogenide glasses and their super-cooled melts.

Using the thermal field, we can cause the substance, which acts well as a transformer, to solve P_4. For example, by freezing water, which transfers pressure well, we can overcome the Pascal law because ice absorbs pressure well:

$$P_4 \equiv F_{ME} \rightarrow S(t_1)$$
$$\uparrow$$
$$F_T(t_1).$$

With S4, one can obtain the needed organization of a system on time (note that all the schemes illustrating the examples involve the symbols of temporal organization of the field). Choosing the corresponding action to provide an after-effect or instantaneous effect, we can control TS functioning.

Certainly, all solutions, Sol_1...Sol_4, are also applicable in combination. In such combination, the most interesting results and highly efficient solutions are often obtained. For example, joint application of Sol_2 and Sol_4 allows an easy change of system organization in space and time, while the combination Sol_3 + Sol_2 + Sol_4 adds the needed properties of good control.

16.4   ENERGY CHAINS

There exists a clear correspondence between EEE and the major functional-energy subsystems of TS: energy source, engine, transmission, working tool, and casting. The last often plays the role of terminating element for energy flows inside and outside the technique (see Chapter 3). The terminating element absorbs energy flux and may coincide not only with the casting but also with the raw object, product, tool, or environment. The energy source corresponds to EEE1, while the engine corresponds to EEE3. The transmission most often corresponds to EEE2 and less often to EEE3. The tool and product can be represented by all four types of EEEs depending on system's or subsystem's purpose. The casting as a terminating element, in its turn, corresponds to EEE4. Therefore, *energy chains* can be constructed from EEEs. These energy chains provide for energy completeness and conductivity of a technique (see Chapter 7). Individual EEEs in the energy chains are joined to each other by input and output fields in accordance with the rule of TS transparency (see Chapters 3 and 7). In the general case, the complete energy chain may look as shown below:

$$S_1 \to F_1 \to S_2 \to F'_2 \to S_3 \to F''_2 \to S_4 \to F'''_2 \to [\text{Spr.}]$$
$$S_5$$

ENERGY SOURCE (EEE1)	ENGINE (EEE3)	TRANSMISSION (EEE2)	WORKING TOOL (EEE2)	CASTING/ PRODUCT (EEE4)

The symbol [Spr.] is used here for a product, although for some TS and TP a product can be a field, e.g., radio transmitter.

Certainly such energy chains are not always present in technical systems. They correspond to energy-independent TS, which are needed in all the above-indicated energy transformations. In other cases, links of the energy chain may either coincide with each other (i.e., be joined) or be absent. For example, when the source provides energy in the already needed form, the engine can be excluded or joined with the energy source. When a system or subsystem does not include an energy source, it is considered energy-dependent. In this case, energy comes to the input of the energy chain as a field from other TS, TP, subsystems, or the environment. The transmission also can be excluded if the energy comes directly to the tool from the engine or the energy source. Therefore, the minimal energy chains $S \to [\text{Spr.}]$ and $[\text{Spr.}] \to S$ consist of only a tool and a raw object/product. Minimal energy chains are always preferable because they, as a rule, provide minimum energy loss. On the other hand, the minimal Su-Field model includes only a tool, product, and an interaction (field/energy) between the two.

Energy chains in which the energy comes from the tool to the product are called *changing* ones. They correspond to TSs whose purpose is to change the state of the product: $S_1 \to S_2 \to S_3 \to [\text{Spr.}]$ and $S_1 \to S_2 \to [\text{Spr.}] \to S_3$. The product in such chains is, as a rule, the terminating element or is immediately followed by the terminating element.

The energy chains in which the energy comes from the product to the tool are called *measuring* ones. They correspond to TSs whose purpose is to obtain information about the state of the product. Usually measuring chains are classified by type of input energy or by type of sensor that can be considered their working tool.

Examples of measurement energy chains are

$$[\text{Spr.}] \rightarrow S_1 \rightarrow S_2 \rightarrow F \rightarrow S_3$$

$$S_1 \rightarrow [\text{Spr.}] \rightarrow S_2 \rightarrow F \rightarrow S_3$$

$$S_1 \rightarrow S_2 \rightarrow F \rightarrow S_3$$
$$\uparrow$$
$$[\text{Sc}]\ [\text{Spr.}]$$

In such chains the product is, as a rule, located at the beginning or in the middle of the energy chain. Moreover, the product is the energy source or energy transformer. It also may control one of the subsystems of the energy chain, indicated above by the symbol [Sc], although the latter chain should be present in "full" form for some problems.

In measuring energy chains, the tool transforms the energy flux into a well detectable or well measurable field (usually electric or electromagnetic field). The terminating element in such energy chains may be a man, logic device, element of another energy chain, or the environment.

In examples of measuring energy chains, the lower chain consists of two chains. The energy flux directly transforming into the TS's or subsystem's useful product traverses one of these chains. The energy flux influencing one of the elements of the first chain and controlling the former flux traverses the other [Sc] chain. Such interaction of energy fluxes indicates existence of two more types of energy chains. The first type, called *manipulated* (*controlled*), incorporates energy chains. The energy flux, passing through these chains, most often supplies the product with energy and/or passes through the product. The energy flux in such manipulated chains is directly connected with obtaining a useful result. The second type incorporates *controlling* energy chains. These chains provide a directed change of the manipulated energy flux by supplying one of its elements with energy. The basic subsystem — *varier* — should be able to provide linear or nonlinear control depending on the actual situation.

Classification of energy fluxes into manipulated and controlling is, certainly, relative. A flux considered the controlling flux relative to one energy flux may be considered the manipulated flux relative to other fluxes. Moreover, in self-controlling systems or subsystems, one energy flux may execute functions of both the manipulated and controlling fluxes. The relativity of this classification is illustrated by the following structure:

$$F$$
$$\downarrow$$
$$S_1 \rightarrow S_2$$
$$[\text{Sc}] \downarrow [\text{Sc}]$$
$$S_4 \rightarrow S_3 \rightarrow [\text{Spr.}]$$

As with the classification of energy chains into changing and measuring, considering the classification of manipulated and controlling energy chains allows us to understand the specificity of their construction.

In order to control the manipulated energy flux, the corresponding energy chain should include at least one well controllable element, which easily changes its state under the action of a controlling field. Changes of this element should also influence the manipulated energy flux in the corresponding way. There are many known "substance — field" pairs providing controllability of a system, such as ferromagnetic substances (often particles) with magnetic or electromagnetic fields, and electric field and electrorheological liquids (mixtures which change their viscosity under the action of the electric field). If an image is an object under control, luminescent substances and UV radiation can be used. Application of substances which experience physical transformations under the action of controlling fields usually gives good results. Possible transformations include, for example, phase transitions: change of the aggregate state of a substance, change of magnetic properties upon passage through the Curie point, etc.

At the same time, substances of the manipulated energy chain may not be sensitive to the fields that are not used in the manipulated and controlling energy fluxes, in particular, to ambient fields.

The controlling energy chain which controls other chains or subsystems by a preset program without feedback is the changing energy chain. Some element of the manipulated chain serves as the terminating element in the controlling energy chain. This role can be played by any element of the manipulated chain, except for, as a rule, the product and the casting (terminating element in the PF chain). In the controlling energy chain itself, some elements or the input field should be organized in space and time in accordance with a preset program (that is, they must change in the corresponding way).

In the case of control with feedback, the "changing" part of the controlling energy chain is supplemented with a "measuring" one, which inputs information from the manipulated energy flux into the controlling one:

$$S_1 \to S_2 \to S_3 \to S_4 \to [\text{Spr.}]$$
$$\uparrow \qquad\qquad \downarrow$$
$$[\text{Sc}][S_6 \longleftarrow S_5]$$
$$F \to S_1 \to S_2 \to S_3 \to [\text{Spr.}]$$
$$\uparrow \qquad \downarrow$$
$$[\text{Sc}][S_6 \leftarrow S_5 \leftarrow S_4\,(x,t)]$$
$$S_1 \to F \to [\text{Spr.}]$$
$$[\text{Sc}]\uparrow \longrightarrow|$$

The measuring and controlling energy fluxes have one more common property: their intensity must be as low as possible. Satisfaction of this requirement corresponds to increasing Ideality of a system.

The above described procedure of energy chain construction from EEEs is valuable only if we succeed in finding the corresponding natural phenomena and effects for the structure to be constructed. A "seeking routine" for the array of effects has been developed in the framework of TRIZ by V. N. Glazunov, M. F. Zaripov, and others. Every physical effect is presented as a transformer of energy flux in the type or by a program. For energy emission or absorption, the concept of the inner

TABLE 16.2
Generic Parts for Energy Transformation/Conversion

Energy Parts	Description	Examples
Conductor	Direct energy flow in space.	Water pipe, fiber optics
Switch	Conduct/stop energy flow in time.	Power MOSFET, valve
Varier	Energy flow controlled by another energy.	Variable electrical capacitor
Resistor	Varier subclass: decreases a value of energy over time.	Electrical resistor, hydraulic filter
Amplifier	Varier subclass: increases an energy value due to an external energy.	Transistor, photomultifier
Converter	Change one type of energy into another type.	Thermal expansion bi-metal, photodiode

energy of a substance is used (at the input for EEE1, at the output for EEE4). The search comes by typical energy fluxes (fields) at the input and output with regard to control over parameters of a physical effect. The found physical effect for energy conversion and transformation should be transferred in the basic energy parts, presented in Table 16.2.

16.5 A CASE STUDY

Let us consider the approach to structure-and-energy synthesis of a system's operating principle using as the example the problem of sorting pellets.

> A product is processed with a jet of steel pellets 5 mm in diameter to form a surface rivet which increases fatigue strength of the product. The pellet surface must be free of cracks, as well as of blebs and lugs, which form surface roughness with the radius of curvature less than 1.5 mm. Sharp edges are especially problematic and unacceptable. Pellets not satisfying these requirements must be sorted out before processing with the pellet jet. Sorting is a manual process with all the attendant human-related troubles, penalties, and consequences. We need either to increase the productivity of sorting or fully automate the process.

The formulation of the function "sorting (separation) of pellets" involves transporting pellets in space. That is, there exist flows of substance (pellets) and flux of mechanical energy which coincide with the former ones. To determine the operating principle of this system, we should find the source of this energy and determine in what way the energy fluxes can be made sensitive to the shape of pellets.

It is easy to find that the Earth's gravity field is the energy source for the mechanical movement of the pellets. However, to make the pellet movement sensitive to its shape is much harder, especially in view of the extremely small deviations in the pellet surface from the ideal geometric sphere. Sorting into two fractions — good (S'pr.) and bad (S"pr.) pellets — consists of two operations: detection and separation. Detection is also of two types: by signal and by behavior. Detection by signal involves reception of information flow (signal) from the object to be detected. This information flow is used for control over another energy flux which separates the detected object from others. (A human sorting pellets works by this principle only.) A technical system of sorting by signal should first have the constructed measuring energy chain sending the needed signal. Since pellets themselves do not emit or adsorb anything, energy transformation is needed (i.e., EEE2 or EEE3 is needed):

$$F' \rightarrow [Spr.] \rightarrow F'' \text{ or } F_1 \rightarrow [Spr.]' \rightarrow F_2$$

It is desirable to have some well measurable and/or well controllable field (usually an electromagnetic one) at the output (F' or F''). The transformation must also be sensitive to the shape of the product [Spr.], the pellets.

Referring to the list of physical effects (see Appendix 3), we find the suitable phenomenon: sensitivity of electrical and magnetic fields to the shapes of bodies forming these fields. In particular, we can use the effect of increase of field strength at decreasing curvature radius of the surface of an electrode or magnetized body.

In the case of detection by behavior, objects are involved in a process and interact with other objects or the environment. Different objects interact differently and manifest themselves through different behavior. Naturally, the objects' interaction that is to be detected must correlate with criteria by which they are sorted. For example, if objects are sorted by density, then they may be placed in a liquid with a density average between the densities of the objects to be separated; the behavior to be detected is dropping or rising in the liquid. In the case of sorting by size, it would be best to pass the objects through a sieve, in which case the behavior to be detected is falling through or being caught by the sieve. As seen from these examples, the process of detection by behavior, as a rule, coincides with separation of one fraction from another. Therefore, sorting with detection of just this type is usually more productive than detection by signal. For sorting by behavior, the changing energy chain should be constructed.

The interaction suitable for our problem may be, for example, pellets rolling down an inclined plane because this movement depends on the shape of pellets:

$$F_1 \rightarrow [Spr.]' \rightarrow F_2'(S'pr.); \quad F_1 \rightarrow [Spr.]'' \rightarrow F''_2([Spr.]'')$$
$$\downarrow\uparrow \qquad\qquad\qquad\qquad \downarrow\uparrow$$
$$S_1 \qquad\qquad\qquad\qquad\qquad S_1$$

Here, S_1 is the inclined plane; F_1 is the gravity field energy; and F_2 is the energy of mechanical movement of the pellets [Spr.].

In order to implement the planned structure, we should strengthen the difference in interaction of the good (S'pr.) and bad (S''pr.) pellets with the inclined plane. Toward this end, we should use generic structure solutions (Table 16.1). Since we cannot replace pellets or separate them into parts, we have only solutions Sol_3 (attachment of a substance) and Sol_4 (action with a field), as well as their combination. We can use the effect of increased strength of the magnetic field with decreased radius of curvature. Then the following principle to strengthen differences in the shape of pellets becomes obvious: magnetize the pellets and sprinkle them with ferromagnetic particles. The particles stick mainly to places with increased magnetic field strength, which would be the bad pellets. The real technical system that was constructed from the basic energy parts, presented in Table 16.2, works very well.

16.6   SYNTHESIS OF TECHNICAL SYSTEMS — RECOMMENDATIONS

We should refine the character of technique organization in space, as well as in time, in which the required actions should take place. In addition, we should find what characteristic spatial and temporal rhythms should be inherent in subsystems or in their elements, in accordance with the principles of Ideality: introduce nothing unnecessary, coordinate rhythms of subsystems or their elements.

Determining the type of the technique and revealing its organization in space and time allow us to outline the preferable fields at the input and output. This is already a large step toward the sought solution, especially for a single-purpose technical system (i.e., TS without secondary UF). With determined fields at the input and output, it is far easier to determine the physical principle of functioning of the subsystems.

Fields at the technique's output are determined by the peculiarities of the output's use and sensitivity of the environment. If the output energy is processed by another TS which is already constructed, then the field at the output should correspond to the type of energy used by this other system. *This is the main synthesis principle for technological processes.* If the subsystem is intended to control some energy flux, then the field at the output should correspond to the well controllable element, being a part of the controlled energy chain. The fields at the technique's output can be given or determined often as preferable from the conditions of functioning. If the technique's output must act upon a person, then limited human abilities should be taken into account. We should also take into account the capability of the environment to respond to the technique's output.

When inventors faced the problem of developing a freon-free aerosol can, it was decided to create the required increased air pressure in the can using human energy. However, in order not to alienate customers, a normal human action was to be selected. The inventors selected shaking of the can ("shake before use").

The output field may also be the field controlling this system. On the other hand, input fields/energies predetermine the TS for their storage, as shown in Table 16.3, adapted from Gilbert Kivenson [2]. When fields are not determined, the following suggestions are worthwhile.

With nothing against it, it is preferable that the field at the input of the energy flux is the same as the field at the output, because transformation by the program is, as a rule, simpler than transformation in the type and converters are more expensive than other basic energy parts.

If the system already includes energy flux sufficient for operation of the given subsystem, a part of this energy flux should become input. For example, in a car

TABLE 16.3
TS for Storing Energy

Type	Technical Systems
Mechanical	Water reservoirs, flywheels, springs
Acoustic	Resonating chambers, strings
Thermal	Heated objects, melted solids, condensed gases
Electrical	Condensers, rechargeable batteries, electrets
Chemical	Fuels, exo- and endothermic reagents
Optical	Phosphors, glowing substances, fuels
Nuclear	Radioactive materials

engine, the energy of crankshaft rotation is used to set into motion all auxiliary subsystems.

The energy flux present in the environment and having the same organization in space and time as the TS to be constructed should be used in a similar way and also become the subsystem's input as either the manipulated or controlling energy flux. For example, to drive the reciprocating valve of the steam engine, the energy of the similarly reciprocating piston is used. If there is no suitable energy flux (field) in the environment to operate the subsystem, but there is a substance that changes its properties in space or time in the suitable way, then the field emitted or changed by this substance should become the system's input. In such a case, the substance becomes a subsystem of the TS, and the energy flux having passed through it is used for organization in space and time of the subsystem.

The product or the raw object itself should also be used to control the technical system. For example, temperature intervals in which zinc-aluminum and molybdenum alloys take the property of super-plasticity are different, rather narrow, and not always constant. To catch this interval and turn on a stamping press in the needed time, we can take advantage of the fact that as the alloy transits into the super-plastic state, its structure transforms and its magnetic permeability changes. (See also the case study in Chapter 18.) Typical controlled properties for different forms of energy are listed in Table 16.4 adapted from Kiverson [2]. A similar result can also be obtained as the system processes the energy flux, especially acting periodically with increasing or decreasing intensity. In such cases, the energy flux itself can prepare the subsystem for action by changing the state of the corresponding elements of the energy chain.

There is one more recommendation for selecting the field to act in the technical system. The field should maximally fit the capabilities of the given subsystems to respond to it or to affect it. In "measuring" subsystems, capabilities of the product or raw object are decisive. If a field's changes are indicative of the state of the raw object or product, and it passes through such a subsystem, then the field type should correspond to the product's properties that distinguish the state to be detected or measured from other states. To detect the starting time of electrolyte boiling, the field passing through the electrolyte should be sensitive to the main factor that

TABLE 16.4
Typical Controlled Properties for Different Energy Forms

Energy Form	Symbol	Typical Controlled Property (Measurements)
Mechanical	ME	Position, dimension, weight, volume, pressure, speed, tension, torsion
Acoustic	A	Power, absorption, transmission, wave shape, spectral content, phase, noise level
Electrical	E	Voltage, current, resistance, inductance, capacity, transconductance
Chemical	C	Equivalents between substances (acid for base, etc.), chemical potential, composition, level of polymerization
Nuclear	N	Particle type and identity, particle energy, mass, velocity, and momentum, quantum emission
Optical	O	Color, absorption, transmission, wavelength, phase, power spectral content, stability of frequency
Thermal	T	Temperature, radiation, conduction, convection, thermal capacity, heat transfer properties

distinguishes this time from others: appearance of steaming bubbles. Such a field may be electrical or acoustic.

In order to create links in the energy chain, it is necessary to consider their characteristics (see Chapter 3). Structure synthesis (typical links and generic solutions) can be applied sufficiently formally.

Lastly, to adequately search for the physical operating principle of a TS and to increase the probability that the obtained technical system will be sufficiently efficient, it is important to make an accurate Su-Field Analysis of TS, account for the operational conditions of the system, make a detailed and unambiguous determination of its inputs and outputs, and fully use the principles of Ideality.

16.7 CONCLUSION

Structure-And-Energy Synthesis works rather well at the conceptual design stage of various technical systems, as it allows us to formulate a "skeleton" of a TS that could be implemented at other design stages. The revealed trends in technique evolution significantly restrict the range of variability of conceptual designs and allow synthesis of a prototype of an Ideal technique.

REFERENCES

1. Goldovsky, B. I. and Vaynerman, M. I., *Rational Creativity,* Moscow, Rechnoj Transport, 1990.
2. Kivenson, G., *The Art and Science of Inventing,* Van Nostrand Reinhold, New York, 1982.

17 Agents Method

17.1   HISTORICAL NOTES

This chapter discusses a problem-solving instrument based on solving Agents. In contrast to all other TRIZ heuristics, this is the first TRIZ method to have been developed almost independently in numerous countries — Russia, Israel, and then US. Agents in TRIZ originally were "small smart people" that could do anything a problem solver needed to do in the problem-to-solution transition. They were derived by Altshuller at the end of 1960s from Synectics, the American method of creativity activation. Ten years earlier, William Gordon, the author of Synectics, had suggested using personal analogy or empathy in the solving process [1]. The essence of empathy is that a person "enters" into the object to be improved and tries to imagine the action required by the problem. During TRIZ classes, Altshuller realized that the weak point of empathy is the strong tendency to reject any action that is unacceptable to the human organism. This drawback is overcome with the help of "small smart people" in modeling [2]. A transition from the "small smart people" to "inanimate particles" was proposed by Solomon D. Tetel'baum about fifteen years ago [3], but the idea was not supported by other TRIZniks who often used teams of boys and girls during their lessons. Due to emigration of some TRIZniks from USSR to Israel in the 1980s, this methodology became popular in the Middle East. In the Israeli teaching experience, it was found that students did not always use small smart people effectively. It seems that subconsciously some students were reluctant to place these small smart people in situations that would be life threatening to humans, such as in strong acids or extreme fields. Therefore, Genady Filkovsky, Roni Horowirz, and Jacob Goldenberg from the Open University in Israel replaced small smart people with inanimate particles [4].* This particles method is now used actively in the Israeli derivative of TRIZ simplification named SIT, where it represents almost half of these problem-solving activities [4, 5]. However, some of the author's students have argued that they can more easily imagine various actions performed by the "small smart people" than by inanimate particles. This is all a matter of semantics and the term itself is not as important as the method. But we will use the neutral term *agents*. The experience of Russian, Israeli, and American specialists is summarized and generalized in the Agents Method described in this chapter.

17.2   INTRODUCTION TO THE AGENTS METHOD

The Agents method (AM) helps solvers find conceptual solutions to technical problems by forward, backward, or bidirectional searching (see Chapter 5), usually with graphical support in the form of simple sketches (see Chapter 10). Agents play the role of the "universal ideal tool" for fulfilling the main function of the system. They can carry out any necessary action; any physical, chemical, biological, or geometric effect can be embodied by Agents, irrespective of whether this effect has originally any technical sense. AM builds the bridge between the two endpoints in TRIZ: the correct statement of the problem (CSP) and the ideal final result (IFR). It offers a structured process for identifying the critical steps of solutions, thereby filling the gap between the problem's situation and its goal.

The development of the decision within the framework of AM follows this hierarchical process:

1. Precise formulation of the CSP and/or IFR. The main question here is *What is the CSP and IFR?*
2. Choose the starting point between CSP and IFR. The main question here is *What is important?* If you cannot answer this, use the bidirectional search.

---

* In general, this idea is not new in problem solving; even the famous antique Greek philosopher Demokrit used small particles for explanation of natural phenomena. The famous physicist James Clerk Maxwell used small demons (human-like beings) for resolution of scientific problems.

3. Create a graphical solution. The main question here is *What does a solution looks like?* Note that it is possible to create the graphical solution starting from CSP, from IFR, or from both ends.
4. The fourth phase is a development of structure logic for the Agents' actions. The main question here is *What are the Agents doing?*
5. Prepare the logic list of Agent properties in relation to actions that were previously decided. The main question here is *What properties must the Agents have?*
6. Generation of the solution. The main question here is *How can Agents be transformed into a technical system or technological process?*

AM helps in finding the conceptual solution of a problem; solvers can then use other TRIZ heuristics to check the engineering validity and/or the consequences of the solution. The experts should usually confirm the idea of the solution during or after the last phase.

17.3 AGENTS

A solver should remember that Agents predetermine the solving process. Let us discuss some major features of Agents before detailing AM itself.

17.3.1 PREFERRED SOURCES

Agents are a sort of tool in relation to a raw object and product.* (Keep in mind that the terms "tool" and "product" are not absolute in TRIZ). Both tool and product can be neighboring subsystems of the same technique. Agents should be a part, a subsystem, of a tool or a product, or they should be available from whole TS or TP, external resources, super-system, or environment, in that order of preference.

Frequently Agents emerge from modifying the tool or product through phase transitions (plasma-gas-liquid-solid states, paramagnetic-ferromagnetic states, etc.), chemical reactions (oxidation-restoration, electrolysis, etc.), splitting, change of concentration, flexibility, etc. On the other hand, they can be created indirectly in TS or TP. Altshuller's Standards 5.5.1 and/or 5.5.2 can help in this process (see Chapter 16 and Appendix 5).

17.3.2 PREFERRED LOCATIONS AND TIMES

The Agents should be placed in the CSP, IFR, or sketch of the intermediate condition in order to specify where Agents should be used, which is usually the operation zone of the tool-product interaction (see Chapter 14). The purpose of Agents is to ensure necessary physical, chemical, biological, or geometrical properties and functions in the operation zone, so a solver should understand what the Agents do in this zone. The Agents are placed in different locations of the operation zone or their work is updated at each stage of the graphical solving process. This process helps to specify where or what kind of changes should be reflected in the subsequent sketch.

---

* For simplicity a product and a raw object are not distinguished in AM.

It is possible that the problem exists only in the operation zone, so it is important to define the Agents' work directly before, during, and immediately after that period. For some problems, the time factor is not important at all, in which case we can eliminate time from our consideration. AM does not require analysis of the Agents' work in time intervals other than the operation. Thus, time, as a factor in the problem, can be removed from many intermediate sketches and also from sketches of the IFR and CSP.

The simplicity of the analysis and orderliness of the solving process are two reasons why it is not necessary to work outside the limits of an operation zone. The simplicity helps make the solving process clearer, definite, and robust, and the orderliness allows the solver to find the best solutions quickly. These reasons are well known as the Ockam's razor* or KISS (Keep It Simple Stupid) principle.

17.3.3 REQUIREMENTS OF AGENTS

Though the various aspects of "life" of Agents will be discussed below in this chapter, the following six remarks seem appropriate here:

1. Agents should be a part of a tool, raw object, and/or product or should be available from resources.
2. Agents can be delivered to the operation zone (in a minimal dose, and gain the effect through better distribution) and exist there during the operation period (or a part of T1 when it is actually necessary).
3. Each type of Agent can carry out only a few specific actions, and usually only one action.
4. It is necessary to keep the number of Agent types as small as possible, in keeping with Ockam's razor principle. It is better to start with the single type and increase the number of types gradually if the problem cannot be resolved with the minimal set of types.
5. Agents of each type always work equally under identical conditions. This action can be altered only if external conditions change.
6. Often Agents can be presented as active materials (substances). Sometimes it would be easier to enter a field instead of substances. In that situation, use knowledge and hypotheses from natural sciences for representation fields through specific that are summarized below:
    Electric — electrons, ions, positrons
    Magnetic — Dirak's monopoles
    Chemical — atoms, molecules, ions, radicals, etc.
    Gravity — gravitons, super-heavy grains
    Olfactory — molecules
    Light — photons
    Radiation — alpha, beta,... -particles
    Thermal — phonons

---

* William of Ockham (1280-1349, Franciscan theologian) proposed the logical and philosophical principle, "It is useless to use greater means to achieve a goal attainable by fewer means." This principle is the ground of mathematics and all natural sciences.

A solver should choose the most convenient way from these two options; our experience shows that it is easier to operate with the material-like Agents than with field-like Agents.

17.3.4 INITIATION/TERMINATION OF AGENTS

It may be necessary to determine how these Agents are *initiated*, introduced into the locations where they are needed, or how they are *terminated*, removed after their actions are complete. Common initiation and termination methods are listed below [5]:

Usual Agents Initiation	Usual Agents Termination
getting there	being annihilated
being present	being removed
being put there	leaving (of their own volition)
created there	remaining
pinning	repinning

All aspects of this list may not be needed for a given problem analysis. Although initiation and termination of Agents have routine forms, their details in a specific application depend on the situation of the Agents being analyzed. If a few types of Agents are used in solving a problem, their initiation and termination should be considered separately.

17.4 GRAPHIC PROCEDURE

Usually AM is used as a visual instrument, working with simple but informative sketches which play an important role in this method. These sketches have all (or at least the majority of) the essential details of the subsystems where the problem appears and omit unnecessary details of the subsystems. The subsystems' and technique's functions usually determine what is essential for the sketches. All aspects that are desirable for problem solving should be reflected in these sketches.

AM set of sketches includes a sketch of the initial CSP that is usually drawn first and as the leftmost sketch, and the IFR that is usually drawn second on the far right. These sketches determine what actions the Agents should take, i.e., what allows them to carry out the requirements specified in the problem. If these two sketches differ considerably, one or several additional sketches are drawn to mirror details of conceptual steps between CSP and IFR. These intermediate sketches describe a process by which a solution is to be developed.

It is not recommended to guess whether it is possible to reach IFR or not after you prepare the IFR sketch. At this phase a solver should not think about how IFR will be achieved or how to get from the CSP to IFR.

It is necessary to enter Agents into sketches within the framework of AM. Agents are represented by simple symbols such as **X, +,** or *****. Different Agents should have different symbols or different colors. Sometimes the Agents can complicate a sketch; therefore, it is possible to represent the CSP and/or IFR sketches twice — with Agents and without them, but the Agents are always necessary on the intermediate sketches.

17.5   SEARCH OF GRAPHICAL SOLUTION

Fortunately, AM accommodates various ways of problem solving. A solver can start from CSP, IFR, or both ends. The possibility to perform a bidirectional search of a formal solution is the most important and attractive feature of AM. Although when doing a forward search it is not necessary to know beforehand if IFR is technically realizable, for backward and bi-directional searches* the solver should know that IFR is technically possible and probable.

In the process of searching for a graphical solution, one first sketches CSP and IFR. Then the solver develops conceptual steps between them and defines how to carry out graphic changes in intermediate sketches so as to build a bridge between CSP and IFR. A solver can draw the intermediate sketches in order to illustrate process of transition from CSP to IFR, or vice versa, or make a few sketches from each initial point for a bidirectional search. Forward, backward, and bidirectional AM processes are fruitful and can provide a solution in the framework of TRIZ. If both IFR and CSP are known, the bidirectional search usually requires less effort than the forward or backward search because it allows the solver to work in another direction when a few equal possibilities to continue the solution search appear at some step of the problem solving. Distinctions between the intermediate sketches specify changes that occur at the given stage, i.e., what Agents should do in order to push the system to the state that is shown on the IFR or CSP sketches. Each intermediate sketch has to bring graphic representation of new information in the problem solving. Otherwise, it has to be removed from consideration. The drawing process does not require consideration of real physical or chemical effects or energies at this stage. Frequently the construction of intermediate sketches represents a quasi-automatic process because any interpretation of sketches is not yet required here. The interpretation as well as the physical, chemical, or mathematical details involved in the given sketches arise with further serious analysis. In the best case, a complete reasoning for transformation between CSP and IFR is built, but often each search direction may have unknown steps. A solver should not try to investigate all oppor-tunities at once but instead concentrate efforts on one, proceeding to the end, even if the complete reasoning cannot be built. Analysis of one direction can be exhaustive, so search of other ways is often not required. With certain skills, the analysis is clear and can be done in all three ways in a rather short time.

17.6   THE AND/OR TREE

The Agents themselves are not necessary; we need their "work" and their properties. Therefore, the following step of AM is the determination of the Agents' character-istics. This step can be performed with the aid of the so-called AND/OR tree, well known in logic. Both IFR and CSP blocks are shown in any AND/OR tree and are the kernels of such trees; hence, in a single-directional search they are shown at the highest tier. All other tiers correspond to small steps in the Agents' prospective actions in the operation zone. Actions are shown inside the blocks, which represent

---

* The criteria for these searches are outlined in Chapter 5.

an answer to the question about what the Agents are doing. Any subsequent (lower) tier connected by branches with the preceding tier stands for a combination of several AND/OR conditions expressed in the positive form. Any set of branches of the AND/OR tree represents a gradual transition between IFR and CSP. Each condition is developed independently on the consecutive tree's tier; i.e., each branch symbolizes the independent analysis. The number of sets should be as large as possible, because each set represents opportunities for an autonomous solution of the problem. The reduction of sets is performed later in AM with a special algorithm.

Action and counteraction is one of the major principles used during the development of the AND/OR tree. For each action there may be a useful opposite action that should possibly be included in the AND/OR tree. Subsequent analysis of Agents carrying out both actions and counteractions can assign situations that would have been encountered directly with other Agents' arrangement or location in an operative zone and period. The potential conflicts between the requirements of Agents existing on various branches are ignored at this stage.

In the case of the bidirectional search, the indicator of a solution is the appearance of identical actions at the branches that grow from the CSP and IFR parts of the tree.

The AND/OR tree is completed by addition of lists of known properties for fulfillment of required Agents actions. The main question here is *How can the Agents of a particular type carry out required actions*? *What* Agents do is established in the previous stage, but at this stage we want to know precisely *how* they do it. Therefore, it is necessary to understand the properties of each type of Agent that allow them to carry out these actions. It is not important to know what the Agents themselves represent to answer both these questions about Agents' actions and properties. Thus, the lists of properties (as detailed and long as possible) can be attached to each action of each branch of the AND/OR tree (see Figure 17.1). The various lists at different branches of the same tree can have identical properties or can be in conflict with each other. A solver should pay steadfast attention to such similarities and conflicts.

It is also possible to add the particular Agents' initiation and termination (see Section 17.3.4) to the AND/OR tree kernels or to include these features in the auxiliary tables. In a forward search, the Agents' initiation and termination can be attached to the CSP or IFR kernel, respectively; in a backward search, they are added to the IFR and CSP kernel, respectively. The initiation/termination tables can be attached to the blocks of the AND/OR tree when a few types of Agents are involved in the solution.

A solver can use hypertext (commonly associated with the WWW) instead of an AND/OR tree. Regardless of the method, a knowledge-intensive approach should be involved at this step of AM which requires a solver to first define the necessary useful actions that Agents could carry out in an operation zone, and then the Agents' properties that allow them to carry out these actions. Actions of Agents and their properties noted in the AND/OR tree or mentioned in the hypertext can embody the solution represented on the sketches. Actions and the properties of the Agents become the functions, characteristics, and parameters of the technique or its subsystems, entered for the realization of the problem solutions.

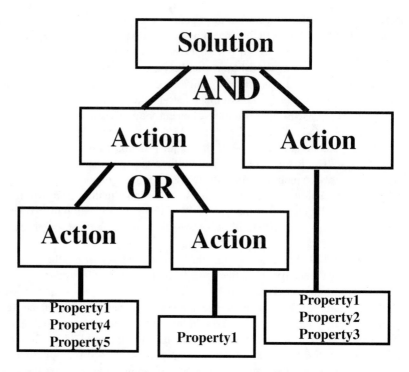

FIGURE 17.1 Schematic diagram of a completed backward AND/OR tree. The top box contains the IFR statement, and the next tier boxes identify Agent actions along the solution path. The lowest tier properties boxes can be linked with lists of known methods to achieve the desired Agents property. (Note that Property 1 appears three times, indicating that redundancy is allowed in the AND/OR tree.)

17.7   ACTIONS AND PROPERTIES OF AGENTS

Actions and properties of Agents are in fact the pointers to the function and resources for problem solutions. They depend on the problem, so it is impossible to predict *a priori* what type of Agent is required. Usually a solver describes the Agents' action by the appropriate verb and the properties by nouns.

17.7.1   AGENT ACTIONS

The purpose of this stage of AM is to generate as many useful actions as possible. Koller's elementary functions [6] can serve as the initial set of possible actions:

- emitting
- conducting
- collecting
- guiding
- transforming
- enlarging

- direction-changing
- directing
- coupling
- connecting
- adding
- storing

Some common actions can be found in the bank of transformation heuristics collected by A. I. Polovinkin and his co-workers [7].* Note that the Agents' actions should not span the problem's boundaries. All actions assume functions and Agent properties, assisting them in carrying out a particular action, and the effect of an action at one tier of the AND/OR tree can differ from the effect of the same action at another tree's tier.

17.7.2   AGENT PROPERTIES

As soon as all details of Agent actions are determined, it is necessary to define which Agent properties should carry out the specified actions. The solver should list all known properties of Agents with which they carry out actions. This list for Agents can be deduced from the Lists of Effects that are part of the knowledge base of TRIZ (see Chapter 9 and Appendix 3) and detailed descriptions of particular effects that can be found in the scientific books and journals for the particular branch. The subsequent list of properties will depend upon the specific situation being analyzed.

For example, consider the action "To come into a pore." The Agents should have appropriate properties, such as:

- size
- electric charge
- flexibility

Such lists of properties belong only to those actions to which they directly correspond. Any property concerning an action is acceptable even if it contradicts the requirements of Agents on other branches of the AND/OR tree. Each list of properties should be examined only in relation to the relevant action.

Due to the Lists of Effects and the algorithm for Agents designed in the framework of TRIZ (see Appendix 3 and the next section) a solver can determine the Agents quite easily.

17.8   ALGORITHM FOR AGENTS

AM includes an algorithm for choosing Agents based on their actions and properties. The base of this algorithm was created by A. M. Pinyayev [9] during his analysis of Altshuller's Standards and developed later for AM by the author. The initial data

---

* This part of the knowledge base of modern TRIZ is presented in Appendix 8.

for this algorithm are Agent properties, useful and harmful functions, and resources as defined in TRIZ. The algorithm is the following:

A. Specify which properties the entered Agents should have in order to accomplish the useful actions. This step results in list A.

B. Formulate the opposite of those Agents' properties specified in list A. Avoid using the word "not" or negative prefixes such as "im-," "non-," and "un-." This step results in list B.

C. For each property on list B, define whether the useful action will be carried out, and if the entered Agents have one of the properties on list A and others from list B.

D. Delete from list B those properties for which the answer "not present" was received at step C. Delete the appropriate properties from list A. The result is list D.
    Note: List D will be shorter than the previous two.

E. For each kind of Agent on list D, determine if any harmful function exists or is performed.

F. Delete from list D those Agents for which in step E the answer is "yes." The result is list F.

G. If list F is *empty* go to step K. If list F is *not empty* go to the next step.

H. Define which properties of the Agents on list F already have
    • Tool or other subsystem of the technique
    • Product or its derivatives
    • Super-system
    • Resources and environment

I. If none of the objects mentioned in the previous step has the necessary properties, formulate the inventive problems to adjust for the missing properties to these subsystems or resources.

J. If you cannot resolve the problems in the previous step, use Agents from outside sources.

K. Change all properties on list D for which the answer was "yes" in step E to supplemental properties. The result will be list A1. If, during this step, all supplemental properties coincide with the properties in list B, you must resolve the physical contradictions. The result will be list A2.

L. Return to step B with either list A1 or A2.

This algorithm makes it possible to determine the most promising types of Agents for technical problems in various engineering fields.

17.9   REALIZATION OF SOLUTION(S)

Three essential components of AM have already been established:

1. Branches of the AND/OR tree between a certain item in the list of properties and the kernel(s) (at least one CSP or IFR),
2. Agents actions and properties lists,
3. Ways of Agents occurrence and disappearance.

It is next necessary to organize them into conceptual solutions. The following four major approaches lead to disclosure of the conceptual solutions of the problem based on previous AM results:

1. Choose Agent properties from pairs of branches, connected by the AND conditions. These "AND branches" interest us because they frequently offer inconsistent conditions. The resolution of this technical contradiction can bring a strong solution.
2. Find the common properties of Agents in different branches that "grow" from the same kernel, which can indicate the possibility of effective solutions.
3. Find the common and consistent properties (combinations of such properties) of Agents in pairs on different branches that "grow" from opposite kernels, which indicates a strong conceptual solution that often can be easily implemented because it meets requirements of both ends of the problem, IFR and CSP.
4. Choose opposite properties/actions of a single type of Agent. Sometimes they reproduce the physical contradictions of the tool. The resolution of this contradiction can bring a strong solution.

If a few real solutions are obtained, it is sometimes possible to choose the best by using the decision aids in Chapter 1.

17.10   A CASE STUDY

The following case study is based on the author's project on self-assembly of quantum dots for nanotechnology applications.

Let us consider a microstructure creation on the surface of semiconductor wafers. This process, known as photolithography [8], is very important to the electronic industry. Photolithography is one of the key limiting factors of yield in the modern semiconductor industry, and it is also the main barrier for future miniaturization of semiconductor devices. Consequently, the search for photolithography alternatives is very important. Our goal here is to show how AM works rather than actually to design a new technological process.

17.10.1   PROBLEM AND SOLUTION CONCEPT

As part of the case study, we need to outline how microscopic patterns are created on the surface of semiconductor wafers with diameters to 65 centimeters (the current limit achieved by the author's team in another project). This surface can be the semiconductor (Si, GaAs, Ge), oxide (SiO_2), some metals (Al, Au, Ti/Ni), etc. All these materials cannot themselves create the patterns, with sizes from a few microns down to 0.1 μm (the current limit for semiconductor devices), so they can be considered all together as a passive substrate. That is why a thin active layer is deposited on a wafer (substrate) surface. Currently special liquids are used which can polymerize after deposition. The resulting material, called photoresist, can change its native chemical resistance against special etching liquids and gases under light

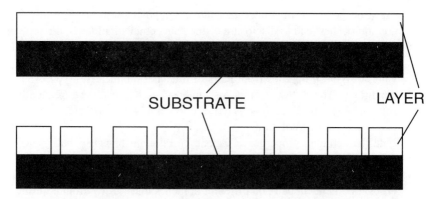

FIGURE 17.2 Changes of a substrate surface due to photolithography.

radiation. The photoresist must have very uniform thickness (±1 μm) along the
substrate surface, and expensive machinery is used in this process.

The common photolithography process is illustrated in Figure 17.2.

Because numerous semiconductor devices can be produced on a single wafer,
the pattern has some symmetry; the devices are symmetric elements of the wafer.

Let us consider a more general problem — creation of a symmetric pattern on
a condensed matter's surface. First of all, we need to remove special terminology;
"photoresist" changes to the more universal word "layer" and "wafer's gas environment" becomes "air." The term "substrate" is already pretty general and need not
be changed.

17.10.2 KERNELS SKETCHES

The subsystems and resources of this problem are material's volume and surface
and the air around the substrate. The initial situation, CSP, is shown in Figure 17.3a
and the final situation, IFR, is shown in Figure 17.3b.

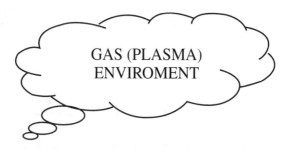

FIGURE 17.3a The initial situation (CSP).

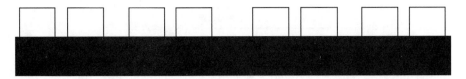

FIGURE 17.3b   The final situation (IFR).

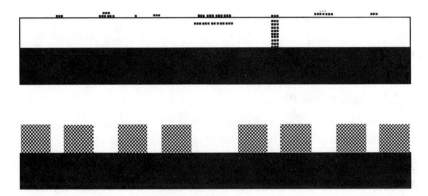

FIGURE 17.4   Placement of Agents (small points).

17.10.3   INTERMEDIATE SKETCHES

Agents are shown in Figure 17.4, an intermediate sketch, to indicate that some changes are needed to achieve the condition shown in the IFR sketch. The placement of Agents in the sketch indicates that something must happen in the zones near Agents. The solution sketch indicates that

  a layer has been built by the Agents (unusual process for the semiconductor industry)

or

  a layer has been removed, and the Agents are gone and no longer have their useful function (as in common photolithography process)

The Agents should be on or in the substrate or layer during the process (the intermediate sketches) and should be removed when the process is done (in the solution sketch).

17.10.4   AND/OR TREE

We will discuss in this subsection aspects of the Agents initiation and termination, as well as the properties of Agents that can be useful for the layer problem.

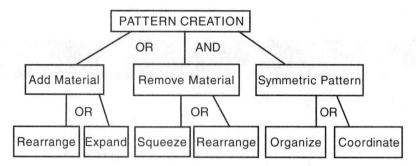

FIGURE 17.5   Solution statement and actions' levels of the AND/OR tree for the layer problem.

17.10.4.1   Agents Actions

Figure 17.5 shows the solution statement with its AND/OR tree and two regular forms for addressing Agent initiation and termination. At the top of the AND/OR tree is the solution statement "Pattern Creation"; subsequent tiers have the following AND/OR statements: Agents are adding the material (substrate?) on or in selected sites; OR Agents are reducing the layer at selected sites; AND Agents are creating a symmetrical pattern. (A well-defined IFR allows work back toward the problem situation sketch.)

The next layer includes the following possibilities, connected only by the OR logical operator with the corresponding previous statements: Agents add new material to the surface; Agents expand existing material of the substrate; Agents rearrange material; Agents remove material from the surface; Agents condense (squeeze) existing material; Agents organize otherwise randomly located activities into symmetric-pattern producing activities; Agents coordinate movement of the material.

17.10.4.2   Agents Properties

Now we need to construct lists of known methods to achieve the properties required of the Agents in this problem:

- For *adding (growing, expanding) material* in or on selected sites, Agents should be sticky for building up material; charged; magnetic (so that they are applied or attracted to the substrate); dispersible for nonuniform application; expandable; reactive to produce lower density material; capable of rearrangement by diffusion; mobile; reactive; etc.
- Properties of Agents for *removing a material* are similar to the properties of Agents for adding material. In addition, Agents can be hot enough to melt or evaporate unwanted portions of material.
- Properties for contracting material are similar to those for expanding it.
- Properties for rearranging material are the same as for nonrearranged material.
- For *organizing actions*, Agents should be replicable in the sense of a template; capable of communication, or monitoring, in order for changes in each area; in step with the others; attractive or repulsive so that they

automatically position themselves equidistant from each other, forming a symmetric pattern; etc.

These lists can be expanded with more detailed consideration of the problem.

17.10.4.3 Agents Initiation and Termination

The pattern layer is much thicker than the substrate, so it is possible to consider only the substrate surface. Some particular ideas can be generated from the common methods of the Agent's initiation and termination on a surface. For demonstration purposes here, even the absolutely crazy (for semiconductors) ideas are not omitted (as they would be in real practice with AM).

Initiation of the Agents' functions is listed below:

already there — part of layer surface and/or core
created there from existing materials — part of substrate, part of air
appearing there without an external force — mobile, self-guided, attractive
being brought there by an external force — mechanical, magnetic, charged

Expansion of the Agents' initiation branch with lists of known properties leads to the following:

- Agents already there: they need to be part of the core material, the surface material, or components of air (including gases, bacteria, dust, pollen, moisture, other chemicals, etc.). If Agents are part of the core (substrate or layer), they should have additional properties enabling them to get to the substrate surface where they are needed. Unfortunately, such a way is not possible in semiconductor manufacturing.
- Agents being created there from existing materials: they need initially to be part of the air, surface, or core and then to be transformed and moved to the surface region to the appropriate chemical state by chemical or nuclear reaction or by biological effects. Unfortunately, such a way is not possible in semiconductor manufacturing.
- Agents being put there because of an external force: they need to have properties that enable them to be "entrained" or "grasped" for moving to the right place. This can be accomplished by mechanical means or by application of fields. Agents can be dispersible and sprayed on as a mist or applied by dipping a substrate into their suspension. If attachable and gaseous, they can be evaporated on the substrate surface. If reactive, they can be applied as a reaction with a substrate surface. Hence, the Agents can have attributes of mass, shape, volume, stickiness, electrical charge, magnetization, roundness (for rolling), etc. Most of these ideas are known and already implemented in the semiconductor industry.
- Agents appearing there without external energy: they must have properties of mobility, self-guidance, and attraction. This idea was new for the semiconductor industry.

Similar lists can be developed for other branches.

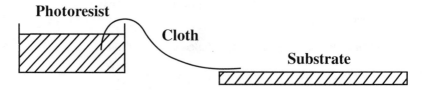

FIGURE 17.6   Photoresist coater.

The consideration of the last new idea led to the following solution for the photolithography process proposed by TRIZ Experts. The solution (see Figure 17.6) replaces expensive equipment for coating wafers by photoresist with an inexpensive and reliable coater. A coater for photoresist uses a special cloth in contact with the photoresist bath on one side and touching the substrate on the other. It evenly coats (thickness 1 μm) the wafer surface and has produced enormous savings of the quite expensive photoresist in semiconductor plants in Russia, Ukraine, and Byelorussia.

Termination of the Agents' functions include the following:

remains — nonadhesive, divisible, soluble, magnetic
is removed — mobile, chemically inert or active
removes itself — reactive, radioactive, edible, mutable
self-annihilate — inert, much smaller than smallest pattern feature, assimilative into layer/substrate materials

The known properties for Agent termination listed above include the following:

- Agents remain: they should be chemically inert, smaller than the smallest pattern feature, or able to assimilate into the layer material and lose their original identity. Unfortunately, such a way is not viable in semiconductor manufacturing.
- Agents are removed: they should be nonadhesive, soluble, divisible into particulates or flakes, charged, or magnetic in order to be easily removed. Some of these ideas are already known and implemented in the semiconductor industry.
- Agents leaving on their own accord: they should be mobile, inert, and want to leave; inertness might be wanted to assure that the Agents will not interact with anything to delay or prevent their leaving. Most of these ideas are already known and implemented in the semiconductor industry.
- Agents self-annihilate: they should be reactive in order to become some other species, radioactive, edible (e.g., by bacteria), or capable of mutation to an innocuous form. This idea seems new for the semiconductor industry.

17.10.5 ALGORITHM APPLICATION

We now apply the Agent algorithm (Section 7.8).

List A
The useful action is listed in the left headers of the following table, and the properties are listed in the left column.

Agents Should Be	How Agents Are Able to Act

Agents' Properties to Add a Material on a Surface

Agents Should Be	How Agents Are Able to Act
Adhesive for accumulating material	Adhesion; they adhere to the surface
	Agents are "woven" in
Reactive with the layer material to change its density or volume	Density (volume) can be changed by nonuniform pressure
Mechanical for manipulation	Rotates itself and drills the micro-voids
Charged	Voltage source
	Electrostatic charging, such as by rubbing
Magnetic (so they can be applied to the layer or are attracted to it)	Agents themselves are magnetic and the layer provides ferromagnetism
Dispersible for nonuniform application	Agents are microscopically small
Small enough to get into cracks, pores, or interstices of the layer's material	Dust Agents
	Molecules
	Bacteria, viruses
Reactive with the material to cause swelling	Chemical reaction
Small enough to be able to diffuse into the material	Atoms, ions
	Gaseous agents
Capable of rearrangement by diffusion	Follows a gradient in diffusion
Reactive to produce lower density material	Radioactive
	Foams
	Mixture
Expandable	Inflatable device
Mobile	Variable concentration both in time and space to cause a variable gradient
	Thermal mobility (Brown's molecular movement)
Reactive	Chemical reaction
	Nuclear reaction

Agents' Properties to Reduce a Material from a Surface

Agents Should Be	How Agents Are Able to Act
Properties for removing material are similar to properties for adding material. In addition, they can be hot enough to melt or evaporate unwanted portions of material.	Agents transform a portion of the waste energy from other manufacturing operations of the semiconductor process into heat.
	Exothermic reaction
	Melting point is quite low, such as room temperature
Properties for contracting material are similar to those for expanding material.	Inflatable device

Agents' Properties to Organize Actions at a Surface

Agents Should Be	How Agents Are Able to Act
Replaceable in the sense of a template	A source delivers new agents after they were consumed
	Return to initial position, e.g., by magnetic field
Capable of communication or monitoring, in order for changes in each area of the surface to be in step with the others.	Electric or magnetic interaction
	Diffusion interaction
	Standing waves
Attractive or repulsive so they automatically position themselves equidistant from each other, forming a symmetric pattern.	Electric or magnetic interaction
	Diffusion interaction
	Standing waves

List B
Agents' properties for adding a material on a surface

Should leave the volume of a substrate.
Inert and keep the density as it is.
Neutral/inert to keep density or reactive to increase density
Para- or diamagnetic.
Fixed for uniform application.
Large enough to stay on the surface.
Inert and keep the surface as it is.
Retain size, shrink, or change size.

Agents' properties to organize the surface and actions on it

Cold.
Retain size, format, weight, etc.
Isolated.
Ignore other Agents.

Step C and List D

Can be performed automatically. Agents not present are listed in *italics*; potentially present Agents are in **bold**:

1. **sticky and fixed**	**UF: adhesion**
2. *sticky and cold*	*UF: could gather other Agents to make new surface*
3. *reactive and leave*	*UF: could be the result of a chemical reaction*
4. *reactive and change size*	*UF: resizing could assist reactivity*
5. *reactive and fixed*	*UF: could increase reactivity*
6. **charged and leave**	**UF: removed Agents could change the surface details**
7. *magnetic and cold*	*UF: potential phase transition*
8. **disperse and neutral**	**UF: Agents may be neutral after dispersion; neutral Agents may be dispersible**
9. **small and resize**	**UF: size variation could move Agents up and down the surface**
10. **small and ignore**	**UF: could assist moving into pores**
11. **swell and increase**	**UF: enlarge surface**
12. *lower density and leave*	*UF: leaving could reduce density*
13. *lower density and resize*	*UF: foaming could reduce density*
14. **mobile and leave**	**UF: increases mobility**
15. **hot and shrink or resize**	**UF: reduces surface**

Step E
In the interest of simplifying presentation of this case study, only the following list is presented:

Agents' properties to reduce the surface	Harmful action that exists or is performed
Properties for removing material are similar to properties for adding material. In addition, they can be hot enough to melt or evaporate unwanted portions of material. Properties for contracting material are similar to those for expanding material.	One could be hurt by the heat. One could be hurt by evaporation. One could be hurt by chemicals.

Steps F and G
Can be performed automatically.

Step H
Product (Surface) — The semiconductors (including substrate surface) are not active to reorganize the material themselves. Shockwaves, which appear when the substrate is hit, could be used to temporarily create the desired pattern, as in a global earthquake. To make it symmetrical throughout the surface, one needs to introduce scatter-centers.

Tool — Available in semiconductor production: heat and other energy wastes, chemicals (liquid, gas, plasma), electricity, ultrasound, etc. Derived resources as well as other subsystems are skipped here for simplicity.

Step I
Possible new idea: *incorporate scatter-centers in the layer's surface to transform an initial shockwave pattern into a random pattern with sufficient amplitude/altitude.*

The analysis of the layer problem is complete. A new idea is obtained and ready now for consideration using other TRIZ heuristics.

17.11 CONCLUSION

AM presented here can be used for solving various technical problems. A solver enters Agents in the CSP, IFR, and intermediate sketches, develops logical and graphical concepts of the decision, and then creates lists of Agents' actions and properties with the help of the AND/OR tree. Often Su-Field Analysis (Chapters 12 and 15) can be done using the discovered Agent actions and properties. At the final stage of AM, a solver should replace the Agents with a real technical subsystem.

The biggest advantage of AM is its flexibility as an instrument for solving problems: a problem solver can use a forward or backward search depending upon the CSP and IFR. However, even if there is no reason to prefer the backward search in a technical problem, a solver should always consider its use if there is no reason to prefer a forward search. For many problems it may not be obvious *a priori* which search, backward or forward, is superior; in such cases, a problem solver might well try bidirectional searching that is also possible in the framework of AM. On the other hand, AM is used as a part of ARIZ (actually Agents replace the X-element and "small smart people" which "participate" in Altshuller's versions of ARIX-85). We will discuss how AM incorporates in ARIZ in the next chapter.

REFERENCES

1. Gordon, W. J. J., *Synectics, The Development of Creative Capacity*, Harper, New York, 1961.
2. Altshuller, G. S., *... And Suddenly the Inventor Appeared,* Detskaya Literatura, Moscow, 1984, 1987, 1989 (in Russian); TIC, Worcester, 1996 (in English).
3. Tetel'baum, S. D., *Su-Field Analysis and Models of Inventive Problems,* OPI, Odessa, 1984 (in Russian).
4. Horowirz, R. and Maimon, O., SIT — A Method for Creative Problem Solving in Technology, in *Proc. 7th International Conference on Thinking*, Singapore, 1997.
5. Sickafus, E. N., *Unified Structured Inventive Thinking — How to Invent*, Ile, Grosse, 1998.
6. Koller, R., *Konstruktionsmethode für Maschinen-, Gerate- und Apparatebau,* Springer, Berlin/Heidelberg, 1979 (in German).
7. Polovinkin, A. I., *The ABC of Engineering Creativity*, Mashinostroenie, Moscow, 1988 (in Russian).
8. Wolf, S. and Tauber, R. N., *Silicon Processing for the VLSI Era: Process Technology,* Lattice Press, Sunset Beach, 1986.
9. http:/www.jps.net/triz/pinyayev1paper.htm

18 ARIZ

18.1 INTRODUCTION

The Algorithm for the Solution of Inventive Problems (or ARIZ from the Russian abbreviation) is the most recognized TRIZ instrument that includes many of the TRIZ heuristics discussed in the previous chapters. ARIZ is a multi-step program of actions that an inventor should make while resolving technical problems with a high degree of difficulty. Genrich S. Altshuller, who created ARIZ and developed various versions of the algorithm for about 30 years, thought that ARIZ

    i. serves as the main instrument for technical problem solving, and
   ii. is the appropriate name for the methodology of technical problem solving known now as TRIZ.

His goal was to organize the thinking of an inventor with a method based on experience extracted from many high-level patents, some psychological research, and heuristics of TRIZ.

There are four essential points to note:

1. ARIZ is specially aimed at solving nontypical inventive problems by elaborating various TRIZ heuristics;
2. fewer than 1% of all technical problem require modern ARIZ (most inventive problems can be solved by individual TRIZ heuristics and/or instruments);
3. some TRIZniks think that work using ARIZ is too austere and exhausting for inventors, especially when they do not have much time for solving a problem;
4. nevertheless, knowledge of ARIZ and the ability to solve problems using it are signs of a good TRIZ expert.

18.2 SHORT HISTORY OF ARIZ DEVELOPMENT

The history of ARIZ is useful for understanding possible ways for establishment of younger TRIZ instruments, such as the Agents Method presented in the previous chapter. Although TRIZniks do not use the early versions of ARIZ anymore, those early versions have many interesting concepts. Moreover, they can perhaps serve as prototypes for possible TRIZ heuristics and/or instruments not yet developed sufficiently. Note that each version of ARIZ has two digits corresponding to the year of its appearance. Some versions from the 1980s have a letter after the digits indicating the modification of ARIZ in that particular year.*

The ARIZ idea appeared in the first paper about TRIZ published by Altshuller and Shapiro [1]. At that time the 1956 List (below) was quite progressive in comparison with the state-of-the-art in the field of problem solving [2], e.g., checklists and questionnaires (see Appendix 1). Ideas of the list were based on biographies of great inventors, polling of inventors, authors' own inventions, and histories of various techniques. Analysis of patent resources was not the main mechanism of the list's construction. It is interesting that all (known so far) research in Western countries following this pattern was established in the 1920s: to extract some typical tricks from previous innovative designs or interviews with inventors until very recently, when the expert systems concept, popular in Artificial Intelligence, showed its weakness. Only in the 1990s did TRIZ experts from the former Soviet Union demonstrate to the Western world the efficiency of analyzing patents and technical information for growth of the problem-solving methodology.

1956 List

I. **Analytical Stage**
   1. Choose the problem
   2. Determine the main part of the problem.
   3. Discover the important contradiction.
   4. Determine the direct reason of the contradiction.

II. **Operation Stage**
   1. Research examples of typical solutions (prototypes) in nature, the technique, and the environment.
   2. Search for the solution by a change in the system, sub- or super-systems, or the environment.

III. **Synthesis Stage**
   1. Introduce changes in the system stipulated by [new] functions.
   2. Introduce changes to methods of the system using (application) stipulated by [new] functions.
   3. Check applicability of principle to solve other technical problems.
   4. Evaluate the solution.

---

* Perhaps new numbering of ARIZ should be adopted, indicating the year of the version's creation and initials of ARIZ authors. Because each version of ARIZ should be stringently tested with known and new problems, it seems impossible that few versions could be created in a single year. So, ARIZ-96SS would reflect that it was created in 1996 by Semyon Savransky.

Altshuller's first book about TRIZ was published in 1961 and 50,000 copies were sold in a few months [3]. As a result, TRIZ became available for engineers in the former Soviet Union. The 1961 List [3], reproduced below, can be considered the ARIZ prototype.

1961 List

1. **Analytical Stage**
   1. Establish the problem.
   2. State the Ideal Final Result (IFR).
   3. Define contradictions that prevent achieving IFR.
   4. Define the causes of the contradictions.
   5. Define the conditions to eliminate the contradictions.

2. **Operation Stage**
   1. Test all possible changes in the parameters and characteristics of the technique itself.
   2. Test the possibility of technique division on independent subsystems:
      - isolate a "weak" subsystem and/or a "necessary and sufficient" subsystem,
      - divide object on equal subsystems and/or on different functional subsystems.
   3. Test possible changes in external (for the given technique) environment and its parameters
      - divide environment on some parts [with different properties],
      - use external environment to perform useful functions.
   4. Test possible changes in techniques working together with the given one:
      - determine connection between prior independent TS functioning in one TP,
      - eliminate one technique by transferring its functions to another one,
      - increase the number of technique samples functioning simultaneously on a definite area using the free reverse side of this area.
   5. Investigate examples from other branches of science and technology (see how the given contradiction is eliminated in other branches).
   6. Return to the initial task and expand its conditions; i.e., transfer to another more common task (if all previously mentioned steps did not work).

3. **Synthetic Stage**
   1. Change shape of the given TS.
   2. Change the other techniques connected with the given one.
   3. Change the methods of using the technique.
   4. Test the use of a found principle for solution of other technical problems.

The 1961 List was not yet a system; "steps" can be rearranged. The word "algorithm" first arose only in 1965 in a paper written by Altshuller. It was an indication of the main aim of TRIZ development at that time. The first popular version of ARIZ (ARIZ-68) appeared in Altshuller's 1969 book, *The Algorithm of Inventions* [4]. This ARIZ-68 was described in common terms of modern TRIZ, compared with George Polya's Schema for solving specifically mathematical problems [5]. Readers can find this common terminology of the early 1970s in Altshuller's book [4]. (Although ARIZ is more detailed and works with engineering problems more concretely than does Polya's Schema, some promising aspects in Polya's Schema were used in versions of ARIZ developed in the 1990s.)

ARIZ was improved in the 1970s because of valuable suggestions for TRIZniks. It became more rigorous and definitive, and it made it possible to solve highly difficult problems. The best ARIZ version of the 1970s — ARIZ-77 — is still useful and recommended to newcomers in TRIZ during their learning curve. ARIZ-77 was also the first version to be presented in English [6] with a few case studies. ARIZ-77 is presented in the next section in terms of modern TRIZ.

In the 1980s new TRIZ heuristics were actively developed and included in ARIZ-81 (modifications A, B, C, D) and ARIZ-85 (modifications A, B, C). Although the number of TRIZniks who helped Altshuller improve the methodology was growing, the creator of TRIZ asked the others not to develop ARIZ. He kept this privilege until 1986 when he decided that other TRIZ experts could also directly participate in ARIZ improvement.* So, ARIZ-85C is the last version of ARIZ created solely by Altshuller [7] who died in 1998. It is currently the last public version of ARIZ.**

Altshuller: ARIZ-68	Polya Schema
1. Selection of the Problem	**Understanding the Problem**
1. Define the final goal of problem solution.	What is the unknown? What are
What is the technical goal of this problem solution (what parameters of a technique should be changed)?	the data? What is the condition? Is it possible to satisfy the
What parameters of a technique cannot be changed while solving the problem?	condition? Is the condition sufficient or insufficient to
What is the economical goal of the problem solution (what expenses will be lower while solving the problem)?	determine the unknown? Or redundant? Or contradictory?
What are the estimated expenses?	Draw a figure. Introduce suitable
What main technico-economic parameter should be improved?	notation. Separate the various
2. Test the possibility to achieve the same goal by solving a round-about problem.	parts of the condition. Can you write them down?

---

* The idea was that Altshuller would supervise the research. A few round-table workshops of ARIZ developers took place in Russia at the end of the 1980s and at the beginning of the 1990s. Later the economic crisis in Russia slowed TRIZ activities overall in addition to specifically ARIZ research and development. Unfortunately, because of the radical change of economics in the country and the illness of Altshuller, TRIZniks stopped sharing their results and Altshuller was unable to supervise ARIZ development.

** The reader can find ARIZ-85C on the Internet (see the author's web site at http://www.jps.net/triz/triz.html or its mirror at http://come.to/triz).

Altshuller: ARIZ-68	Polya Schema

Altshuller: ARIZ-68

In other words, assume that this problem cannot be solved. Then consider what

   a. other problem should be solved in order to achieve the required result,

   b. other technico-economic parameters should be improved in order to solve the round-about problem.

3. Define the solution of what problem — initial or round-about — can be more effective.

In other words, compare the initial and round-about problems with the tendencies of the given and leading branches of technology developments. Compare the results for the initial problem with the round-about one, and then make the selection.

4. Define the required quantitative indices (speed, efficiency, precision, dimension).

5. Make the "correction in time" in the required quantitative indices.

6. Specify the requirements caused by concrete conditions of the problem realization.

Take into consideration the presumable scale of its usage and the peculiarities of its introduction, in particular the permissible degree of solution complexity.

2. Specification of the Problem Conditions

1. Specify the problem using patent literature.

Find how similar or round-about problems have been solved by reviewing patent literature and the leading branches of science and technology.

2. Apply time-dimension-cost operator.

Answer the following questions:

   a. Can the given problem be solved, not taking into consideration the expenses?

   b. How is the problem changed if the required parameter is zero or is ten times as large as normal?

3. How is the problem changed if no specific terms are used?

3. Analytical Stage

1. Define the Ideal Final Result.

Show the technique schematically how it was and how it should be.

Simplify the final schema as much as possible while preserving the working capability of the technique.

2. Define what prevents achieving IFR and why.

3. Define the conditions under which we can get IFR (under what conditions the preventive factor will disappear).

Can the preventive factor disappear or be preserved yet also be not preventive?

Polya Schema

Devising a Plan

Have you seen it before? Or have you seen the same problem in a slightly different form? Do you know a related problem? Do you know a theorem that could be useful? Look at the unknown. Try to think of a familiar problem having the same or a similar unknown.

There is a problem related to yours and solved before. Could you use it? Could you use its method? Should you introduce some auxiliary element in order to make its use possible? Could you restate the problem? Could you restate it still differently? Go back to definitions.

If you cannot solve the proposed problem, try to solve first some related problem. Can you imagine a related problem that is more accessible? A more general problem? A more specific problem? An analogous problem? Can you solve part of the problem? Keep only a part of the condition, drop the other part; how far is the unknown then determined and how can it vary? Can you derive something useful from the data? Can you think of other data appropriate to determining the unknown? Can you change the unknown, the data, or both, if necessary, so that the new unknown and the new data are nearer to each other? Did you use all the data? Did you use the whole condition? Have you taken into account all essential notions involved in the problem?

Altshuller: ARIZ-68	Polya Schema
4. What mechanism should eliminate the prevention? Define the state of this mechanism and how it changes during the working process.	**Carrying Out the Plan** Carrying out your plan of the solution, check each step. Can
5. If necessary, the analysis is repeated.	you see clearly that the step is correct? Can you prove that it is
4. Operation Stage	correct?
1. Test the possibility of eliminating the technical contradiction using the Matrix.	**Examining the Solution Obtained**
2. Test the possibility of environmental change of a technique.	Can you check the result? Can you
3. Test the possibility of changes in the techniques that function together with the given one.	check the argument? Can you derive the result differently? Can
4. Test the possibility of time changes:	you see it at a glance? Can you
Is it possible to eliminate the contradiction by prolonging the ongoing action according to the condition of the problem?	use the result of the method for some other problem?
Is it possible to eliminate the contradiction by shortening the ongoing action according to the condition of the problem?	
Is it possible to eliminate the contradiction by performing the required action in advance, before operation of the technique?	
Is it possible to eliminate the contradiction by performing the required action after the technique finishes the operation?	
If, according to the conditions of the problem, the action is continuous, test the possibility of transforming it to periodic action.	
If, according to the conditions of the problem, the action is periodic, test the possibility of transforming it to continuous action.	
5. How are similar problems solved in nature (organic and inorganic)?	
What are the tendencies of development?	
What changes should be made while considering the materials used in nature?	
5. Synthetic Stage	
1. Define the changes that should be made to other subsystems of the technique if we change one subsystem of it.	
2. Define the changes of the technique's objects functioning together with the given one.	
3. Test whether the changed technique can be used as a new one.	
4. Use the found technical idea while solving other technical problems.	

A few new versions of ARIZ have been proposed in the 1990s. These versions have new stages and sub-algorithms as well as a larger number of steps, which is not very good because the Ideal Algorithm is, perhaps, an equation for solving all technical problems. They also improve the degree of algorithmization of the first part of ARIZ-85C (which was weaker than that in ARIZ-77). In addition, they include more comprehensive and deeper functional analysis of the technique and its objective evolution trends and more accurately work with resources and some new TRIZ heuristics. Later versions are supported by some TRIZ companies and considered proprietary information, so the possibility for careful comparison of these versions does not yet exist.

18.3   PRE-ARIZ

The following few cautions were made by Altshuller [7] for using the so-called pre-ARIZ presented later in this section. They are relevant for all versions of ARIZ created in the last quarter of twentieth century and some are also significant for other problem-solving instruments.

1. ARIZ is an instrument for systematic thinking, but it does not replace thinking. The important part of ARIZ is removing psychological inertia which depends on an engineer's experience, skills, and cultural background. Do not hurry when you solve a problem with ARIZ. You should make a detailed consideration of each step. In addition, it is necessary to note all ideas arising while solving a problem.
2. The process of solving inventive problems subsequently requires good knowledge of various TRIZ heuristics, and ARIZ-85 combines many of them. Learn the TRIZ concepts and heuristics that are incorporated in ARIZ. When you have enough experience with ARIZ you can replace the notes like "a-*.* Read chapter N…" to "Refresh your memory about…"; later mind-maps of TRIZ parts will be enough and eventually you can remove such notes at all.
3. ARIZ is designed for solving unusual problems. Possibly your problem can be solved with less complicated TRIZ heuristics and instruments. (Modern TRIZ heuristics and instruments are so effective that ARIZ is required for fewer than 1% of problems.)
4. Mini-problems and maxi-problems are distinguished in ARIZ (see also Chapter 10). The *mini-problem* occurs when the system remains unchanged or is only simplified, but disadvantages disappear or a desirable improvement is achieved. The *maxi-problem* does not intrude any constraints on the possible changes; in this case, a new system based on a novel principle of functioning can be developed. Mini-problems in the technique usually correspond to line two in the table below, while maxi-problems in the technique correspond to line three:

Possible TS alterations	Possible TP alterations
1. TP of TS production without changes in TS itself	1. TS performs TP without changes in TP itself
2. TS or subsystems without changes in mode of PF (UF) performance	2. TP or subsystems and operations without changes in mode of PF action
3. PF of TS (revolutionize TS)	3. PF of TP (revolutionize TP)

5. Because implementing mini-problem solutions is usually cheaper and simpler than maxi-problem solutions, ARIZ focuses on mini-problems first. The mini-problem can be obtained from the invention situation that introduces restrictions, i.e., that everything must remain unchanged or become simpler, but simultaneously the required action appears or the harmful action disappears. Focusing on the mini-problem does not mean that we try to solve a small problem. On the contrary, introducing additional requirements (the result should be obtained "with nothing added") sharpens the conflict and cuts routes to compromise solutions. In general, ARIZ requires sharpening of the contradiction rather than smoothing for both mini- and maxi-problems.

6. When solving a problem according to ARIZ, the solution is formed gradually, in a way similar to how the sun rises above the horizon, illuminating all dark sides of the initial situation with the technical problem. It is important not to stop halfway through and not to interrupt the process with the first signs of a solution and "fix" rather than a completely finished solution. The solution process must be finished according to ARIZ. Don't stop at the first "weak" solution; a new, "stronger" solution should still be sought, even if it takes time. A problem's solution is accompanied with the breaking of old concepts and emergence of new ones. It can be difficult to express these new concepts with words or pictures; use any suitable representation for them. Specialized terms should be replaced with simple or general terms in order to avoid or overcome psychological inertia (see Chapter 10).

7. ARIZ has been tested in solving many problems, practically the entire pool of problems used in TRIZ training. Forgetting this, some people propose "hot" improvements based on the solution of only one problem. Although these improvements may be useful for this one problem, as a rule they hamper the solving of all other problems. Consequently, any proposed improvements should be first tested out of ARIZ (as was the case with the Agents method).

   (A new heuristic should always be compared with existing TRIZ instruments; if this "newcomer" is good, it will be appreciated by the TRIZ community.)

8. After being added to ARIZ, each suggestion must be tested in solving at least 20–25 sufficiently complicated problems that cannot be resolved by other TRIZ tools. ARIZ is constantly improving, so it needs new ideas; however, these ideas must first be carefully tested.

   (Note that some of the latest ARIZ versions have not passed this last criterion yet.)

9. ARIZ is a sophisticated instrument. Do not use it for solving new technical problems if you have not passed at least the 80-hour training course.

   (ARIZ is not a simple heuristic that can be taught in a single training session. It is complex and difficult and requires much training to be used effectively, but it offers good students a great source of inventive power, as proved by the thousands of people who have studied it sufficiently. If you have read all previous chapters carefully and did not have difficulty in solving the problems presented in previous chapters, you are ready to use ARIZ.)

18.4   ARIZ-77

ARIZ-77 consists of the seven parts presented below.

Part 1: Selection of the Problem

1.1. Determine the final aim or ultimate objective of the solution of a problem:
   a. What features of the technique need to be changed?
   b. What features of the technique cannot be changed after the solving of the problem?
   c. What is the main technical and/or economic indicator that should be improved if the problem is solved?
   d. What approximately is the permissible expenditure?
1.2. What other problems must be solved in order to obtain the requisite final result if the initial problem is fundamentally insoluble? Check the deviation route.
   a. Transform the problem by switching to the super-system or subsystem level.
   b. On three levels (super-system, system, and subsystem) transform the problem, having replaced the requisite action (or property) by its reverse.
1.3. Determine which problem's solution is more purposeful — the original one or one of the deviations. Compare two possibilities of solving the problem directly or indirectly; choose the better way.
   Note that one should take into account objective factors (what are the reserves of development of the system given in the problem?) and subjective ones (is one oriented toward which problem? the maximum or the minimum?).
1.4. Determine the required qualitative parameters (indicators) describing your technique.
1.5. Increase the requisite quantitative indicators, taking into account the time necessary for implementing the solution.
1.6. Determine general manufacturing requirements. Specify exactly the requirements necessitated by the specific conditions in which the realization of the solution is proposed.
   a. Take into account the features of implementation, in particular the admissible degree of complexity of the technique itself and the solution.
   b. Take into account the proposed scale of implementation.
1.7. Check whether the solution is solved by a direct application of Standards. If an answer is obtained, go to 5.1. If not, go to 1.8.
1.8. Specify the problem by using patent information.
   a. What answers can be drawn from patent information on problems which are similar to or the opposite of the one in question in the same

and leading branch of technology? This information helps make your problem more precise and general.

1.9. Use the Size-Time-Cost operator.

Part 2: Constructing a Model of the Problem

2.1. Write down the specifications of the problem using special terminology. Then formulate the problem in simpler and more general terms.

2.2. Isolate and write down the conflicting pair of elements. If only one element is given in the problem specification, go to step 4.2.

Rule 1. It is necessary to introduce a tool and a product on which the tool can act directly into the conflicting pair of elements.

Rule 2. If one element (usually the tool), according to the specifications of the problem, can be in two states, then one should choose that state which ensures the best implementation of the primary function of the technique.

Rule 3. If the problem has a pair of homogeneous, mutually interacting elements (Al, A2... and B1, B2...) it is sufficient to take one pair only (A1 and B1).

2.3. Identify interrelationships between the components found in 2.2 and classify them as useful or harmful functions.

2.4. Write down the standard formulation of the model of a problem, indicating the conflicting pair and the technical contradiction.

Part 3: Analysis of the Model of the Problem

3.1. Select from the conflicting pair one element that can easily be changed, replaced, etc.

a. Tools are easier to change than products. Artificial objects are easier to change than natural ones.

b. If there are no easily changeable elements in the technique, one should indicate an external environment.

3.2. Write down the standard formulation of the IFR:

An element (indicate an element chosen at step 3.1) *itself* removes HF (indicate) while retaining the ability to perform UF (indicate).

a. The IFR formulation should always include the word "itself."

3.3. Identify the element's zone (indicated in step 3.2) that cannot cope with a set of two interactions demanded in the IFR. What in this zone cannot fulfill the imposed functional requirements: a substance, a field? Show this zone on a schematic drawing denoting it in color, lines, etc.

3.4. Formulate antagonistic physical requirements for the state of the isolated zone of the element of the conflicting interactions (actions, properties) as: For UF security (indicate the interaction which should be retained) it is necessary (indicate the state). AND For prevention HF (indicate the harmful interaction or interaction that should be introduced) it is necessary (indicate the state).

a. The state indicated at this step should be mutually opposed, e.g., being heated, mobile, or charged vs. being cold, immobile, or uncharged.
3.5. Write down the standard formulation of the physical contradiction.
   a. The full formulation — The denoted zone of the element (indicate) should (indicate the first state noted at step 3.4), in order to perform UF (indicate), and should (indicate the second state noted at step 3.4) in order to prevent HF (indicate).
   b. The short formulation — The denoted zone of the element (indicate) should be and should not be.

Part 4: Removing the Physical Contradiction

4.1. Examine the simplest transformations of the denoted zone of the element; i.e., separate the contradictory properties
   a. in space;
   b. in time;
   c. by using transitory states in which contradictory properties either coexist or appear alternately;
   d. by rebuilding the structure; particles of the denoted zone of the element are given which possess the property, and all the denoted zones as a whole are marked by the requisite (conflicting) property.
   If an answer is obtained (i.e., it emerges that a physical action is necessary) go to step 4.5; otherwise, go to step 4.2.
4.2. Use Su-Field transformations. If a physical effect is obtained, go to step 4.4. If there is no physical effect, go to step 4.3.
4.3. Use the list of application of physical effects and phenomena. If a physical effect has been obtained, go to step 4.5; otherwise, go to the next step (4.4).
4.4. Use the Matrix of eliminating technical contradictions. If a physical effect has been obtained before this, use the Matrix for the verification of the obtained solution.
4.5. Use the obtained physical answer to solve the technical problem; formulate the method of solving it and develop the concept of a technique which would implement this solution.

Part 5: Preliminary Assessment of Obtained Solution

5.1. Evaluate the initial solution using the following checklist:
   a. Does the solution fulfill the main requirement of the IFR (the element itself...)?
   b. Which physical contradiction has been eliminated (if any at all) by the solution obtained?
   c. Does the solution provide the possibility to control a new technique element? Which subsystem? How is this control implemented?
   d. Is the solution found for a "one-off" model of the problem also suitable in real manufacturing ?

If the solution does not satisfy at least one of the a–d points above, return to step 2.1.

5.2. Check against patent information whether the solution obtained is formally an invention.

5.3. What secondary effects can arise in the technical implementation of the idea so produced? List possible subordinate problems — inventive, design, accounting, administrative.

Part 6: Development of the Obtained Answer

6.1. Determine how the super-system, to which the new technique belongs, will change.

6.2. Check whether the new technique will have to be applied in a new way. For good solutions other possible applications often exist (extra-effect).

6.3. Apply the solution to other technical problems.

    a. Examine the possibility of using the reverse of the idea obtained.

    b. Construct a morphological box and examine possible reconstruction of the solution.

Part 7: Analysis of the Process of Solution

7.1. Compare the real solving process with the theoretical (according to ARIZ scheme). Report any deviations.

7.2. Compare the obtained solution with TRIZ heuristics (Standards, the lists of physical effects, the Principles). Report any deviations.

Altshuller's solution of an illustrative problem with algorithm ARIZ-77 is presented in Section 18.5.1 below. Resolutions of several problems by ARIZ-77 are found in Altshuller's first book published in English in the middle of the 1980s [Ref. 6, 285–298]. Although this book was written more than 20 years ago, it is still a good source of TRIZ.

18.5   ARIZ-85

There are nine basic stages in ARIZ-85C with different aims for each stage [7].

ARIZ-85AS has Stages 1-9 similar to ARIZ-85C and Stage 0 is added and should be performed before starting the problem-solving process.

18.5.1   ARIZ-85AS

In 1996 the author proposed the latest known version of ARIZ. The main advantage of this algorithm is its ability to generate a few strong solutions, not a single one as in previous versions of ARIZ. Unfortunately ARIZ-96SS is too complicated to be presented in this book. Some ideas of ARIZ-96SS were reduced and simplified and included in the steps and notes of ARIZ-85AS, which is given below. ARIZ-85AS is based on the last Altshuller version that was developed and polished by the author based on his practice and suggestions of other TRIZ experts, notably Alex M. Pinyayev,

ARIZ-85C Stages

#	Stage Name	Stage Aim
1	Problem Analysis	Describe the technique where the problem appears and pass from a vague invention situation and administrative contradiction to the evident technical contradiction and then to the distinctly constructed statement of the problem.
2	Analysis of the Problem's Model	Recognize all available resources that can be used in the problem solution, and find the generic structure of the problem.
3	IFR and Physical Contradiction Determination	Formulate IFR that indicates the direction of the strongest solution and the physical contradiction that hampers achieving IFR.
4	Substance-Field Resources Applications	Mobilization and systematic usage of all substance-field resources for the problem solution.
5	Information Resources Application	Utilize TRIZ information resources, i.e., the lists of effects and phenomena, registry of the typical transformations, etc.
6	Problem Replacement	Test the problem essence. The invention problems cannot be set correctly from the beginning. Changes of the essence (lifting the initial restrictions and constraints, removing psychological inertia, etc.) is common for such problems. The solving process is inherently the process of problem correction.
7	Contradiction Elimination Analysis	Physical contradiction must be eliminated practically, Ideally, without anything added. The goal here is to check the quality of the solution found in previous stages.
8	Obtained Solution Application	Employ the obtained solution. A truly good idea not only solves the specific problem but also gives the universal key for many similar problems.
9	Solving Process Analysis	Evaluate ARIZ and your solving process. (NOTE: This stage is for professional TRIZniks only.)

Simon S. Litvin, and Vladimir A. Korolyev. ARIZ-85AS has some elements of ARIZ-96SS and can be considered the best currently available in the public domain for solving technical problems.

Stage 0 — Information

0.1. Collect and classify information about the technique you would like to improve or create and its environment, as well as the initial problem situation.
Notes:
   a-0.1 Read Chapters 1 and 10 about information and presumptions.
   b-0.1 It is often important to know the history of attempted solutions to the problem and of the technique developments and the desired timetable for solving the problem and implementing a solution.
0.2 Determine
   allowed changes in the technique and limitations;
   expected technical and economic characteristics and parameters in the new technique.

Notes:

a-0.2 Rank minimal changes by answering the following questions:
What are the technical and economic goals of solving the problem?
(What characteristics and parameters of the technique should or can
be changed? How will profit be increased after the problem is
solved? Does the additional profit cover expenses for changes in
the technique?)

b-0.2 Rank accepted solution by answering the following questions:
1. While solving a problem, what characteristics and/or parameters
   of the technique must deliberately not be changed?
2. What are the approximately admissible expenses?

Stage 1 — Problem Analysis

1.1. Write the statement about the technique in the following form:
Technique for (*indicate the primary function — PF*) includes (*list the
main subsystems*). The technique and its main subsystems have the following (*indicate UF*), but also have the following (*indicate HF*).
Notes:

a-1.1 Read Chapter 3 for discussion of subsystems and functions of a
technique.

b-1.1 Check the correctness of including the subsystem by answering
the question, "Will the problem disappear if we remove this subsystem?"

c-1.1 If a subsystem has elements that are important for solving the
problem, those elements should be listed independently in the
registry of main subsystems.

d-1.1 Indicate in the scheme not only main subsystems, but also the
environment that interacts with the subsystems and describe the
conditions at which the technique exists.

e-1.1 Use the following syntax, filling in the relevant nouns and verbs,
to extract the PF: Subject (Subsystem) — Predicate (Change of
Subsystem) — Direct Object (receiver of the primary function's
action).

f-1.1 Remember the relativity of UF and HF.

1.2. Determine the structure of the potential problems in the technique.
Notes:

a-1.2 Read Chapter 4 about structure of technical problems.

b-1.2 The problems cannot be structured in the terms presented in Chapter 4 if the answer to one of the following questions is "no":
Does HF increase if UF performance can grow by the method that
is described in the problem conditions or by well-known methods?
Does UF performance decrease if HF can be eliminated or decreased by the method that is described in the problem conditions
or by well-known methods?
If no contradiction exists, you have a *routine* or *synthesis problem*.

b-1.2.1 Routine problems can be solved by common, regular methods, well known in each engineering field. One class of synthesis problems was described in Chapter 16. Some TRIZ instruments and heuristics for synthesis and genesis of a technique were mentioned in Chapters 2 and 7.

1.3. Write statements about the administrative contradiction and conditions of the mini-problem in the following form:

Administrative contradiction — The following result (*indicate*) should be obtained, but…(*indicate the administrative contradiction*).

Mini-problem — It is necessary to remove HF by minimally changing PF or UF in the technique (*subsystem*).

or

It is necessary to get UF without HF with minimal changes in the technique (*subsystem*).

Notes:

a-1.3 Read Chapter 4 for discussion of administrative contradictions.

b-1.3 Solve the maxi-problem if there are no real restrictions at step 0.2.

c-1.3 Reformulate the problem in simple words using TRIZ slang (see Chapter 11), if necessary.

1.4. Present the structure of the mini-problem in a graphical scheme.

1.5. Reduce the structure of the mini-problem to one of the generic problems' structures.

Note:

a-1.5 Unfortunately, "pruning" TRIZ heuristics that help simplify the problem structure are too complicated to be considered in this book. Nevertheless, this step can be done "by hand."

1.6. Attempt to adapt the solutions of a problem similar to this one.

Notes:

a-1.6 The generic structure of the problem allows choosing the right analogies from the fund of solved problems regardless of the solutions' engineering domain.

b-1.6 Although technical problems are infinite in number and variety, the number of generic structures and types of contradiction they involve is comparatively small (see Chapter 4). Therefore, many problems can be solved by analogy with other problems with a similar generic problem structure that reflects the relations between contradictions in a technique. Externally, problems can differ significantly from each other, and analogy can be found only after analysis in this stage of ARIZ.

1.7. If the administrative contradiction remains unsolved, go to Stage 2. If the problem is solved, go to Stage 7, although in this case it is also recommended to continue the analysis (Stages 2–6) during which the problem structure can be reduced to the point contradiction and other strong solutions can be obtained.

Stage 2 — Problem Reduction

2.1. Separate and write down the main conflicting pair of articles: a raw object and a tool. Determine HF and UF that appear during transformation of the raw material into a product.

Notes:

a-2.1 Read Chapter 3 for discussion of raw objects, products, and tools.*

b-2.1 The raw material is the object of UF and HF, and it can usually be deduced from the third term of the syntax given in Note e-1.1. As a rule, the product is mentioned in the problem statement and in the administrative contradiction. The tool performs PF or UF for the transformation of the raw material into a product and can be deduced from the conflicting pair of articles (one of which is the raw object); HF arises in another if its PF or UF increases. The tool can usually also be deduced from the first term of the syntax given in Note e-1.1.

c-2.1 One of the articles of the conflicting pair (tool versus product or raw material) may double. For example, given are two different tools that should simultaneously act upon the raw material, and one tool hinders the other. On the other hand, given are two different raw materials that should be subjected to action of the same tool; one raw object hinders the other during their transformation into products.

d-2.1 At this step we select the most important pair contradictions in the entire structure of the problem.

2.2. Formulate the opposite state for the raw object or for the product that influences TC selection.

Formulate the opposite state for the tool that determines UF or PF increase and HF decrease.

Notes:

a-2.2 If the tool, according to the problem conditions, can be in two states, you should indicate both of them because they predetermine two possible technical contradictions.

b-2.2 If the problem includes many pairs of similar interacting elements, it is sufficient to consider only one conflicting pair.

c-2.2 Usually the tool is changed as a result of the implementation of the solution. Rarely the solution leads to a necessity for modifying the raw object or product. In the framework of a mini-problem, it is possible to change a raw material and/or a product in the following cases:

• if another raw object can be transformed into the requested product (usually this is a raw material with similar but not identical properties);

---

* There is absolutely no difference between a raw material and a raw object; in fact, one word describes both these terms in Russian (the original language of TRIZ and ARIZ).

- if the change is insignificant and does not influence functioning of the product;
- if the change can be made at the various (usually previous) stages of a technique's life cycle.

d-2.2 Sometimes the problem conditions give information only about the raw object and/or the product, not about the working tool or operations, so there is no clear tool. In such cases, the tool can be found by considering conditionally two states of the raw object or product, although one of them is obviously unallowable.

e-2.2 Sometimes it is difficult to determine PF and the working tool in measurement problems. An engineering measurement is almost always performed in order to change, such as to treat a subsystem or to fabricate products. The only exclusion is some problems of measurement for scientific purposes. Usually the entire measuring technique (not only a sensor, i.e., the main measuring subsystem) is needed for PF in measurement problems. Often a measurement can be considered a change of information about the raw object.

g-2.2 Usually the problem itself dictates the principal article for the consideration — the product or the raw object. Only the product is considered below for the sake of simplicity, as is common in previous versions of ARIZ.

2.3. Define at least two technical contradictions (TC-1 and TC-2) between the product and the tool. Construct verbal and graphical schemes of TC-1 and TC-2. In terms of functions, TC-1 and TC-2 can be written as

TC-1: If UF (*indicate*) performance increases, then HF (*indicate*) also increases above a threshold;

TC-2: If HF (*indicate*) decreases, then PF (*indicate*) performance also decreases below a threshold.

Notes:

a-2.3 Read Chapters 4 and 13 for discussion of technical contradictions.

b-2.3 Obtain a general formulation of the problem in terms of technical contradictions.

c-2.3 Both TC give complementary requirements to the tool and product. It is unknown, until later steps, which contradiction is best suited for the problem.

d-2.3 Check if there are discrepancies at steps 2.1 and 2.2. If you find any, eliminate them by correcting the formulation of the administrative contradiction or improving the logic of flow between these three steps.

2.4. Select the technical contradiction, TC-1 or TC-2, that ensures the best performance of the primary function of the technique shown in the problem conditions at step 1.1. Check the possibility of applying the Matrix to the solution of the selected technical contradiction.

Notes:

a-2.4 Read Chapter 13 for discussion of the Matrix for resolution of technical contradictions.

b-2.4 Having selected one of two contradictions, we have selected one of two opposite states of the tool.

c-2.4 If two UF are present in the initial situation, choose that TC which best fulfills the subsystem's PF. Further solution should deal with just this state.

d-2.4 The Matrix presents the set of the typical technical contradictions. TC absent in the Matrix can also be used, if they better reflect the essence of the problem.

2.5. If the problem remains unsolved, go to the next step. If the problem is solved, go to Stage 7, although in this case it is also recommended to continue the analysis by the next step or by Stage 3 where the problem structure can be reduced to the point one.

2.6. Apply the Parameters Operator by indicating the limit state (performance, action) of elements.

Notes:

a-2.6 Read Chapter 11 for discussion of the Parameters (Size-Time-Cost) Operator.

b-2.6 Enforce the contradiction until the qualitative change of the problem appears. A few grades of contradiction strength are possible, and various new problems can be obtained at each such grade.

2.7. Write down the strengthened formulation of the problem model indicating the conflicting pair.

Notes:

a-2.7 The problem model is conditional, and some subsystems of the technique are artificially selected in it. The presence of other subsystems only helps to understand the technique better.

b-2.7 Return to step 2.1 after this one and check the logic of the problem model construction. It often proves possible to refine the selected scheme of the conflict by indicating some change in the subsystem or technique, any unknown characteristic, or parameter.

c-2.7 Enforced formulation of the problem allows a more Ideal solution that can be applied to a wider spectrum of similar problems.

2.8. Construct the Su-Field model of the problem. Check the possibility of applying the Standards to the solution of the problem model.

Notes:

a-2.8 Read Chapter 12 about Su-Fields and Chapter 15 about Standards.

b-2.8 Analyses performed during the first stage and construction of the model make the problem much clearer. In many cases, they allow you to see familiar details in nonusual problems. This makes possible more efficient use of the Standards than in the case when the Standards are applied to the initial formulation of the problem.

2.9. If the problem remains unsolved, go to Stage 3. If you have solved the problem and do not need other solutions, go to Stage 7 of ARIZ, although in such a case it is also recommended to continue the analysis from Stage 3.

Stage 3 — Problem Sharpening

3.1. Define the operative zone and the operative period.
Notes:
a-3.1 Read Chapter 14 about operative zone and period.
b-3.1 Determine the operative zones Z1 where the tool interacts with the product and the operative periods T1 where HF should be decreased or eliminated and UF should be increased or remain constant.
c-3.1 Usually it is enough to consider T1 and Z1, but sometimes consideration of T2 and T3 as well as Z2 could be useful.
d-3.1 Specify the operative period and the operative zone for UF and HF, if possible.

3.2. First formulate IFR on the basis of previous analysis.
IFR-1: Agents, which absolutely do not complicate the system and cause no harm, eliminate (*indicate HF*) during the operative period within the operative zone, keeping the tool capable of performing (*indicate UF*).
Notes:
a-3.2 Read Chapter 5 about IFR and Chapter 17 about Agents Method.
b-3.2 The IFR formulation above is only a sample. In addition to the conflict "HF related to UF," other conflicts are also possible, such as "introduction of a new UF causes technique complication (growth of NF number)" or "one UF is incompatible with another one." The general idea of all IFR formulations is that acquiring a UF (or elimination of HF) should not be accompanied with deterioration of other properties (or appearance of HF).
c-3.2 Agents can resolve the problem at this step if additional substances and/or fields with the required properties can be introduced into the technique.
d-3.2 The best scenario occurs if Agents link the initial problem state with IFR-1 within the operative zone and during the operative period by a forward, backward, or bidirectional method.

3.3. Strengthen the IFR formulation by introducing an additional requirement: introduction of new substances and fields is not allowed; only the existing resources should be used.
Notes:
a-3.3 Read Chapter 6 about resources and find out the available resources in your technique.
b-3.3 During the solving of the mini-problem, consider resources in the following order: tool resources, resources of other subsystems of the technique, environment resources, differential resources, product and raw object resources (if allowed). The potential states of resources are functional (transmitting, repulsive), physical (hot, superconductive), chemical (reactive, heat-generating), geometric (long, round), biological (cardiovascular, fermentive), etc. Here the available resources play the role of a supporting subsystem.

c-3.3   The presence of different resources causes further analysis. In practice, problem conditions usually decrease the number of lines (see
Note d-3.3). Nevertheless, a successive analysis should be replaced
by a parallel one, i.e., by a system thinking operator (see Chapter
11 about the multi-screen approach). When solving a mini-problem,
it is sufficient to perform analysis up to obtaining the solution idea;
if the idea is obtained, for example, at the "tool screen," you can
stop checking other screens. When solving a maxi-problem, it is
worthwhile to check all existing screens for a given case. This means,
for example, that having obtained a solution at the "tool screen," you
should also check the lines of the environment, secondary resources,
and the resources of other subsystems of the technique.

d-3.3   Use the idea of preferable resource to decrease the number of lines
in solving:
> The preferable resource eliminates or prevents HF, while it does
> not cause the destructive phenomena during the operative period
> in the operative zone for HF and does not inhibit the tool from
> performing UF during the operative period in the operative zone.

e-3.3   Formulate the opposite state (requirements) for each resource without proving its possibility. It is desirable to describe the condition
of the preferable resource by various synonyms and to prove the
opposite condition by requiring preservation of effective UF fulfillment or prevention of additional HF.

f-3.3   Exclude the "unreasonable" requirements to the condition of the
preferable resource. Define which condition should have the preferable resource during the operative period and in the operative
zone to ensure or supply the requirement of the strengthened IFR
on elimination or prevention of HF.

g-3.3   Analysis of available free resources is somewhat preliminary here.
More detailed consideration is done at Stage 4.

h-3.3   Usually the product or raw object is unchangeable, and its resources
are very limited. Nevertheless, some problems could be solved
solely by the resources of the product, as is employed in Stage 4.

i-3.3   Present the available resources as Agents and try to solve the
problem with AM (see Chapter 17). Available resources here play
the role of a supporting tool or additional raw material.

3.4.   Formulate the physical contradiction (PhC) at the macro-level.
Macro-PhC: The operative zone must contain the preferable resource
(indicate the physical state or actions) in order to ensure (indicate the
macro-state) and must contain no preferable resource (or must contain the
preferable resource with opposite state or actions) in order to ensure
(indicate another macro-state).

or

The preferable resource during T– in the limits of Z– should be (*first
state*), to remove or prevent HF (*indicate*), and to be (*second state opposite
to the first*), that during T+ in the limits of Z+ UF (*indicate*) is fulfilled.

Notes:

a-3.4 Read Chapters 4 and 14 about physical contradictions.

b-3.4 The preferable resources can resolve this problem if they can act as necessary in the existing or available states of the conflicting pair introduced to solve the problem (what should it keep, eliminate, improve, ensure, etc.).

c-3.4 The preferable resource must have or be the first state during $T+$ within $Z+$ in order to provide the requirement of the strengthened IFR formulation on elimination or prevention of HF. The preferable resource must have or be the second (opposite) state during $T-$ within $Z-$ in order to reserve the efficient execution of PF (UF) or prevent appearance of any additional HF.

d-3.4 Existence of a few formulations for the physical contradiction testifies for the usage of different synonyms at the description of the condition or tells of the presence of different problems which demand separate solving.

e-3.4 Emergence of several formulations of the physical contradictions indicate either the use of different synonyms when describing the subsystems, the states of the resource (synonymous physical contradictions), or the presence of different problems which should be solved individually.

3.5. Formulate the physical contradiction at the micro-level.

Micro-PhC: The operative zone during the operative period must (*physical micro-state*) in order to perform (*one of the conflicting actions*) and may not (*opposite physical micro-state*) in order to perform (*another conflicting action or requirement*).

Notes:

a-3.5 If it is difficult to construct the complete formulation of the physical contradiction, you may construct a brief formulation:

Subsystem (or an operation for TP) in the operative zone/period (*specify*) must be, in order to (*indicate*), and may not be, in order to (indicate).

b-3.5 If a problem can be solved only at the macro-level, you may fail to complete this step. However, even in this case the attempt to construct a micro-PhC is useful because it gives additional information about how the problem can be solved at the macro-level.

c-3.5 The transition from macro- to micro-level often enables the solver to avoid psychological inertia (functional and terminological) and to find previously missed attributes of the technique.

3.6. Formulate the second IFR.

IFR-2: The operative zone (*indicate*) during the operative period (*indicate*) must ensure itself (*indicate the opposite physical macro- or micro-states and their properties*).

Notes:

a-3.6 The first three ARIZ stages significantly reconstruct the initial problem. This step summarizes the results of this reconstruction.

By constructing the IFR-2 formulation, we simultaneously obtain a new problem — a more difficult one. In the following, you should solve just this problem.

b-3.6 If it is difficult to imagine IFR-2, you may construct the extended formulation:

The substance or the field in the operative zone (*indicate*) during the operative period (*indicate*) must ensure itself (*indicate the opposite physical macro- or micro-characteristics and their properties*).

3.7. Check the possibility of applying the Standards to the problem solution formulated in IFR-2.

Notes:

a-3.7 Define which of two states (required by IFR-2) the preferable resource already has.

b-3.7 Su-Field of this earlier, latent problem always differs from the model of the initial problem (step 2.8), which allows more effective use of the Standards.

3.8. If the problem remains unsolved, proceed to Stage 4. If the problem is solved, go to Stage 7, although in this case it is also recommended to continue the analysis by Stage 4.

Stage 4 — Resources Applications

Note: The available resources were considered in steps 3.3 and 3.4. Using only these resources is not sufficient. The following steps force us to use first derivative or differential resources (see Chapter 6), later the product or raw object resources, and then external substances and fields. External substances and fields can be acceptable only if the available resources (product's, raw object's differential and derivative) did not help solve the problem. Resource mobilization is not aimed at using all the resources for solving a mini-problem. Its aim is to find the most appropriate solution with minimal expense.

4.1. Determine whether the problem can be solved using the mixture of substances' resources.

Notes:

a-4.1 If the substances' resources, as they are, can be used for solving the problem, the problem probably would not have arisen or would have been solved automatically. Usually new substances are needed. At the same time, the use of new substances leads to system complication, additional harm factors, etc. Principal point of resources application is to overcome this contradiction and to introduce new substances, without actually introducing them.

b-4.1 In the simplest case, this step is a transition from two monosubstances to a nonhomogeneous bi-substance (see Chapter 7).

c-4.1 Determine whether the problem can be solved by replacing the resource substances with emptiness (a void) or a mixture of these substances with emptiness.

4.2. Determine whether the problem can be solved using derivative and/or differential resources.

4.3. Determine whether the problem can be solved using product resources.
Note:

a-4.3 As known, the product is an unchanged element of the mini-problem. What resources can the raw object or product contain? Sometimes the raw object or product allows the following:
a. modification of characteristics and/or parameters;
b. expenditure (i.e., change) of some of its elements or subsystems when the raw object or product as a whole is unchangeable;
c. transition to a super-system (a brick remains unchanged, but a building changes);
d. use of multi-level structures;
e. joining with "nothing," i.e., with emptiness;
f. temporal changes.

Thus, the raw object or product can be a part of resources in only such rare cases when it can be easily changed, remaining unchanged.

4.4. Determine whether the problem can be solved using the combination of a field and an additional substance that responds positively to the field.
Notes:

a-4.4 Examples of such combinations (or couples) are magnetic field-ferromagnetic liquid, UV radiation-luminophor, heat field-shape memory metal.

b-4.4 Often usage of an electrical field or two interacting electrical fields instead of a new substance is fruitful for problem solving. Particularly, electrons (or their flow — electrical current) should be used if the conditions of the problem do not allow the use of both available and derivative resource substances. Electrons are always present in any substance and are the source of an electrical field providing a high ability to be controlled.

4.5. Determine whether the problem can be solved using resources from the super-system, in particular the resources of the alternative techniques.

4.6. Use the Agents method to join the problem statement and IFR-2.
Change the scheme in such a way that Agents act without this conflict.
Notes:

a-4.6 Agents were created from the available resources of the technique (e.g., native substances and fields) at steps 3.2 and 3.3 but failed to solve the problem. Here Agents are created from foreign substances and fields and should link IFR-2 with the initial situation.

b-4.6 Agents represent here the additional variable tools of the problem model under consideration.

c-4.6 Sometimes it is possible to execute step b-4.6 by combining two images corresponding to a bad and a good action in one sketch. It is useful to make a series of consecutive sketches for describing Agents' bad and good actions.

d-4.6 The sketch is used to imagine the action of Agents within the operation zone and period and to formulate the requests for resources.

e-4.6 Apply the forward search if the initial conditions of the problem are determined very precisely by the correct problem statement, and the task is to find a method to reach a solution close to IFR-2. Apply the backward search if the correct problem statement is doubtful.

4.7. Consider the possibility of a problem solution (in IFR-2 formulation with regard to resources refined in Stage 4) by the Standards.

Note:

a-4.7 A return to the Standards takes place at steps 2.8 and 3.7. Before this step, the main idea was to use available resources, circumventing wherever possible new fields and substances (see also the note to this stage). If we fail to solve the problem in the framework of available, derivative, or differential resources, new fields and substances have to be added. Many Standards and the Agents Method deal only with the technique of additions introduction.

4.8. If the previous steps lead to the problem solution, go to Stage 7 when other solutions are not desirable. If a solution is not found (i.e., all substance-field resources and Agents are insufficient) or other solutions are desirable, go to Stage 5. By entering Stage 5, the problem becomes much clearer and its direct solution using the knowledge databases becomes possible.

Stage 5 — TRIZ Knowledge Bases Application

5.1. TRIZ knowledge bases (see the appendixes) can be applied early, e.g., at step 1.6. Only the solutions that coincide with IFR-2 or are practically close to it are applicable during Stage 5.

5.2. Consider the possibility of a problem solution (in the formulation IFR-2 with regard to resources refined in Stage 4) by analogy with nonusual problems solved earlier.

Notes:

a-5.2 We already applied the heuristic "solution by analogy" at step 1.6 when the problem had a nonpoint structure. During Stages 3 and 4 we simplified the problem structure at the level of the physical contradiction and complicated the problem structure by introducing resources into consideration. Therefore, we can again return to the fund of solved problems.

b-5.2 Note that many TRIZ experts have their own registries of solved problems, but work should be done to combine these solutions into a single, uniform fund. In the meantime, the fund of patents can be used.

5.3. Consider the possibility of a problem solution with the help of the lists of effects.

Note:

a-5.3 See Appendix 3.

5.4. Consider the possibility of a problem solution using the typical transformations.

Note:

a-5.4 See Appendix 8 for the typical transformations.

5.5. Consider the opportunity to take "one step back" from IFR-2 if the problem is unsolved.

Note:

a-5.5 Sometimes a little further prompting of the solution is necessary by slightly deteriorating the technique. Solving this simple problem quite often provides a solution for the initial problem.

Stage 6 — Problem Replacement

6.1. If the problem is solved, go through Stages 7-9.

6.2. If there is no solution, check whether the conflicting pair of articles is the right combination. If they are not, another product and/or another tool should be chosen at step 2.1.

6.3. If there is no solution, check whether the formulation of the administrative contradiction (step 1.3) is the combination of several different problems. If that is the case, the formulation of the mini-problem (step 1.3) should be changed by separating the different problems for sequential solution (usually it is sufficient to solve only the small number of problems).

Notes:

a-6.3 Estimate the importance level and degree of difficulty (see Chapter 1) of these different problems.

b-6.3 Solve the main problem, if it is impossible to solve the optimal problems.

c-6.3 Use decision aids (see Chapter 1) for choosing the optimal problems (weight of the importance level and degree of difficulty). Start with the problem with a high level of importance and small D).

d-6.3 Check the possibility of solving the main problem after the solution of each additional problem is obtained.

6.4. If there is no solution, change the problem by selecting another TC at step 2.4.

6.5. If there is no solution, return to step 1.3 and repeatedly formulate the mini-problem relating it to a super-system. If necessary, such a return might be done several times, such as with transition to a super-super-system, etc.

6.6. If there is no solution in the previous steps (6.2–6.5), formulate and solve the maxi-problem.

Note:

a-6.6 Maxi-problems can often be solved through synthesis and genesis of a new technique. Unfortunately, a set of TRIZ instruments for these tasks is not considered in this book. Convolution of similar and alternative subsystems is a very popular TRIZ instrument for new technique synthesis. It is presented in Chapters 7 and 16.

Stage 7 — Solutions Analysis

7.1. Check the obtained solution(s).

Consider each added substance and field:

- Can you find the necessary substances and fields in other subsystems of the technique or in other available resources?
- Is it possible *not* to introduce *any* new substances and fields but to use derivative resources?
- Is it possible to use self-regulating or smart substances?

Introduce the corresponding corrections into the solution.

Note:

a-7.1 Self-regulating (under the given problem conditions) substances can change their physical parameters in a certain way with changing external conditions. Application of self-regulating substances allows change of the state of a system or measurements without additional devices.

7.2. Evaluate each obtained solution.

Control questions:

- Does the solution provide fulfillment of the IFR-1 main requirement?
- Is a physical contradiction eliminated, and if so, which one?
- Does the obtained system include at least one well-controlled element? Which one? In what way can it be controlled?
- Is the solution found for the problem model suitable for real conditions?

If the solution does not meet at least one of the four control questions, return to step 1.1.

7.3. Select the best solution from the obtained solutions.

Notes:

a-7.3 Use the evolution trends, corollaries, and paths of the technique and its super-system (see Chapter 7).

b-7.3 Use the answers to questions in step 7.2 as well as decision aids (see Chapter 1) for selecting the solution.

7.4. Check by patent data the formal novelty of the selected solution.

7.5. What sub-problems can arise by technical elaboration of the selected solution?

Notes:

a-7.5 List those possible sub-problems regardless of their nature (design, calculations, management changes, etc.).

b-7.5 Solve the necessary sub-problems (see also Note to step 6.3).

7.6. Proceed from the selected conceptual solution to a technical one: formulate the method and construct the principal scheme of the technique accomplished in this manner.

Stage 8 — Selected Solution Application

8.1. Determine in what way the super-system, including the improved technique, must change.

8.2. Check whether the improved technique or super-system can be used in a new way.

Note:

a-8.2 The legal pattern "Usage of *known* as *unknown application* of *known*" exists in patent practice of some countries.

8.3. Use the selected solution in solving other problems:

a. Formulate the obtained heuristic in a general form;

b. Consider the possibility of direct use of this heuristic in solving other problems;

c. Consider the possibility of using the heuristic opposite to the obtained one;

d. Construct the morphological box and consider the possible reconstructions of the solution by positions of this box;

e. Consider how the obtained heuristic varies by varying each important parameter of the technique (or the main improved technique's sub-systems) applying the Parametric Operator;

f. Consider the possibility of using the obtained heuristic in the multi-screen approach.

Notes:

a-8.3 Read Chapter 1 about the morphological box.

b-8.3 If the work is performed not only to solve one specific technical problem, careful completion of sub-steps 8.3a–8.3e may become the starting point of developing a branch of TRIZ based on the obtained heuristic.

8.4. If you have enough time, perform steps 7.4 and 8.3 for nonselected obtained solutions.

Stage 9 — Solving Process Analysis

9.1. Compare the real progress of this problem solution with the theoretical one (by ARIZ). Report the deviations from ARIZ in a special form. Distribute copies of the report.

9.2. Compare the result obtained with the data from TRIZ heuristics and instruments as well as with TRIZ knowledge databases (effects, transformations, etc.). If this heuristic or effect is unknown in TRIZ, list it in the preliminary collection of TRIZ heuristics. Share the findings with other TRIZ experts.

a-9.2 Check the obtained heuristic with the patent fund in accordance with the procedure described in Chapter 8.

9.3. Estimate the degree of difficulty of the resolved problems and the Level of the obtained solutions. Collect the statistical data in the pattern "TRIZ heuristic (*indicate*) helps to resolve the problem (*indicate the engineering field*) with **D** = ... (*verify the number*) at the Level ... by (*indicate how many specialists participated in the estimation of the Level*)." Share periodically your statistics with other TRIZ experts.

18.6 CASE STUDIES

Primitive technical systems with only a few elements are chosen to illustrate ARIZ. Consequently, some steps and even stages are obvious, so they are omitted here for simplicity.

18.6.1 SOLUTION OF A PROBLEM WITH ARIZ-77

Problem:

A polishing disk inadequately processes a product with a complicated shape, with recesses or protuberances, such as a spoon. Polishing in some other way is inconvenient and complicated. Use of polishing disks made of ice is too expensive in this case. Also unsuitable are elastic inflatable disks with an abrasive surface since they wear down too quickly. What can be done?

1.1–1.9 — Problem already done.
2.2 — The product is a curved surface (e.g., spoon). The tool is a polishing disk.
2.3 —
    a. The disk possesses the ability to polish (UF).
    b. The disk does not possess the ability to adapt itself to curved surfaces (HF).
2.4 — The disk and a product are given. The disk possesses the ability to polish but cannot adapt itself to the curved surface of the product.
3.1 — The shape of a product cannot be changed: a flat spoon would not hold liquid. The disk can be changed (while retaining, however, its ability to polish). These are the specifications of the problem.
3.2 — The disk itself adapts to the curvature of the product while retaining the ability to polish.
3.3 — The external layer of a disk (the outer ring, frame); a substance (abrasive powder, solid body).
3.4 — In order to polish, the outer layer of a disk should be firm (or rigidly fixed to the central part of a disk for transmission of force) AND in order to adapt to the curvature of products, the outer layer of the disk should not be firm (or should not be rigidly fixed to the central part of a disk).
3.5 — The outer layer of a disk should be firm in order to polish a product and should not be firm in order to adapt to the curvature of the product.
The outer layer of the disk should be and should not be.
4.1 — Standard transformations do not produce an obvious solution to this problem.

4.2 — According to the typical solution, the substance S2 should be developed into a Su-Field, introducing field F and adding S3 or dividing S2 into two interacting parts. (The idea of dividing the disk was beginning to take shape at step 3.3. But if one simply divides the disk, the outer part will fly off under the effect of centrifugal force. The central part of the disk should hold strongly to the outer part and at the same time should give it the possibility of being freely exchanged…) Later, according to the typical solution, it is desirable to translate the Su-Field (obtained from S2) into a Su-M_Field, i.e., to use a magnetic field and a ferromagnetic powder. (This enables one to make the outer part of the disk movable and exchangeable, and it guarantees the requisite bond between both parts of the disk).

4.3 — The substitution of field or substance links by means of electromagnetic fields.

Other effects relate to liquids and solid bodies; they call for the introduction of additives or do not ensure automatic self-regulation.

4.4 — According to the specifications of the problem, one needs to improve the ability of the disk to grind products of different shapes. This is an adaptation (line 35 in the Matrix). One way is to use a selection of various disks. On the debit side one loses time when exchanging and selecting disks, thus lowering productivity: columns 25 and 39. The methods according to the table are 35, 28; 35, 28, 6, 37. The recurring, and hence most likely, methods are 35, exchanging the total state (the outer part of the disk is pseudo-liquid, made of moving particles), and 28, a direct indication to switch to a Su-M_Field, which has been done above.

4.5 — The central part of the disk is made from magnets, and the outer layer is of ferromagnetic particles or abrasive particles baked onto the ferromagnetic particles. Such an outer layer will assume the shape of the product. At the same time, it will retain its firmness necessary for polishing.

18.6.2   SOLUTION OF A PROBLEM WITH ARIZ-85AS

Problem:

A punch becomes disabled quickly during a cold punching (cutting out) of molybdenum parts for semiconductor devices. The "evident" source of trouble is the very hard molybdenum. Molybdenum cannot be replaced because it is the only material compatible with the semiconductor (the same thermal expansion coefficient). There are restrictions for the punch and die substitution also. But it cannot be allowed that this technical system does not work because assembly of the semiconductor devices is the main company business.

Let us show how to solve this problem with ARIZ-85A5.

0.1 — The technique for making parts comprises a blank (the raw object), package (the product), punch, and die.

1.1 — The primary function (the punch and die change the blank's shape) is accompanied by the harmful function (the punch and die break during PF performance).

Initial situation: The state of the system elements before the blow:

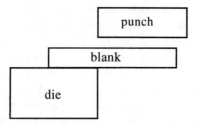

The elements' state at the beginning of a blow (the elements come into contact):

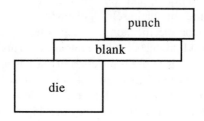

The elements after a blow:

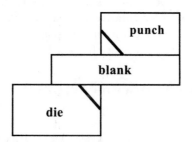

The areas of destruction of the punch and die are marked with bold lines.

1.2 — Because the technique under consideration is very simple, it is extremely easy to determine the structure of the problem as

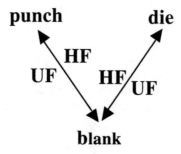

1.3 — The administrative contradiction is:

The high mechanical stability of the punch and die should be obtained at minimum changes in the technique. But the punch and die break during the technique work.

The formulation of a mini-problem:

It is necessary to remove the possibility of being broken for the punch and die, while keeping their ability to change the shape of a blank with minimum changes in the subsystems.

1.4–1.6 — The problem has a known generic structure and can be resolved by analogy with a solution of the same generic problem [8], but for our purposes here (to demonstrate how ARIZ-85AS works), we will forget the solution and assume that a solution for a similar problem is unknown or cannot be adapted to this problem.
    Note: The structure of the problem is linear and reflects HF and UF between 3 elements.
    Because the nature of HF (UF) between the punch and the blank and between the blank and the die is the same, the linear problem can be reduced to the pair problem between the tool (die and punch) and the raw object (blank), as is shown in the figure below.
    2.1 — We already defined the blank as the raw object and the package as the product. The punch and die present the tool. No new UF and HF are found at this step.
    2.2 — UF coincides with PF in this problem. So it is more rational to consider the punch and the die together and consider how this tool interacts with the product or raw object (package/blank).

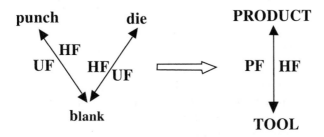

2.3 — TC-1: If the punch hits the blank, then the blank takes the needed shape of the package, but after several blows both the punch and die break.
    TC-2: If the punch does not hit the blank, then the punch and die do not break, but the blank does not take the needed shape of the package.
    The goal is to provide the needed shape of the blank and to keep the punch and die unbroken with minimal changes made in the system.
    Below is a graphical scheme of the conflict:

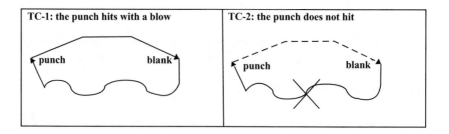

| TC-1: the punch hits with a blow | TC-2: the punch does not hit |

2.4 — PF changes the blank's shape into the shape of the package (punching). Therefore, TC-1 is selected.

2.7 —

1. Strengthen the conflict: the punch and die break after the first blow. If the punch hits the blank with a blow, then the blank takes the needed shape, but the punch and die break after a *relatively small number of blows*.
2. Strengthened formulation of a conflict: If the punch hits the blank with a blow, then the blank takes the needed shape, but the punch and die break after the *first blow*.
3. We should find an unknown sub-subsystem, characteristic, or parameter that would retain the punch's capability to hit the blank with a blow and at the same time ensure the safety of the punch and die.

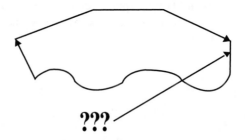

Note: The unknown subsystem, characteristic, or parameter acts on the blank rather than on the punch and die because in this case it *prevents* negative action of the blank on the punch and die rather than protects the punch and die from being broken (it is easier to prevent a disease than to treat).

2.8 — Selection of the Standards was performed using the algorithm presented in Chapter 15. Standard 1.2.4, Transition to the double Su-Field*, was selected. It is proposed to introduce the second field to neutralize the harmful action of the product S_2 on the tool S_1.

---

* If both useful and harmful effects take place between substances in Su-Field and substances which must be in direct contact, the problem is solved by passing to double Su-Field, in which the available field F_1 retains its useful effect, while the new added field F_2 neutralizes (compensates for) the harmful effect (or transforms it into a useful one) [4].

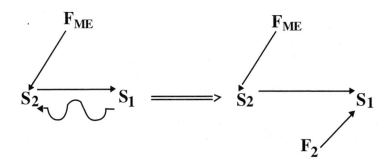

Note: It is a good idea but the nature of the second field F_2 is unclear, so we will go forward without analyzing the different possible fields.

    3.1 — Operative zone: the zone of deformation, which includes the blank, punch, and die (the zone where the blank *must* be shaped plus the zones where the break of the punch and die *takes place*).

    Operative period:

    T1 — the time during which the punch hits the blank with a blow.
    T2 — the time prior to the blow.

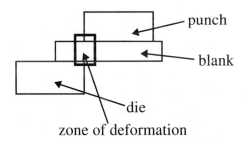

zone of deformation

    3.2 — IFR-1: Agents, absolutely not complicating the system and producing no harmful effects, eliminate the break of the punch and die, while the punch hits the blank within the zone of deformation; the punch capability to hit the blank with a blow and thus to shape it is kept unchanged.

    3.3 — Strengthened formulation of IFR-1: The material of the blank, absolutely not complicating the system and producing no harmful effects, eliminates the break of the tool, while the punch hits the blank within the zone of deformation; the punch capability to hit the blank with a blow and thus to shape it is kept unchanged.

    Note: According to Note b-3.3, we should first consider the use of the tool. However, in this problem we started the analysis from raw object resources. ARIZ-85 does not prohibit it (see Note h-3.3).

    The substances (resource) cannot be changed because only what is mentioned in the problem statement can be used.

    3.4 — Physical contradiction at the macro-level: The material of the blank in the zone of deformation should be able to be easily shaped during the blow so the blank can take the needed shape, and it should be difficult to deform because the blank possesses this property according to the conditions of the problem.

The preferable resource should alter the deformation ability (hard vs. soft) of the blank material. Hypothesis: various chemical bonds in the alloyed molybdenum.

3.5 — Physical contradiction at the micro-level: The zone of deformation should contain loosely bound particles of the substance during the blow for the blank material to be easily shaped, and it should contain strongly bound particles for the blank material to resist being deformed.

or

The chemical bonds of the blank material should be weak in the deformation zone during a blow and the chemical bonds should be strong for the blank material to be hardly deformable.

3.6 — IFR-2: The blank is endowed with a property γ (hard), while its part is endowed with the property anti-γ (the blank is soft in the zone of deformation).

4.1–4.5 — Substance and field resources:
  1. Inside the technique:
      A. resources of the tool — punch and die material,
      B. resources of the raw object/product — blank material.
  2. Outside the technique:
      A. electric current,
      B. air,
      C. lubricant.

The strength of punching can be increased without increasing the strength of the blow, if air under the blank is replaced with a vacuum. This solution fits, perhaps, only for thin parts and cannot improve considerably the assembly of semiconductor devices.

4. 6 — The blank in the deformation zone during the blow should itself provide the weakly and strongly bound Agents.

Hypothesis: Agents help alter the chemical bonds.

The correct statement of the problem (see figure at step 1.1).

The desired solution (IFR): The bold line marks agents, which are loosely bound to each other.

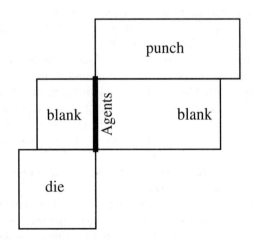

The agents appear *only* at the moment when the tool comes into contact with the raw object.

The state of the elements after the blow: Agents disappear without leaving a trace.

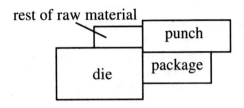

Obviously, the Agents should be a field which can penetrate a substance without leaving a trace, as Standard 1.2.4 told us at step 2.8. Because the blank is made of metal, it is worth first considering the use of magnetic and electric fields.

Solution: Electrodes are applied to the punch and die. As the die comes into contact with the punch, the blank is heated along the line of punching.

Any metal interacts with an electric field. Effects with the electric field should be considered.

4.7 — No additional solutions by Standards are found.

5.3 — Application of the known physical effects: the use of the effect of magneto-plasticity — metals become plastic under the action of a magnetic field. After relieving the magnetic field, the hardness reconstructs.

6.1–6.6 — Analysis by Stage 6 is omitted because the problem is solved.

7.1 — Checking the answer. The solution uses the available resources — electric current.

7.2 — Tentative evaluation of the solution:
  a. The strengthened formulation of IFR-1 is asserted.
  b. Physical contradiction at the macro- and micro-levels is eliminated.
  c. The system includes a well-controllable element — the electric current.
  d. The solution is suitable for conditions with multiple cycles.
Therefore, the evaluation of the solution is positive.

7.3 — The solution corresponds to a known ET path.

7.6 — The solution was implemented in a Russian semiconductor plant and results in large profit.

8.1–9.3 — Stages 8 and 9 were omitted because the solution is not absolutely new (similar idea known in different engineering field).

18.7 CONCLUSION

At first glance, ARIZ seems very complicated (especially its recent versions) and difficult. It is not an equation, but rather a multi-step process of asking a series of questions that move a solver back and forth from the Functional Domain and Physical Domain, integrating different TRIZ heuristics. ARIZ is an innovative instrument developed during the last 50 years for engineers who have been trained in TRIZ and who are able to use properly and creatively its heuristics and concepts. ARIZ's complexity, formalism, and a certain excess of rationalizing and control reflect the socio-psychological environment in which TRIZniks and problem solvers work. Analysis of the algorithm's history increases understanding of its logic and case studies assist in recognizing how ARIZ works.

There is no particular reason to engage in the careful, conscious analysis of a problem with ARIZ when you can immediately get some good ideas on how to solve it. In such a case, just go ahead and solve the problem "naturally." However, after you solve it, or, even better, while you are solving the problem, analyze what you are doing. It will greatly deepen your understanding of problem-solving heuristics, and you might discover a new heuristic or a new application of an old heuristic. Nevertheless, practice in solving problems with ARIZ is very helpful. The objective technique's evolution trends and resolution of contradictions are the most important criteria for estimating a solution obtained in the framework of ARIZ and other TRIZ heuristics and instruments.

I would recommend designating several weeks for solving your first real problem by ARIZ. It is sometimes suggested to start with ARIZ-77, wait a couple of weeks after completing the solution with this algorithm, and then solve the same problem with ARIZ-85AS. We find that it is a good exercise for a few problems, and then you can switch to ARIZ-85AS. Regular practice is quite important, and as a result you will be able to solve very difficult technical problems in a few days.

A few new special algorithms have been developed on the basis of TRIZ in the last several years for solving technical problems in the areas of technique synthesis and diagnostics, yield enhancement, and others. Unfortunately the scope of this work and the level of TRIZ presentation adopted here do not allow inclusion of these results. At the same time, ARIZ has become universal: people are beginning to apply it to nontechnical problems in management, arts, marketing, etc. Various TRIZ experts in the former Soviet Union have developed algorithms for such nontechnical problems. This research is the base of "soft TRIZ," also beyond the scope of this book.

REFERENCES

1. Altshuller, G. S. and Shapiro, R. B., About the psychology of inventivity, *Problems of Psychology*, 6, 37, 1956 (Electronic copy at http://www.jps.net/triz/triz0000.htm).
2. Gordon, W. J. J., *Synectics, The Development of Creative Capacity*, Harper, New York, 1961.
3. Altshuller, G. S., *How to Learn to Invent*, Tambovskoe Knignoe Izdatel'stvo, Tambov, 1961.
4. Altshuller, G. S., *Algorithm of Inventions*, Moskovskij Rabochij, Moscow, 1969, 1973 (in Russian); TIC, Worcester, 1999 (in English).
5. Polya, G., *How to Solve It: A New Aspect of Mathematical Method*, Princeton Univ. Press, Princeton, 1973.
6. Altshuller, G. S., *Creativity as an Exact Science: The Theory of the Solution of Inventive Problems*, Gordon and Breach, New York, 1984.
7. Altshuller, G. S., *To Find an Idea*, Nauka, Novosibirsk, 1986 (in Russian).
8. Yuganson, E., Danilin, E., and Kommel', F., *Tekhnika Molodezhi*, 5, 14, 1969 (in Russian).

19 Conclusion

Norbert Wiener (1894–1964), the famous American mathematician and one of the fathers of cybernetics and computer science, wrote, "We are living in an age that differs from all previous ones by the fact that the invention of new machinery … is no longer a sporadic phenomenon, but has become an understood process to which we resort, not merely to improve the scale of living and the amenities of life, but from a desperate necessity to render any life whatever, and certainly any civilized life, possible in the future" [1]. TRIZ transfers the genesis of inventions from occasional events into regular practice through its unique ability to help obtain strong solutions of the following type of nonroutine problems:

- achieve a specific improvement, enhance quality, and reduce cost of an existing technique;
- forecast a specific technique and create a conceptually new technique.

TRIZ accumulates and condenses all respective human knowledge and then applies it to solving inventive problems. The first aim of TRIZ is to transform current inventive problems into the routine problems of the future. The second is to develop an inventive person who is able to solve technical and nontechnical problems creatively. This book summarizes the results of TRIZ methodology developments in the twentieth century by experts mostly from the former Soviet Union. In the last few years TRIZ popularity has grown exponentially in the US, Japan and the Pacific Rim, and Western European countries. TRIZ will become more versatile in the twenty-first century as its developers continue the search of new heuristics, construction of new powerful instruments for problem solving based on contradiction resolution and technique evolution, and application and modification of known heuristics and instruments to yet untouched fields (e.g., software and genetic engineering).

Although I defined TRIZ as a methodology in this book, one can consider TRIZ as an embryo of a new science that breaks current paradigms in solving nonmathematical problems, engineering design theory, history of technique, and human creativity development. It seems possible to present evolution of a methodology with the same S-curve that often describes the evolution of techniques, biological and economical systems, etc. During the last 50 years, TRIZ achieved the point where these methodologies began to be accepted in academia (see figure). Its early childhood was very difficult because TRIZ was almost forbidden in the USSR, where it was developed, and unknown outside the country. It became very powerful for problem solving because of the enthusiasm of TRIZ experts. It can potentially grow forward, although its progress will slow if TRIZ is not supported as other sciences. It seems that four application streams are possible:

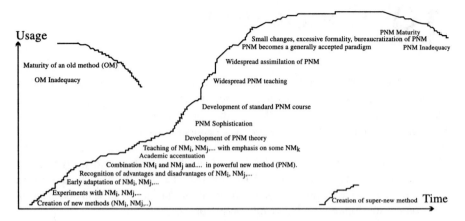

FIGURE 19.1 S-curve for the general trends of method evolution. Currently TRIZ is at the fifth step in this curve.

- Theoretical TRIZ — creation of new ideas and concepts, search and recognition of new heuristics, construction of new instruments, representation and convolution of knowledge of other sciences for solving problems, as well as analysis, development, and mathematical description of already known TRIZ ideas and concepts.
- Experimental TRIZ — test of new heuristics and instruments during the solving of real problems and design of techniques, application of known TRIZ heuristics and instruments in new engineering fields where these heuristics and instruments had not been implemented.
- Applied TRIZ — usage of known TRIZ heuristics and instruments for solving particular problems in various engineering fields where these heuristics and instruments already worked.
- Pedagogic TRIZ — development of instruments and methods for effective teaching of TRIZ, search of known solutions of easily understandable problems and their presentation in TRIZ education, as well as implementation of TRIZ in the teaching of other sciences, liberal arts, and methodologies.

TRIZ experts should develop this scientific methodology along the first two streams; the third is available to any engineer who knows TRIZ, and the last should be developed by TRIZniks and experts in education. TRIZ theoretical and experimental research is the process of discovering a heuristic. Design, problem solving, invention creation (i.e., applied TRIZ) is the process of using that heuristic. The TRIZ expert discovers a class of heuristics — a generalization — and the designer, problem solver, or engineer prescribes a particular embodiment of it to suit the particular result or objects he or she is concerned about. Pedagogic TRIZ allows the designer, problem solver, or engineer to learn TRIZ and then to apply it.

Due to the long standing development by relatively large groups of researchers, TRIZ has become the most powerful methodology for effective and inventive thinking during resolution of nonroutine technical problems. It helps fulfill the main axiom of our society:

> Both the quantity and quality of human requirements and
> needs, as well as requirements for humans, increase with time.

It happens because TRIZ makes it possible to satisfy human requirements by solving various technical and nontechnical problems in a short time and with high quality. TRIZ also provides people (who know this methodology) the ability to develop creativity and innovativeness and to be a specialist-universalist centaur. TRIZ can be considered a basis for developing personal and group creativity and inventiveness.

Everyone who solves technical problems perhaps already knows and uses some of the heuristics and instruments described in this book You may be using the heuristics known to you less effectively than you could be if trained in TRIZ. And if you are not an extremely dexterous and ingenius problem solver, some heuristics and instruments described in this book could be unknown to you, yet very useful.

I hope you learn the powerful concepts of TRIZ and have gained new ideas from this book. And I hope that you will apply them to solve previously insolvable problems. If you do, I consider that I achieved my goal with this book.

REFERENCE

1. Wiener, N., *Invention: The Care and Feeding of Ideas,* MIT Press, Cambridge, 1993.

Appendixes

APPENDIX 1
Popular Checklist and Questionnaire

ALEX OSBORN'S QUESTIONNAIRE FOR NEW IDEAS

Put to Other Uses?
New ways to use as is? Other uses if modified?

Adapt?
What else is like this? What other idea does this suggest? Does past offer a parallel? What could I copy? Whom could I emulate?

Modify?
New twist? Change meaning, color, motion, sound, odor, form, shape? Other changes?

Magnify?
What to add? More time? Greater frequency? Stronger? Higher? Longer? Thicker? Extra value? Plus ingredient? Duplicate? Multiply? Exaggerate?

Minify?
What to subtract? Smaller? Condensed? Miniature? Lower? Shorter? Lighter? Omit? Streamline? Split up? Understate?

Substitute?
Who else instead? What else instead? Other ingredient? Other material? Other process? Other power? Other place? Other approach? Other tone of voice?

Rearrange?
Interchange components? Other pattern? Other layout? Other sequence? Transpose cause and effect? Change pace? Change schedule?

Reverse?
Transpose positive and negative? How about opposites? Turn it backward? Turn it upside down? Reverse roles? Change shoes? Turn tables? Turn other cheek?

Combine?
How about a blend, an alloy, an assortment, an ensemble? Combine units, purposes, appeals, ideas?

APPENDIX 2
Forecast of Technical Systems

Evolution paradigm introduced in biology [1] became popular in social and natural sciences, economics, and liberal arts [2–6]. Although thousands of papers and books are published about evolution of different systems and artifacts, most are not more than a collection of facts and speculations. All attempts to construct a forecasting theory or method for predicting the future of a technique were unsuccessful. Some TRIZ experts have recently claimed the creation of simple, qualitative forecasts for technical systems (TS) and technological processes (TP) based on regularities of technique evolution [7, 8]. *The target of this work is to check the possibility of quantitative forecasts of TS.*

Any forecast must provide the probability of achieving a given level of TS performance by a certain time [4, 5]. To provide the right description for forecasting a technique, we should choose the appropriate mathematical apparatus for its description. It seems that such an apparatus has been developed in statistical physics that describes sets of objects regardless of their nature [9–11]. Statistical physics works with micro-states of objects (analog of the technique states), time, space, and different averages that also are applicable to TS and TP [12]. An extensive number of states G_i are scaled like e^N for a TS that consists of N elements or subsystems in phase space Q of a technique. All states G_i are metastable because of technique changes due to development.

First of all, the question about ergodicity is raised for objects that should be forecasted [13, 5]. Due to ergodicity, the time $\langle Q \rangle_t$ and ensemble $\langle Q \rangle_E$ averages of microstates for a set (or ensemble in the statistical physics language) of the ergodic objects are equal, and, therefore, the behavior of such objects can be predicted if their current state is known [9–11]. The following important necessary conditions for ergodicity must occur in a system [9–11, 14–16]:

    i. Conservation of the Hausdorff dimensions **d** for the invariant ergodic informational measure of a dynamic system.

    ii. Accessibility of almost all phase space with equal probability for the Lebesgue measure.

    iii. Complete nondegeneration of a full system's Hamiltonian and energy equipartition.

    iv. Positive values of the maximal Lyapunov's exponent L_{max}, i.e., the exponential divergence of initially close trajectories in the phase space.

Let us discuss the ergodicity of TS based on modern knowledge about technique. Unfortunately, it is impossible to check any of the necessary conditions of ergodicity in direct experiments with real technical systems because we cannot measure ergodicity metrics such as the maximal Lyapunov's exponent. On the other hand, the numerical experiments at modern superfast CRAY computers for analysis of a set consisting of nonidentical elements require more time than design of a new technique usually takes, so the numerical experiments are ineffective [17]. Hence, for analysis of TS ergodicity, we will use the indirect information that can be extracted from knowledge about technique. A summary of the necessary conditions analyses follows.

i. It is well known that technique evolution leads to a change in the number of elements and links between subsystems; i.e., the Hausdorff dimensions **d** are not conservative metrics of the TS or TP.

iii. Any technique does not yet develop itself. Various factors, such as availability of raw materials, quality of labor forces, necessary knowledge for development, etc. play important roles in the evolution of a technique. Due to the complexity of a real technique, various barriers can separate the transitions between any neighboring states N of the phase spaces. Then the probabilities of the transition $G_{i-1} \to G_i \to G_{i+1}$ and the transition $G_{i+1} \to G_i \to G_{i-1}$ are different, even if the equal "weight" corresponds to the states G_{i+1}, G_i, and G_{i-1}; i.e., the equal accessibility of all states G_i is violated. Each such transition can be considered a step in a technique's development.

iv. If long-term development observed in some techniques meets the change of the phase space ($|dQ/dt| > 0$), then $g_{ij} = G_i - G_j$ decreases with time t and the maximal Lyapunov's exponent

$$L_{max} = \lim_{t \to \infty} \{ \ln[g_{ij}(t)/g_{ij}(0)] \}/(2t)$$

is negative. This slowing or saturation in evolution [2–7] begins only when the technique needs additional sources for its development, which usually lie out of the main knowledge domain of the system resources.

Therefore, the necessary conditions for ergodicity are violated for the technical systems, which is why the application of statistical methods to the design of TS [12] is under question.

Moreover, the technique phase space is not compact and is divided into fragments. Any technique has a finite time-of-life; hence, the average in time can be performed only up to the limit of a TS lifetime. Therefore, some states G_i of the phase space Q are not probed in a finite period because not all possibilities of system development are tried by its designers. It is possible to speculate that these fragments allow existence of competition between various technical systems with the same primary function.

The nonuniform probability P of existence in various states can be used for definition of the nonergodicity parameter F [18] that is the width of the state distribution probability due to the violations of the equality

$$\forall G_i = \forall G_j; \qquad G_i, G_j \subset N.$$

This parameter can be applied to various systems including techniques [18]. In the ergodic system, $F = 0$ and the dependence $P(N)$ is characterized by the delta-function. On the other hand, in the absolutely nonergodic system, $F = 1$ and the uniform distribution is realized for $P(N)$ in intervals from 0 up to 1. In the majority of the nonergodic systems, the values of F are far from the lower and upper limits [18] and $P(N)$ is characterized by nonuniform distribution. Such distribution leads to the well-known S-shape evolution curve [2–8]. This S-curve can be represented by the equation of the so-called inverse Kohlrausch law

$$M \sim \exp \left[(-t/t_0)^{-\delta} \right].$$

Here M is the measure of TS performance, t_0 is the characteristic mean time of system development, and δ is the Kohlrausch exponent that depends on a value of the driven force for some types of the movement of the representative point through the phase space [19]. The comparison of the Kohlrausch law with the evolution curves of various TS (airplanes, semiconductor chips, light sources) leads to the values 0.4–0.8 for the exponent δ. The uncertainty for forecast is in order of the Kohlrausch exponent δ that reflects the nonergodicity parameter F. Unfortunately, δ or F are too big for any reasonable practical predictions about the evolution of a technique.

Therefore, the lack of time dependence of the evolution in nonergodic systems (including TS and TP) forbids a quantitative forecasting for such objects.

REFERENCES

1. Darwin, C., *On the Origin of Species.* Harvard University Press, Cambridge, 1964.
2. Binder, K. and Young, A. P., *Rev. Mod. Phys.* 58 (1986) 801.
3. Marshall, S. and Service, E. R., Eds., *Evolution and Culture*, University of Michigan Press, Ann Arbor, 1960.
4. Arrow, K., Anderson, P. W., and Pines, D., Eds., *The Economy as a Complex Evolving System,* Addison Wesley, New York, 1988.
5. Aoki, M., *New Approaches to Macroeconomic Modeling*, Oxford University Press, New York, 1996.
6. Kuhn, T., *The Structure of Scientific Revolutions,* University of Chicago Press, Chicago, 1962.
7. Altshuller, G. S., *To Find an Idea,* Nauka, Novosibirsk, 1991.
8. Altshuller, G. S., Zlotin, B. L., Zusman, A. V., and Filatov, V. I., *Search of New Ideas,* Kartya Moldovenyaske, Kishinev, 1989.
9. Toda, M., Kubo, R., and Saitf, N., *Statistical Physics I. Equilibrium Statistical Mechanics,* Springer, Berlin, 1983.
10. Balescu, R., *Equilibrum and Nonequilibrum Statistical Mechanics*, John Willey & Sons, New York, 1975.
11. ter Haar, D., *Rev. Mod. Phys.,* 27 (1955) 288.

12. Yoshikawa, H., Extended General Design Theory, in *Proc. IFIP W.G. 5.2: Design Theory for CAD*, North-Holland, Amsterdam, 1985.
13. Durlauf, S. N., Nonergodic Economic Growth, *Rev. Economics Studies,* 60 (1993) 349.
14. Shurenkov, V. M., *Markov's Ergodic Processes,* Nauka, Moscow, 1989 (in Russian).
15. Petersen, K., *Ergodic Theory,* Cambridge University Press, New York, 1983.
16. Kornfel'd, I. P., Sinai, Ya. G., and Fomin, S. V., *Ergodic Theory,* Nauka, Moscow, 1980 (in Russian).
17. Thirumalai D. and Mountain, R. D., *Phys. Rev.,* E47 (1993) 479.
18. Savransky, S. D., Quasielastic Neutron Scattering QENS'93 Singapore, *World Scientific,* (1994) 194.
19. Palmer, R. G., Stein, D. L., Abrahams, E., and Anderson, P. W., *Phys. Rev. Lett.,* 53 (1984) 958.

APPENDIX 3
Databases of Effects

List of Physical Effects

Required Action	Physical Effects
1. Measure temperature	Thermal expansion and its influence on natural frequency of oscillations
	Thermal-electrical phenomena
	Radiation spectrum
	Changes in optical, electrical, and magnetic properties of substances
	Transition over the Curie point; Barkhausen and Seebeck effects
2. Reduce temperature	Phase transitions
	Peltier, Seebeck, and Thomson effects
	Magnetic calorie effect
	Thermal-electrical phenomena
3. Increase temperature	Electromagnetic induction
	Eddy current
	Surface effect
	Dielectric heating
	Electronic heating
	Electrical discharge
	Absorption of radiation by substance
	Thermal-electrical phenomena
4. Stabilize temperature	Phase transitions, including transition over the Curie point
	Evaporation
5. Locate an object	Introduction of marker substances which are capable of transforming an existing field (such as luminophores) or generating their own (such as ferromagnetic materials) and therefore are easy to detect
	Reflection and emission of light
	Photo-effect
	Deformation
	Radioactive and Xray radiation
	Luminescence
	Changes in electrical or magnetic field
	Electrical discharge
	Doppler effect
6. Move an object	Applying magnetic field to influence an object or magnet attached to the object
	Applying magnetic field to influence a conductor with DC current going through
	Applying electrical field to influence electrically charged object
	Pressure transfer in liquid or gas
	Mechanical oscillations

List of Physical Effects

Required Action	Physical Effects
	Centrifugal force
	Thermal expansion
	Pressure of light
7. Move a liquid or gas	Capillary force
	Osmosis
	Toms, Bernulli, Weissenberg effects
	Waves
8. Move an aerosol	Electrolyze
(dust particles, mist,	Applying electrical or magnetic field
smoke, etc.)	Pressure of light
9. Formate mixtures	Ultrasonics
	Cavitation
	Diffusion
	Applying electrical field
	Applying magnetic field in combination with magnetic material
	Electrophoresis
	Solution
10. Separate mixtures	Electric and magnetic separation
	Applying electrical or magnetic field to change pseudo-viscosity of a liquid
	Centrifugal force
	Sorption
	Diffusion
	Osmosis
11. Stabilize an object's	Applying electrical or magnetic field
position	Holding in a liquid that is hardening under electrical or magnetic field influence
	Gyroscope effect
	Reactive force
12. Generation and/or	Generating high pressure
manipulation force	Applying magnetic field through magnetic material phase transitions
	Thermal expansion
	Centrifugal force
	Changing hydrostatic forces via influencing pseudo-viscosity of an electroconductive or magnetic liquid in magnetic field
	Use of explosive
	Electro-hydraulic effect
	Optical-hydraulic effect
	Osmosis
13. Change friction	Johnson-Rabeck effect
	Radiation influence
	Abnormally low friction effect
	No-wear friction effect
14. Destroy object	Electrical discharge
	Electro-hydraulic effect
	Resonance

List of Physical Effects

Required Action	Physical Effects
	Ultrasonic
	Cavitation
	Use of lasers
15. Accumulate a mechanical and/or thermal energy	Elastic deformation Gyroscope Phase transitions
16. Transfer energy (mechanical, thermal, radiation, energy transfer)	Oscillations Alexandrov effect Waves, including shockwaves Radiation Thermal-conductivity Convection Light reflection Fiber optics Lasers Electromagnetic induction Superconductivity
17. Influence on a moving object	Applying electrical or magnetic fields (no-contact influence instead of physical contact)
18. Measure a dimension	Measuring oscillations' natural frequency Applying and detecting magnetic or electrical markers
19. Change a dimension	Thermal expansion Deformation Magneto-striction Piezoelectric effect
20. Detect surface properties and/or conditions	Electrical discharge Reflection of light Electronic emission Moire effect Radiation Auger spectroscopy
21. Vary surface properties	Friction Absorption Diffusion Bauschinger effect Electrical discharge Mechanical or acoustic oscillation Ultraviolet radiation
22. Detect volume properties and/or conditions	Introduction of marker substances which are capable of transforming an existing field (such as luminophores) or generating their own (such as ferromagnetic materials) depending on considered material properties Changing electrical resistance depending on structure and/or properties' variations Interaction with light Electro- and/or magneto-optical phenomena

List of Physical Effects

Required Action	Physical Effects
	Polarized light
	Radioactive and Xray radiation
	Electronic paramagnetic or nuclear magnetic resonance
	Magneto-elastic effect
	Transition over the Curie point
	Hall, Moessbauer, Hopkins, and Barkhausen effects
	Ultrasonics
23. Vary volume properties	Varying properties of liquid (pseudo-viscosity, fluidity) via influence by electrical or magnetic field
	Influence by magnetic field via introduced magnetic material
	Heating
	Phase transitions
	Ionization by electrical field
	Ultraviolet, Xray, or radioactive radiation deformation
	Diffusion
	Electrical or magnetic field
	Bauschinger effect
	Thermoelectric, thermo-magnetic, or magneto-optical effect
	Cavitation
	Photo-chromatic effect
	Internal photo effect
24. Develop certain structures, structure stabilization	Interference
	Standing waves
	Moire effect
	Magnetic waves
	Phase transitions
	Mechanical and acoustic oscillation
	Cavitation
25. Detect electrical and/or magnetic fields	Electrization
	Electrical discharge
	Piezo- and segneto-electrical effects
	Electrets
	Electronic emission
	Electro-optical phenomena
	Hopkins and Barkhausen effect
	Hall effect
	Nuclear magnetic resonance
	Gyromagnetic and magneto-optical phenomena
26. Detect radiation	Optical-acoustic effect
	Thermal expansion
	Photo-effect
	Luminescence
	Photo-plastic effect

List of Physical Effects

Required Action	Physical Effects
27. Generate electromagnetic radiation	Josephson, Cherenkov, Gunn effects
	Induction of radiation
	Tunnel effect
	Luminescence
28. Control electromagnetic field	Use of screens
	Changing properties (for example, varying electrical conductivity)
	Changing object shapes
29. Control light, light modulation	Refraction and reflection of light
	Electro- and magneto-optical phenomena
	Photo-elasticity
	Franz-Keldysh, Gunn, Kerr, and Faraday effects
30. Initiate and/or intensify chemical reactions	Ultrasonics
	Cavitation
	Ultraviolet, Xray, and radioactive radiation
	Electrical discharge
	Shockwaves

LIST OF CHEMICAL EFFECTS

TRANSFORMATION OF SUBSTANCE

Carry in space: transport reactions, thermo-chemical method, in hydrate/hydride condition, in compressed gases, as a part of the future alloy, in adsorbents, as explosive mixes, molecular self-assembly, complexons, liquid membranes.

Change of mass: transport reactions, thermo-chemical method, translation in chemically connected species, translation in hydrate/hydride condition, in exothermic reactions.

Change of concentration: transport reactions, translation in chemically connected species and allocation, translation in hydrate/hydride condition, in compressed gases, displacement of chemical balance, adsorption-desorption, semitight membranes, complexons, liquid membranes.

Change of specific weight: translation in chemically connected species, translation in hydrate/hydride condition.

Change of volume: translation in chemically connected species, transport reactions, translation to hydrate/hydride condition, dissolution in compressed gases, in exothermic or thermo-chemical reactions, dissolution, at explosion.

The changed forms: transport reactions, thermo-chemical processing, gas hydrates, compressed gases, hydrides, melting-freezing.

Change of electrical properties: process of hydride synthesis (hydrogenation), restoration oxide, dissolution of salts, neutralization of electrical charges, at displacement of chemical balance, electrification due to oxidation, gases at radioactive radiation, electro-chrome, wetting ability of a layer, complexons.

Change of optical properties: restoration oxide, color, generation of light, change of light transmission, in mono-molecular layers.

Change of magnetic properties: hydrogenation, oxidizers, clusters.

Change of biological properties: translation in chemically connected species, influence by ozone, water ability/repellency, complexons.

Change of chemical properties: transport reactions, thermo-chemical processing, chemical linkage of gases, gas hydrate, compressed gases, hydrogenation, restoration oxide, exothermic reaction, thermo-chemical reaction, melting-freezing, dissolution of salts, displacement of chemical balance, influence by ozone, in photochrome, ability to repel water, translation in a microcondition, complexons, liquid membranes.

Change of a phase condition: transport reactions, thermo-chemical processing, chemical linkage of gases, gas hydrates, compressed gases, hydrides, melting-freezing, dissolution of salts, allocation from solutions, adsorption-desorption, photochrome.

Disposal (destruction): translation in chemically connected species, translation in hydrate condition, in compressed gases, hydrogenation, exothermic or thermo-chemical reaction, dissolution, influence by ozone, complexons, liquid membranes.

Stabilization (temporary reduction of activity): chemical linkage of gases, translation in hydrate condition, in compressed gases, in hydrides, melting-freezing, in adsorbent, complexons.

Transformation of two and more substances into one: transport reactions, thermo-chemical method, chemical linkage of gases, gas hydrates, compressed gases, hydride, oxidation-restoration, exothermic reaction, thermo-chemical reaction, dissolution, connection of mutual-active substances, influence by ozone, photochrome, complexons.

Protection of one substance from penetration by another: by chemical linkage of one of them, protection by hydrates, dissolution in compressed gases, protection by hydride, incineration, oxidation, from oxidizers, water ability/repellent, semitight membranes, liquid membranes.

Drawing one substance on a surface of another: transport reactions, in hydrate condition, with the help of hydrides, oxidation-restoration, connection of mutually-active substances, photochrome, electrochrome, molecular self-assembly, water ability/repellent, liquid membranes.

Connection of diverse substances (condensation, congestion): with the help of hydrates or hydrides, welding, melting-freezing, molecular self-assembly.

Division of substances (allocation of one from another): transport reactions, allocation of chemically connected gases, from compressed gases, from hydrides, restoration from oxide, displacement of chemical balance, from adsorbent, from ozonides, water ability/repellent, semitight membranes, complexons, liquid membranes.

Destruction of substance: transport reactions, thermo-chemical method, destruction of chemically connected substances, allocation from compressed gases, saturation by hydrogen, destruction oxide, incineration, dissolution, displacement of chemical balance in mixes, connection of mutually-active substances, oxidation, explosion, complexons.

Accommodation of one substance in the friend: transport reactions, chemical linkage of gases, gas hydrate, in compressed gases, in hydrides, in adsorbent, dissolution, complexons, molecular self-assembly, liquid membranes.

Reception of new substances (synthesis): transport reactions, thermo-chemical method, chemical linkage of gases, gas hydrates, hydride, restoration from oxide, exothermic or thermo-chemical of reaction, connection of mutually-active substances, at displacement of chemical balance, influence by ozone, oxidizers, super-oxidizers, ozonides, molecular self-assembly, complexons.

Organization of a closed cycle on substance (absorption-allocation): transport reactions, chemical linkage-allocation of gases, dissolution in compressed gases, hydride, adsorption-desorption, with the help of ozonides, in electro-chromes, complexons, liquid membranes.

Assembly of substance from atoms: transport reactions, allocation from chemically connected species, allocation from compressed gases, from hydrides, restoration from oxide, connection of mutually-active substances, molecular self-assembly, semitight membranes, transition molecules unit, complexons, liquid membranes.

Reception of substances with well organized structure (reception of pure substances): transport reactions, in chemically connected species, allocation from compressed gases, from hydrides, molecular self-assembly, complexons, liquid membranes.

Transport of one substance through other: transport reactions, thermo-chemical method, in chemically connected species, in compressed gases, in hydrides, hydrogen through metals, in thermo-chemical reactions, with use of phase transition, during the displacement of chemical balance, in adsorb species, semitight membranes, complexons, liquid membranes.

TRANSFORMATION OF SOME FORMS OF ENERGY

Reception of heat (input of thermal energy in system): incineration gas of hydrates, incineration of hydrogen, with the help of hydrides, power-intensive substances, exothermic reaction, with use of strong oxidizers, ozone decomposition.

Reception of cold (conclusion of thermal energy from system): decomposition gas hydrates, with the help of hydrides, endothermal reaction, during solution.

Reception of mechanical pressure: decomposition of gas hydrates, decomposition hydrides, loss of strength of metals at hydrogen penetration, swelling of metals, at decomposition of liquid ozone.

Generation of light radiation: chemo-luminescence.

Storage of heat: in chemical reactions, at phase transitions.

Storage of cold: in hydrides.

Storage of light energy: photochrome effect.

Transport of thermal energy: transport reactions, in hydride accumulators.

Transport (drain) static electricity: metalization of fabrics, processing ozone, covering by material with wetting ability.

Regulation of light energy: photochrome effect.

Power influence on substance: crown charge, radioactive radiation, cavitation, UV-light (5), electrical field, electrical current, electromagnetic field, IR-light, microwaves category, visible light, thermal energy.

See also the **List of Energies.**

INFORMATION TRANSFORMATION

Indication of the current information about substance: hydrogen, metal-organic impurity in gas, ozone, chemoluminescence in oxidation reactions, fluorescence, hydro-photo, hydrodynamics of flows.

Indication of the information about energy: thermal at phase transition, thermal in thermo-chromes, crown of the category on formation ozone, radioactive radiation on ozone formation, radioactive radiation in radio-chromes, visible radiation in photochrome, UV-radiation in photochrome.

LIST OF GEOMETRICAL EFFECTS

Required Action (property)	Geometric Effects
1. Reduce or increase the volume of a body at constant weight	Dense packing of elements. Gaufres. Hyperboloid of revolution of one nappe.
2. Reduce or increase the area or length of a body at same weight	Multistory configuration. Gaufres. Use figures with variable section. Mobius strip. Use the surrounding areas.
3. Transform one kind of a movement into another	Triangle of Relo. Conic-like beetle. Crank-rod transfer.
4. Concentrate energy/particles flow	Paraboloid, ellipse, cycloid.
5. Intensify process	Transition from processing on a line to processing on a surface. Mobius strip. Eccentricity. Gaufres. Screw. Brushes.
6. Decrease loss of energy or substance	Gaufres. Change of section of working surfaces. Mobius strip.
7. Increase process accuracy	Special selection of forms or trajectories for a movement of the processing tool. Brushes.
8. Increase control	Brushes. Hyperboloid. Spiral. Triangle. Use objects of an impermanent form. Change the progressive movement to rotary. Nonalignment screw mechanism.
9. Decrease control	Eccentricity. Replace round objects on multi-coal.
10. Increase lifetime, service reliability	Mobius strip. Change of the contact area. A special choice of the form.
11. Decrease expenses	Principle of similarity. Conformity (orthomorphism) reflections. Hyperboloid. Use a combination of simple geometric forms.

After you choose the appropriate effect, it is a good idea to read books and papers about the particular branch of science or to consult with a specialist in the field. Of course, not all effects are mentioned here. The Database of about 6000 Effects is available in TRIZ software, but, unfortunately, the software is not free from mistakes, so double-check an effect with scientists if you are going to use it in your practice.

The statistical analysis performed by Russian TRIZniks (without examination of patents in "biology-reach" classes, such as biotechnology and genetic engineering) in the 1980s shows that frequency of application of the physical, chemical, and geometric effects is 9:0.8:0.2. Perhaps this ratio will change soon due to the growing importance of genetics and bioengineering.

APPENDIX 4
Energies List

Energy is constantly used to meet human technical and social needs, and it is an important problem in scientific and technical progress. To systematize the information on energy transformation, the following list can be used. It is organized according to Type of energy, Kind of substance motion, and Type of interaction.

1. Mechanical energy.
    1.1. Motion of solid bodies.
        1.1.1. Movement.
        1.1.2. Rotation.
        1.1.3. Oscillation.
    1.2. Directed motion of bodies and particles.
        1.2.1. Motion of liquid.
        1.2.2. Motion of gas.
        1.2.3. Motion of molecules.
        1.2.4. Motion of atoms.
        1.2.5. Motion of elementary particles.
    1.3. Turbulent motion of particles.
        1.3.1. Motion in liquids.
        1.3.2. Motion in gases.
        1.3.3. Motion in plasma.
    1.4. Wave perturbations.
        1.4.1. Acoustic waves.
        1.4.2. Surface waves.
    1.5. Elastic deformation of bodies.
        1.5.1. Bending of solid bodies.
        1.5.2. Tension of solid bodies.
        1.5.3. Torsion of solid bodies.
        1.5.4. Compression of solid bodies.
        1.5.5. Compression of liquids.
        1.5.6. Compression of gases.
        1.5.7. Compression of plasma.
    1.6. Molecular and atomic interaction.
        1.6.1. Molecular forces of surface tension at media interface.
        1.6.2. Atomic forces of surface tension at media interface.
        1.6.3. Molecular capillary forces.
        1.6.4. Atomic capillary forces.

        1.6.5. Molecular sorption forces.

        1.6.6. Atomic sorption forces.

2. Thermal energy.

  2.1. Chaotic thermal motion of molecules and atoms.

        2.1.1. Displacement of molecules.

        2.1.2. Displacement of atoms.

        2.1.3. Rotation of molecules.

        2.1.4. Rotation of atoms.

        2.1.5. Vibration of molecules.

        2.1.6. Vibration of atoms.

  2.2. Thermal motion of molecules at mixing of bodies.

        2.2.1. Mixing of solid bodies.

        2.2.2. Mixing of liquids.

        2.2.3. Mixing of gases.

        2.2.4. Mixing of solid bodies and liquids.

        2.2.5. Mixing of liquids and gases.

        2.2.6. Mixing of solid bodies and gases.

        2.2.7. Mixing of solid bodies, liquids, and gases.

  2.3. Thermal motion of electrons.

        2.3.1. Excitation of electronic orbits in solid bodies.

        2.3.2. Excitation of electronic orbits in liquids.

        2.3.3. Excitation of electronic orbits in gases.

  2.4. Thermal motion of charges.

        2.4.1. Mixing of charges in equilibrium plasma.

        2.4.2. Mixing of charges in nonequilibrium plasma.

        2.4.3. Mixing of charges in electron-positron gas.

3. Energy of phase transitions.

  3.1. Phase transition of the first type.

        3.1.1. Change of aggregate state.

        3.1.2. Change of crystal structure.

        3.1.3. Dissolution of crystals.

        3.1.4. Evaporation of crystals.

  3.2. Phase transition of second type.

        3.2.1. Transition into and from the superconducting state.

        3.2.2. Transition of a ferromagnetic into paramagnetic and vice versa.

        3.2.3. Transition of helium-1 into helium-2 and vice versa.

        3.2.4. Polarization and depolarization of dielectrics.

4. Chemical energy.

  4.1. Chemical reaction of a compound.

        4.1.1. Oxidation.

        4.1.2. Recombination.

        4.1.3. Neutralization.

        4.1.4. Polymerization.

        4.1.5. Hydration.

        4.1.6. Formation of hyperfragments (positronium, ionium, etc.).

        4.1.7. Ionization with addition of electron or ion.

4.2. Chemical reactions of decomposition.
    4.2.1. Decomposition of molecules.
    4.2.2. Decomposition of polymers.
    4.2.3. Dissociation.
    4.2.4. Ionization in the process of decomposition.
    4.2.5. Dehydration.
4.3. Chemical reactions of substitution.
    4.3.1. Substitution in electrolytes.
    4.3.2. Substitution in crystals.
    4.3.3. Substitution at media interface.
4.4. Chemical reactions of transition from an excited state to the ground one.
    4.4.1. Transition of solid bodies.
    4.4.2. Transition of liquids.
    4.4.3. Transition of gases.

5. Nuclear energy.
  5.1. Nuclear synthesis.
    5.1.1. Spontaneous nuclear synthesis.
    5.1.2. Controlled nuclear synthesis.
    5.1.3. Pulse nuclear synthesis.
    5.1.4. Combined nuclear synthesis.
  5.2. Nuclear fission.
    5.2.1. Spontaneous nuclear fission.
    5.2.2. Controlled nuclear fission.
    5.2.3. Pulse nuclear fission.
    5.2.4. Combined nuclear fission.
  5.3. Radioactive transformations.
    5.3.1. Transformation at trapping of an electron with K-shell by nucleus.
    5.3.2. Transformations at isomeric transitions.
    5.3.3. Transformations at gamma-quantum emission.
    5.3.4. Radioactive decay.
  5.4. Annihilation.
    5.4.1. Annihilation of leptons.
    5.4.2. Annihilation of bosons.
    5.4.3. Annihilation of hyperfragments.
    5.4.4. Annihilation of atoms.
    5.4.5. Annihilation of molecules.
    5.4.6. Annihilation of macro-systems.

6. Energies of standard physics fields.
  6.1. Gravitational interaction.
    6.1.1. Gravitation of heavy masses.
    6.1.2. Gravitational waves.
    6.1.3. Gravitons (theoretical).
    6.1.4. Transmutations (transitions).
    6.1.5. Collapse.

APPENDIX 5

Altshuller's Standard Solutions of Invention Problems

Class 1. Construction and Destruction of Su-Field Systems

1.1. Synthesis of Su-Fields
    1.1.1. Making Su-Field
    1.1.2. Inner complex Su-Field
    1.1.3. External complex Su-Field
    1.1.4. External environment Su-Field
    1.1.5. External environment Su-Field with additives
    1.1.6. Minimal regime
    1.1.7. Maximal regime
    1.1.8. Selectively maximal regime
1.2. Destruction of Su-Fields
    1.2.1. Removing of harmful interaction by adding a new substance
    1.2.2. Removal of harmful interaction by modification of the existing substances
    1.2.3. Switching off harmful interaction
    1.2.4. Removal of harmful interaction by adding a new field
    1.2.5. Turn-off magnetic interaction

Class 2. Development of Su-Fields

2.1. Transition to complex Su-Fields
    2.1.1. Chain Su-Field
    2.1.2. Double Su-Field
2.2. Forcing of Su-Fields
    2.2.1. Increasing of field's controllability
    2.2.2. Tool fragmentation
    2.2.3. Transition to capillary-porous substances
    2.2.4. Dynamization (flexibility)
    2.2.5. Field organization
    2.2.6. Substances organization
2.3. Forcing of Su-Fields by fitting (matching) rhythms
    2.3.1. Field-Substances frequencies adjustment
    2.3.2. Field-Field frequencies adjustment
    2.3.3. Matching independent rhythms

2.4. Transition to Su-M_Field systems
   2.4.1. Making initial Su-M_Field (or "proto-Su-M_Field")
   2.4.2. Making Su-M_Field
   2.4.3. Magnetic liquids
   2.4.4. Capillary-porous Su-M_Field
   2.4.5. Complex Su-M_Field
   2.4.6. Environment Su-M_Field
   2.4.7. Usage of physical effects
   2.4.8. Su-M_Field dynamization
   2.4.9. Su-M_Field organization
   2.4.10. Matching rhythms in Su-M_Field
   2.4.11. Su-E_Fields
   2.4.12. Electrorheological suspension

Class 3. Transition to Super-System and to Microlevel

3.1. Transition to bi-systems and poly-systems
   3.1.1. Creation of bi-systems and poly-systems
   3.1.2. Development of links
   3.1.3. Increase of difference between system's elements
   3.1.4. Convolution
   3.1.5. Opposite properties
3.2. Transition to micro-level
   3.2.1. Shift to micro-level

Class 4. Standards for System Detection and Measurement

4.1. Roundabout ways to solve problems of detection and measurement
   4.1.1. Change instead to measure
   4.1.2. Copying
   4.1.3. Sequential detection
4.2. Synthesis of Su-Field measurement systems
   4.2.1. Creation of measurable Su-Field
   4.2.2. Complex measurable Su-Field
   4.2.3. Measurable Su-Field at environment
   4.2.4. Additives in environment
4.3. Forcing of measuring Su-Fields
   4.3.1. Physical effects applications
   4.3.2. Resonance
   4.3.3. Resonance of additives
4.4. Transition to Su-M_Field systems
   4.4.1. Measurable proto-Su-M_Field
   4.4.2. Measurable Su-M_Field
   4.4.3. Complex measurable Su-M_Field
   4.4.4. Environment measurable Su-M_Field
   4.4.5. Physical effects related to magnetic field

4.5. Direction of measuring system evolution
    4.5.1.   Measurable bi- or poly-systems
    4.5.2.   Evolution line

Class 5. Standards for Using Standards

5.1. Adding substances at construction, reconstruction, and destruction of Su-Fields.
    5.1.1.   Round-about ways:
        5.1.1.1.   "Emptiness" instead of substance
        5.1.1.2.   Field instead of substance
        5.1.1.3.   External addition instead of internal one
        5.1.1.4.   Particularly active addition in very small doses
        5.1.1.5.   Substance in very small doses
        5.1.1.6.   Addition is used for awhile
        5.1.1.7.   A copy instead of a subsystem
        5.1.1.8.   Chemical compound
        5.1.1.9.   Addition is obtained from the subsystem itself
    5.1.2.   Substance(s) separation
    5.1.3.   Substance(s) dissipation
    5.1.4.   Big additives
5.2. Adding fields at construction, reconstruction, and destruction of Su-Fields
    5.2.1.   Using existing fields
    5.2.2.   Fields from environment
    5.2.3.   Substances as fields sources
5.3. Phase transitions
    5.3.1.   Change of the phase state
    5.3.2.   Second type phase transition
    5.3.3.   Phenomena coexist with phase transition
    5.3.4.   Two-phase state
    5.3.5.   Interaction between phases
5.4. Application peculiarities of physical effects
    5.4.1.   Self-driven transition
    5.4.2.   Increase of output field
5.5. Creation of particles
    5.5.1.   Substance destroying
    5.5.2.   Integration of particles
    5.5.3.   How to use Standards 5.5.1 and 5.5.2

APPENDIX 6
Relations between TRIZ Heuristics

I. Altshuller's methods to resolve the physical contradiction vs. Standards:

Method	3	4	5	6	7	8 and 9	10	11
Standard	3.1.1	3.1.3	3.1.5	3.2.1	5.3.1	5.3.2	5.3.4 and 5.3.5	5.4.1

II. Altshuller's Principles to resolve the technical contradiction vs. Standards:

Principle #	Standard #
1	2.2.2, 2.2.4, 3.2.1, 5.1.2
3	1.1.8.2, 1.2.5, 2.2.6, 5.1.1.5
4	2.2.6
5	1.1.2–1.1.5, 3.1.4
11	1.1.8.1
13	2.4.6
15	2.2.4, 2.4.8
16	1.1.6, 5.1.4
18	2.3.1, 2.4.10, 4.3.2
19	2.2.5, 2.4.10
20	2.3.3
22	1.2.2
23	2.4.8, 5.4.1
24	1.1.7, 2.4.9, 2.4.5, 1.1.2–1.1.5, 4.1.2, 5.1.4
25	2.4.8, 5.4.1
26	4.1.2, 5.1.1.7
28	2.2.1, 2.4 (all, especially 2.4.11), 4.2 (all), 5.1.1.2
29	2.4.3, 5.1.1.1, 5.1.4
30	2.2.6
31	2.2.3, 2.2.6, 2.4.4, 5.1.1.1
32	4.1.3, 4.3.1
34	5.1.3
35	5.3.1, 1.1.2-1.1.5, 2.4.12
36	2.4.7, 4.1.1, 4.3.1, 5.3.2, 5.3.4, 5.3.5
37	4.1.1, 4.3.1
38	5.5 (all), 5.1.1.4
39	1.1.3, 1.1.5
40	5.1.1.1

III. Altshuller's methods to resolve the physical contradiction vs. Principles to resolve the technical contradiction:

Methods	Principles
1	1,2,3,7,17,24,26
2	15,9,19,21,32,34
3	3,5,22
4	13
5	13
6	1,28
7-10	36
11	27

IV. Separation heuristics for resolution of the physical contradiction vs. Altshuller's Principles.

Separation in space	1, 2, 3, 4, 7, 17, 24, 26, 30.
Separation in time	9, 10, 14, 15, 16, 18, 21, 27.

The original Altshuller numbering for all types of heuristics is used in all tables above.

APPENDIX 7
A Dozen Frequently Used Hints

1. FIELD-SUBSTANCE INTERACTION

HINT 1. DUALISM

A. If the specific field is needed for required action realization and it is absent in the technique, use the substance for transforming existing field(s) into the required action.
B. If the specific substance is needed for required action realization and it is absent in the technique, use the field for transforming existing substance(s) into one needed for the required action.

List of energies presented in Appendix 4 gives the idea of "active" fields; lists of phenomena (see Appendix 3) include some "transforming" substances.
All fields can be easily transformed into the thermal field.

HINT 2. TURN INTO FERRO...

A. Substitution of the substance–function carrier by ferromagnetic substance which can transform the magnetic field energy into the desired action.
B. Switch from the rigid ferromagnetic substance to ferromagnetic particles and/or to a magnetic liquid.

2. MEASUREMENTS AND CONTROL

HINT 3. DENIAL FROM MEASUREMENTS

A. Change the system so that it is not necessary to hold measurements at all.

HINT 4. SUBSTITUTION

A. Substitute the direct operations under the measured subsystem with operations under its model or picture.
B. Use the optical combination between the subsystem's image and its gauge for difference detection.

Hint 5. Replacement

A. Replace the measurement process by consistent discovery of changes.

Hint 6. Synthesis

A. Omit some field flux through the technique, which might be detected easily, and then make a decision about modification in the technique by the output change of this field.
B. Use easily detected additions — the reformer substances (ferromagnetic particles, luminophores, bubbles, foam, chemical indicators, etc.) or the source of some easily detected field (smell, luminescent, electric, magnetic, etc.).

Hint 7. Increase control ability

A. Introduce a new field into the technique, which can be managed more easily than the existing fields, necessary for realizing the primary and secondary functions.
B. Apply averaging procedure to the parameter distribution in space or in time (using the inertia of the impact).
C. Use small differences of large values.
D. If the given maximum cannot be achieved, make slightly more and remove the excess.
E. Rely on a slightly different parameter value if the given "flat" optimum cannot be achieved.

Hint 8. Increase manipulation ability

A. Incorporate the field and the energy reformer substance, which can realize the control functions under the substance.
B. If the whole cannot be modified, change the most sufficient substance.
C. Use small differences of large values.
D. Move manipulated subsystem close to controlling one in the energy chain.
E. Influence the environment instead of influencing the technique.

3. ELIMINATION OF HARMFUL INTERACTION

Hint 9. Overcome the harm

A. If the HF depend on the UF, and the growth velocity for UF is higher, then the process should be limited in time or compressed into periodic action.
B. Modify a substance to resist HF (e.g., coating).

HINT 10. COMPENSATING THE HARM

A. Use various HF possibly existing in a technique to compensate each other.
B. Compensate for the negative results of HF by previously introduced changes or by something for neutralizing the effect.
C. Allow HF choosing cheap unstable substance instead of durable expensive substance.
D. Modify a field existing in a technique to resist HF (e.g., screening).
E. Use the neutral functions possibly existing in a technique to counteract HF.

HINT 11. DESTRUCTION OF HARMFUL INTERACTION
BETWEEN SUBSTANCES

A. Incorporate a new substance between the existing substances. As a rule, this new substance may be either the variation of some existing substance or a mixture of the existing substances (especially when substances are involved in variation or motion, such as change of substance condition, decomposition, division, breaking, chemical reaction).
The most common modification is:
   A part of the product acting as the tool or the tool made partly of the same substance as the product.
B. Introduce a substance to absorb HF actions, close the harmful energy chain (i.e., HF will be directed to this substance).
C. Incorporate the field that would neutralize negative result of HF.

HINT 12. DESTRUCTION OF HARMFUL INTERACTION BETWEEN
SUBSTANCE AND FIELD.

A. Incorporate a new field which would neutralize HF of the existing field on the substance.
B. Incorporate substance which would neutralize the negative result of HF.

Notice: These hints, collected and selected by B. Goldovsky, G. Frenklach, and S. Savransky in the 1980s and 1990s, are based mostly on Altshuller's Standards.

APPENDIX 8
Polovinkin's Heuristic Expedients for Systems Transformations

1. SHAPE TRANSFORMATIONS

1.1. To use circular, spiral, treelike, spherical, or other compact shapes.

1.2. To carry out apertures or cavities in a system. Inversion of expedient.

The instruction "Inversion of expedient" is given at the end of the description of many expedients with the purposes of text reduction. Generally it is recommended also to make the converse transformation or to search in the opposite direction. Hence, the expedient 1.2 would have the following complete description: "To make an aperture and cavity in a system, OR to exclude an aperture and cavity from a system."

1.3. To check the system shape's conformity to symmetrical patterns. To proceed from the symmetric shape and structure to asymmetric. Inversion of expedient.

1.4. To proceed from rectilinear parts, flat surfaces, cubic, and multilateral shapes (especially in places of interfaces) to curve, spherical, and streamline shapes. Inversion of expedient.

1.5. To give convex (or more convex) shape to a system working under a load.

1.6. To compensate the undesirable shape by accretion with the opposite outline form.

1.7. To carry out a system in the form of
 • another technical system with a similar function or purpose
 • a human body
 • an animal, plant, or their bodies.

1.8. To adapt a system to the shape of parts of a human body.

1.9. To use a natural principle of formation in live or inanimate nature in similar conditions of work.

1.10. To separate a raw material (sheet or volumetric) rationally (optionally); to change the shape of details for more complete use of a raw material.

1.11. To design the shape of details as close as possible to the shape and sizes of cutout parts.

1.12. To find global-optimum shape of a system.
1.13. To find the best integral shape of a system (visual allocation of the main subsystem or functional element, elimination or screening of most unimportant and auxiliary subsystems or details, etc.).
1.14. To use various kinds of symmetry, asymmetry, dynamic, and static properties of the form, rhythm (alternation of identical or similar (sub)systems or elements), nuances, and contrast.
1.15. To carry out harmonic coordination of the shapes of various (sub)systems or elements (choice of scales and ratio between a system and its environment, use aesthetically of preferable portions).
1.16. To choose (or to create) the most beautiful shape of the system and its casing.

2. STRUCTURE TRANSFORMATIONS

2.1. To exclude the most intense (loaded) (sub)system.
2.2. To exclude a (sub)system while preserving all former functions of the system. Because one subsystem carries out a few functions, the necessity of other subsystems disappears. To remove excessive details even at the loss of one percent of effect.
2.3. To attach a new (sub)system to the system by rigid or knuckle-connected plate (core, pipe, etc.) located in operative zone or in contact with this working zone or in the environment.
2.4. To attach the additional specialized tool to the base system.
2.5. To replace communications (way or means of connection) by (sub)systems; to carry out rigid connection by flexible parts. Inversion of expedient.
2.6. To replace a source of energy, type of a drive, color of cable, etc.
2.7. To replace the mechanical circuit by an electrical, thermal, optical, or electronic one.
2.8. Essentially to change configuration of (sub)systems; to reduce layout expenses.
2.9. To concentrate controls in one place.
2.10. To consolidate (sub)systems by the uniform case, or to make the system integral.
2.11. To enter a uniform drive, control system, or power supply.
2.12. Connect homogeneous subsystems or subsystems intended for adjacent operations.
2.13. To merge subsystems with independent functions into one system which retains all these functions.
2.14. To use an aggregation. To create a base design (e.g., a uniform frame) on which it is possible "to hang" various (or in various combinations) subsystems, instruments, etc.
2.15. To combine obviously or traditionally incompatible (sub)systems by removing the contradictions.
2.16. To choose a raw material, ensuring the minimum labor input during the processing or manufacturing of subsystems (and/or elements).

2.17. To use sliding, folding, modular, inflatable, and other subsystems (elements), ensuring greatly reduced size during switching of the technical system from a working condition to nonworking.

2.18. To find global-optimum structure.

3.   TRANSFORMATIONS IN SPACE

3.1. To change traditional orientation of system in space: horizontal to vertical or inclined; to put on its side; to turn upside down; to rotate.

3.2. To use "an empty space" between subsystems (one subsystem can pass through a cavity in the other).

3.3. To merge separate subsystems by location one inside the other (as in the Russian nested dolls, *Matrioshka*).

3.4. To change the settlement on one line to several lines or planes. Inversion of expedient.

3.5. To move from one plane to several planes or in three-dimensional space; to proceed from one-layer configurations to multi-layer. Inversion of expedient.

3.6. To change the direction of action of labor or environment.

3.7. To proceed from contact on a point to contact on a line; from contact on a line to contact on a surface; from contact on a surface to volumetric (spatial). Inversion of expedient.

3.8. To carry out interface on several surfaces.

3.9. To move instruments closer to the operation zone (the place of fulfillment of tool functions) without moving other subsystems or the whole system.

3.10. To place subsystems in the most convenient place for them to take effect without time or energy expenses for their delivery.

3.11. To proceed from sequential connection of subsystems to parallel or mixed. Inversion of expedient.

3.12. To divide a subsystem in order to move each resultant part to the operative zone.

3.13. To divide a subsystem into two parts, volumetric and nonvolumetric; to move the first part for a boundaries-limiting volume.

3.14. To move subsystems that can be affected by harmful factors far from a zone of the harmful factors' action.

3.15. To transfer the system or its subsystems to another environment that excludes action of the harmful factors.

3.16. To overcome traditional spatial restrictions or overall dimensions.

4.   TRANSFORMATIONS IN TIME

4.1. To transfer fulfillment of action to another operation time. To carry out required action prior to the beginning or after conclusion of (sub)system's work.

4.2. To proceed from continuous submission of energy (substance) or continuous action (process) to periodic or pulsing. Inversion of expedient.

4.3. To proceed from a permanent-in-time regime to a nonstationary mode.

4.4. To exclude useless (harmful) time intervals; to use a pause between pulses (periodic actions) for realization of other action.

4.5. To provide a continuous useful action to carry out work of the system permanently without single courses. All (sub)systems of the system should work all the time with a complete load.

4.6. To change the sequence of fulfillment of operations or functions.

4.7. To proceed from sequential realization of operations to parallel (simultaneous). Inversion of expedient.

4.8. To combine technological processes or operations. To merge homogeneous or adjacent operations. Inversion of expedient.

5. TRANSFORMATIONS OF MOVEMENTS AND FORCES

5.1. To change a direction of rotation.

5.2. To replace forward (rectilinear) or reciprocal movement by rotation. Inversion of expedient.

5.3. To remove or to reduce single, return, and intermediate movements.

5.4. To change essentially the direction of a movement, including changing to the opposite direction.

5.5. To replace a traditional complex trajectory by a movement along a line or circle. Inversion of expedient.

5.6. To replace a bend by stretching or compression. To replace compression by stretching.

5.7. To divide a system into two parts, heavy and light; to move only the "light" part.

5.8. To change conditions of work so that it is not necessary to move a subsystem against a gravitational force.

5.9. To replace friction of sliding with friction of rolling. Inversion of expedient.

5.10. To proceed from a constant physical field to a changeable one. Inversion of expedient.

5.11. To divide a (sub)system into parts capable of moving each other.

5.12. To change conditions of work so that dangerous or harmful movements occur at high speed. Inversion of expedient.

5.13. To use magnetic forces.

5.14. To compensate action of weight of a (sub)system by connecting it with the (sub)system having elevating force.

5.15. To make a shifting (sub)system motionless. Inversion of expedient.

6. TRANSFORMATIONS OF A MATERIAL

6.1. To carry out the (sub)system and (sub)systems cooperating with it from the same material or from a material with similar properties. Inversion of expedient.

6.2. To carry out a (sub)system or its surface from "porous" material. To fill holes by a substance.

6.3. To divide a (sub)system into parts so that each part can be made from the most suitable material.

6.4. To remove a superfluous material (one not carrying a functional load).

6.5. To change surface properties of a (sub)system; to strengthen a surface; to neutralize properties of a material on a surface, etc.

6.6. To replace a rigid part with malleable (sub)systems; instead of rigid volumetric designs, to use flexible shells and films. Inversion of expedient.

6.7. To change physical properties of a material, for example by changing its state.

6.8. To replace some environment systems with systems having other physical and chemical properties.

6.9. To use other material (cheaper, newer, etc.).

6.10. To use details from a material with subsequent solidification.

6.11. To separate a harmful or undesirable impurity from a substance.

6.12. To replace a traditional environment. To consider an opportunity to use a strongly rarefied, inert, aquatic, liquid, or other environment.

6.13. To replace subsystems by their optical copies (images); to change the scale of the image. To proceed from visible optical copies to infrared, ultraviolet, and other images.

6.14. To replace an expensive, durable (sub)system with a cheap, short-lived one.

6.15. To replace a diverse material and shape of a (sub)system by a unified or standard (sub)system.

6.16. To carry out (sub)systems from materials with differing characteristics producing the necessary effect (e.g., materials with different thermal expansion).

6.17. Instead of firm parts, to use liquid or gaseous (inflatable, water or air pillows, hydrostatic, etc.). Inversion of expedient.

6.18. To choose materials ensuring waste reduction during the manufacturing of details (e.g., to proceed from application of details made by cutting to details made from plastic or metal powder).

6.20. To proceed to waste-free technologies (e.g., to result in a more useful substance that can be used for manufacturing of other details).

6.21. To consolidate materials by mechanical, thermal, electro-physical, electrochemical, laser, and other kinds of processing.

6.21. To use materials with higher specific characteristics (corrosion resistance, electrical, etc.).

6.22. To use reinforced, composite, porous, and other new prospective materials.

6.23. To use a material with characteristics changeable in time (rigidity, transparency, etc.).

7.   EXPEDIENTS OF DIFFERENTIATION

7.1. To divide a driven flow (substance, energy, information) into two or more.

7.2. To divide a friable, liquid, or gaseous (sub)system into parts.

7.3. To section a (sub)system, easily separated.

7.4. To differentiate a drive and other sources of energy; to move them to the instruments (executive bodies) and operation (working) zones.

7.5. To give each (sub)system independent control, management, and drive.

7.6. To split traditional system into homogeneous (sub)systems, carrying out similar functions. Inversion of expedient.

7.7. To divide a (sub)system into parts, then to produce, process, load, etc. each part separately, and then to carry out assembly.

7.8. To divide system into parts: hot and cold; to isolate one from the other.

7.9. To present a system as a compound construction; to construct it from separate (sub)systems and parts.

7.10. To give a block structure to a (sub)system, in which each block accomplishes an independent function.

7.11. To allocate in a system the necessary (sub)system (necessary property) and to strengthen it and/or to improve conditions of its work.

8. QUANTITATIVE CHANGES

8.1. To change parameters of a (sub)system (its elements) or environment.

8.2. To increase or to reduce the number of identical or similar (sub)systems in the system. To change the number of simultaneously working or processable subsystems (e.g., engines, instruments, etc.).

8.3. To change overall dimensions, volume, or length of a (sub)system while switching it from a working or nonworking condition.

8.4. To increase (or to reduce) the degree of splitting of the (sub)system.

8.5. To insignificantly reduce the required effect.

8.6. To use the superfluous decision concept (if it is difficult to receive 100% of required effect, set the purpose to receive a little more).

8.7. To change harmful factors so that they cease to be harmful.

8.8. To reduce the number of functions of a (sub)system so that it becomes more specific, appropriate only to its main functions and requirements.

8.9. To exaggerate considerably the sizes (or other parameters) of a (sub)system and to find application for this. Inversion of expedient.

8.10. To increase the intensity of technological processes by making an operational working zone in the shape of a platform or closed volume.

8.11. To create local quality; to carry out local concentration of forces, pressure, voltage, etc.

8.12. To find global-optimum parameters of a technical (sub)system on various criteria of development.

8.13. To proceed to other physical principles of action with cheaper or accessible sources of energy or higher efficiency.

8.14. After constructive improvement of any (sub)system, to define how other (sub)systems should be changed, such that the efficiency of the system as a whole increases.

9.   TRANSFORMATIONS IN ACCORDANCE WITH THE EVOLUTION TRENDS

9.1.   To turn from homogeneous or disordered substances or fields to inhomogeneous and ordered (in space and in time).

9.2.   To coordinate (or vice versa) the action of the substance carrier of function with the eigen-frequency of the substance-object of function.

9.3.   To fill the pause of one kind of action by another action.

9.4.   To change the rigid structure of the substance-function carrier into a soft, dynamic structure according to the trend: Solid $\rightarrow \ldots \rightarrow$ Liquid $\rightarrow$ Gas $\rightarrow$ Field.

9.5.   To turn from the field of direct action into the changing field, then to the impulse field.

9.6.   To switch the substance that works on the macro-level to another that works on a micro-level.

9.7.   To switch from mechanical fields to acoustic, electric, chemical, and magnetic fields.

9.8.   To combine (temporarily or permanently) into a super-system, the subsystems for performance of similar functions first, then the subsystems for different functions, and finally the subsystems for opposite functions.

REFERENCES

Polovinkin, A. I., *Laws of Organization and Evolution of Technique,* VPI, Volgograd, 1985 (in Russian).

Polovinkin, A. I., *Theory of New Technique Design: Laws of Technical Systems and Their Applications,* Informelektro, Moscow, 1991 (in Russian).

Polovinkin, A. I., *The ABC of Engineering Creativity,* Mashinostroenie, Moscow, 1988 (in Russian).

Index

inventive, 4, 23
linear, 65, 67, 68
maxi-, 309
mini-, 309, 310, 322, 333
network, 65, 67
nonroutine, 4, 18
pair, 65, 66
point, 64
reduction, 318
replacement, 327
routine, 4
selection of, 306, 311
sharpening, 321
simple classification of, 6
-situations, 184
solution, 146, 326
space, 5
star, 65, 68
structures of, 27
synthesis, 316
technical, standard solutions of, 259–265
Problem solving, 3–20, 125
    back method to technical, 78
    decision aids, 15
    difficulty of problem, 5
    information and, 15
    methodology of inventive, 22
    methods of creativity activation, 8–15
        checklists and questionnaires, 9–10
        morphological box, 11–14
    nonroutine, 29
    problem solving and information, 15–17
    psychological inertia, 5–7
    requirements, 17–18
        necessary qualities for solver of nonroutine
            problems, 18
        requirements for inventive problem solving,
            17–18
    routine and inventive problems, 3–5
    trial-and-error method, 7–8
Product(s), 34, 325
    collateral, 92
    resources, 325
Productivity, 203
Psychological inertia, 5, 6, 7, 161
Psychologists
    American, 166
    characters of people counted by, 169
Pulsation, 13

Q

QFD, see Quality Function Deployment
Qualitative technological forecasting, 123

Quality Function Deployment (QFD), 28, 98, 126
Quantification, 21
Questionnaires, 9

R

Radioactive materials, 281
Raw object, 34
Reaching, 74
Readily available field resources, 86
Readily available functional, 87
Readily available matter resources, 86
Readily available resources, 85
Readily available time resources, 87
Real-Ideal Transition method, 182
Realization, fee for, 76
Recognizable substance, 189
Recycling, 103
Reduction, 109
Removal/extraction, 204–205
Repairability, 202
Resistor, 278
Resonating chambers, 281
Resource(s), 102
    applications, 324
    availability of, 89
    collateral, 90
    derivative energy, 86
    derivative field, 86
    derivative functional, 87
    derivative information, 86
    derivative matter, 86
    derivative space, 87
    derivative time, 87
    derived, 85
    detail analysis of, 85
    energy, 92
    energy-field, 84
    environmental, 83
    external, 85
    functional, 84
    harmful, 90
    information, 84
    internal, 85
    limitations, overcoming of, 90
    mind-map of, 94
    mobilization, 324
    natural, 83
    product, 325
    readily available field, 86
    readily available matter, 86
    readily available space, 87
    readily available time, 87